Applied Mathematics in Energy and Mechanical Engineering

Editor
Leonid Plotnikov
Department of Turbines and Engines
Ural Federal University
Named after the First
President of Russia B.N. Yeltsin
Yekaterinburg
Russia

Editorial Office
MDPI
St. Alban-Anlage 66
4052 Basel, Switzerland

This is a reprint of articles from the Special Issue published online in the open access journal *Axioms* (ISSN 2075-1680) (available at: www.mdpi.com/journal/axioms/special_issues/1K40HD3JFV).

For citation purposes, cite each article independently as indicated on the article page online and as indicated below:

Lastname, A.A.; Lastname, B.B. Article Title. *Journal Name* **Year**, *Volume Number*, Page Range.

ISBN 978-3-7258-0346-0 (Hbk)
ISBN 978-3-7258-0345-3 (PDF)
doi.org/10.3390/books978-3-7258-0345-3

Applied Mathematics in Energy and Mechanical Engineering

Editor

Leonid Plotnikov

MDPI

Basel • Beijing • Wuhan • Barcelona • Belgrade • Novi Sad • Cluj • Manchester

Contents

About the Editor

Leonid Plotnikov

Leonid Plotnikov is a professor at the Department of Turbines and Engines of the Ural Federal University named after the first President of Russia B.N. Yeltsin. His areas of scientific interests are gas dynamics and heat exchange of gas flows in various technical applications, piston engines, gas exchange systems and experimental studies. Leonid Plotnikov has an H-index of 10 according to the Scopus database. He is a reviewer for the journals *Energy, Thermal Science and Engineering Progress, International Journal of Heat and Mass Transfer, International Journal of Engine Research, Thermal science, Axioms, Energies, Mathematics* and others.

Preface

Today, the development of many branches of science, engineering and technology directly depends on the level of the mathematical apparatus used. This is due to the rapid adoption of mathematical modeling, digital twins, machine learning and artificial intelligence algorithms. Therefore, new achievements in applied mathematics are a driving force in the development of energy and mechanical engineering.

The purpose of this Special Issue was to publish modern results in the field of applied mathematics in relation to energy and mechanical engineering. The scientific works of the Special Issue are devoted to modern and current achievements in mathematics as applied to various technical areas. Thus, the special issue contains mathematical articles on the following topics: thermal engineering, gas dynamics, materials science, electrical engineering, 3D printing, mechanical engineering technology, digital twins, optical measurements, vibration reliability, electric vehicles, etc. Overall, the authors presented innovative ideas, methodologies and techniques that can contribute to the development of mathematics, energy and mechanical engineering, covering a large number of mathematical problems for various technical areas. Moreover, it can be assumed that these are not all relevant and important problems in applied mathematics, energy and mechanical engineering. Therefore, future Special Issues of *Axioms* will undoubtedly be very popular.

High-quality articles were prepared by specialists working at universities from different countries, including China, Australia, Bangladesh, India, Italy, Palestine, Russia, Taiwan, Turkey and the UK. This reaffirms the importance and relevance of applied mathematical problems in energy and mechanical engineering.

It is hoped that this Special Issue will be useful for specialists from various fields of science and technology, and will inspire scientists and specialists to pursue further research in mathematics, energy and mechanical engineering.

The statistics of the Special Issue are as follows:

Submissions – 28.

Publications – 15.

Rejections – 13.

Article types: Research Article – 15.

Thanks to all the authors, Assistant Editors, editorial staff and peer reviewers for their valuable contributions to the Special Issue.

Leonid Plotnikov
Editor

axioms

Article

Nonlinear Tolerancing: Variation Simulation and Assembly Analysis with Regard to Contact Interaction of Parts

Sergey Lupuleac *, Margarita Petukhova *, Julia Shinder, Maria Titova, Nadezhda Zaitseva and Maria Churilova

Institute of Physics and Mechanics, Peter the Great St. Petersburg Polytechnic University, 195251 Saint Petersburg, Russia; zajtseva_ni@spbstu.ru (N.Z.); churilova_ma@spbstu.ru (M.C.)
* Correspondence: lupuleac@spbstu.ru (S.L.); margarita@lamm.spbstu.ru (M.P.)

Abstract: The variation analysis is a key tool for ensuring the high quality assembly in the process of developing the technology for manufacturing of aircraft parts. One of the main factors in variations is the deviations in the positioning procedure. This paper is devoted to the development of an approach that allows taking into account the variations during positioning and merging it with the special algorithm of contact problem solving. The impact of varied boundary conditions is incorporated into an additional vector of forces that can be interpreted as reactions to the shift of supports. The obtained results are illustrated with a case of wing-to-fuselage assembly.

Keywords: variation simulation; assembly analysis; contact problem; method of influence coefficients; mathematical modeling

MSC: 74M15; 65K10

check for updates

Citation: Lupuleac, S.; Petukhova, M.; Shinder, J.; Titova, M.; Zaitseva, N.; Churilova, M. Nonlinear Tolerancing: Variation Simulation and Assembly Analysis with Regard to Contact Interaction of Parts. *Axioms* **2024**, *13*, 67. https://doi.org/10.3390/axioms13010067

Academic Editors: Hans J. Haubold and Leonid Plotnikov

Received: 1 November 2023
Revised: 12 January 2024
Accepted: 16 January 2024
Published: 20 January 2024

1. Introduction

As it was mentioned in the review by Ceglarek et al. [1], in the aircraft manufacturing and automotive industries, approximately 70% of all technological changes are carried out at the production stage, since at the design and implementation stages there is not yet enough information to accurately predict the results of technological processes. The development of digital technologies makes it possible to predict the quality of production and the probability of production defects using mathematical modeling of the assembly process. The main difficulty in predicting the outcome of a production process is the inevitable deviation of all real components in practice from the ideal constructions. Hu et al. [2] admit that the source of these assembly deviations is both manufacturing defects of individual parts and possible inaccuracies of each step of the assembly process. The cumulative effect of these deviations can lead to the fact that the assembled structure will not meet the technical requirements for reliability and functionality, as was noticed by Söderberg et al. [3]. Therefore, in the mathematical modeling of the assembly process, it is important to take into account these deviations and analyze their influence on the final deviations in the assembled structures.

The most common type of part assembly in aircraft manufacturing is riveting, which ensures the required reliability and durability of the structures. One of the important stages of riveted assembly is assembly with the use of temporary fasteners. These fasteners are installed in a specified set of assembly holes and clamp the parts together during further assembly operations.

In mass production conditions, the arrangement of temporary fasteners cannot be adjusted to the individual characteristics of incoming parts. Therefore, a single template for installing temporary fasteners is developed for each joint. Despite possible random assembly deviations, this template should eliminate the gap between parts. As admitted by Weber [4], the development of effective templates for temporary fasteners is one of the actively used methods for optimizing the assembly process in aircraft manufacturing.

The main approach to handling problems of analyzing the assembly process, taking into account the deviations, is statistical modeling. The general solution scheme in this case is based on the Monte Carlo method and consists of cycled modeling of the outcome of the assembly process for random initial deviations. Based on the data obtained, the required probabilistic characteristics of the analyzed process are determined using statistical methods. It is necessary to have a large set of implementations of assembly deviations to carry out the statistical analysis. Additionally, since most of the parts used in aircraft manufacturing are compliant structures, it is important to take into account the deformation of the parts during their contact interaction.

The standard method that allows taking into account assembly deviations is the direct Monte Carlo method, as described by Gao et al. in [5]. The realizations of the initial deviations are obtained by generating random numbers from given distributions, and the modeling of the assembly process is carried out using the classical finite element method. However, the use of the finite element method in the series of calculations for large and complex structures is a very labor-intensive approach.

In 1997, Liu and Hu [6] proposed the method of influence coefficients (MIC), which makes it possible to establish a linear relationship between the initial deviations and the characteristics of the analyzed assembly process. This method significantly reduces the time required for statistical modeling. In this case, however, it is assumed that contact between the assembled structural elements occurs only at predetermined points (for example, at welding points), which makes MIC a purely linear method of analysis and significantly reduces the reliability of results in many application areas. However, the simplicity and speed of calculations have made MIC an extremely popular research method in many fields, including aircraft manufacturing [7–12].

In 2008, Warmefjord et al. [13] proposed the development of MIC by taking into account the possibility of contact between assembled structural elements at arbitrary points in the junction area. In this work, the method of influence coefficients is combined with methods of direct search for contact points. This approach is quite widely used. For example, articles [3,14–18] describe MIC application to the analysis of various assembly processes. However, this approach is difficult to implement and does not allow working with sufficiently detailed models.

In the approach proposed in 2010 by Lupuleac et al. in [19], the variational formulation of the contact problem (e.g., see the book of Wriggers [20]) is reduced to a quadratic programming problem (QPP). The most common methods for QPP were investigated, and specialized solvers have been developed for handling problem features. The dual active set method proposed by Goldfarb [21] is proven to be fast and efficient for small to medium-sized problems (up to ten thousand unknowns, see Burton and Toint [22]). The problem requires focusing on the modification made by Powell [23] that is preferable for ill-conditioned problems. The warm start is easy to implement for active set methods, and one can make use of previously found solutions for subsequent computations. The next one, the interior-point method, is a polynomial-time algorithm that allows for solving a wide range of optimization problems, such as linear, convex quadratic programming, etc. [24]. This method is known to be one of the main methods used for solving large sparse problems and is implemented in many commercial optimization software packages (CPLEX, IMSL, MATLAB, and others). The appropriate choice of a starting point and the advances in the solution of the linear system for determining the Newton direction make the interior-point method applicable to the types of problems considered. Reformulating the QPP in a dual and special "relative" form [25] provides the simple box constraints and leads to the Newton projection method [26], which has a quadratic convergence rate. As a result, the contact problem for deformable parts is solved quickly and without loss of accuracy. This methodology was significantly expanded in subsequent works devoted to modeling the assembly process in the aircraft industry [27–30].

This study attempts to combine the approaches proposed in [9] and [19]. Thus, the approach presented here combines the simplicity and speed of MIC with the accuracy of contact analysis.

Note that the methods taking into account the deviations of parts are widely used for the optimization and development of assembly technology for aircraft structures. In particular, the optimization of the wing-to-fuselage assembly process for Airbus A350 aircraft is considered in [27]. The optimization of the wing assembly process for Airbus A320 is discussed in [28]. The procedure for optimizing the S19 splice joint assembly in the tail section of A350 is described in [29].

The remainder of the article is organized as follows: Section 2 is devoted to the mathematical approaches used in the work. Sections 2.1 and 2.2 describe a methodology that allows positioning deviations to be easily taken into account when solving contact problems in assembly modeling for compliant structures. Section 2.3 describes special methods that allow one to quickly solve the numerous quadratic programming problems that arise when modeling assembly processes. Section 2.4 discusses the statistical analysis of the results. Section 3 describes the results of mathematical modeling. The surrogate model of the wing-to-fuselage upper joint that is used for the numerical experiments is presented in Section 3.1. Section 3.2 illustrates the influence of deviations in the positions of support points on the gap between assembled parts. The choice of the most appropriate numerical method for the considered problem is discussed in Section 3.3. The main modeling results on the influence of positioning deviations on the assembly quality and resulting deviations are provided in Section 3.4. The emulation of positioning deviations by setting the initial gaps between assembled parts is discussed in Section 3.5. The main conclusions of the work are contained in Section 4.

2. Methodology Concept

2.1. Problem Statement

This section is devoted to the problem statement and description of the numerical procedure.

Let us consider the finite element model (FEM) of two panels to be assembled, as shown in Figure 1a. The parts may come into contact in the area where finite element nodes are marked by colored balls; further on, this area is referred to as the junction area. The panels are fixed along the edges (clamps are drawn in Figure 1a). According to the FEM, it is possible to split the nodes into three groups: nodes that can be brought into contact—the computational ones; nodes where fixations are set—nodes to be varied and all the rest. The degrees of freedom (DOF) in computational nodes are referred to as $U_C = (U_{Ci})_{i=1}^{N_C}$, the DOF in nodes to be varied—as $U_V = (U_{Vi})_{i=1}^{N_V}$, the DOF in the rest of nodes—as $U_R = (U_{Ri})_{i=1}^{N_R}$. All the groups of DOF are represented in Figure 1b.

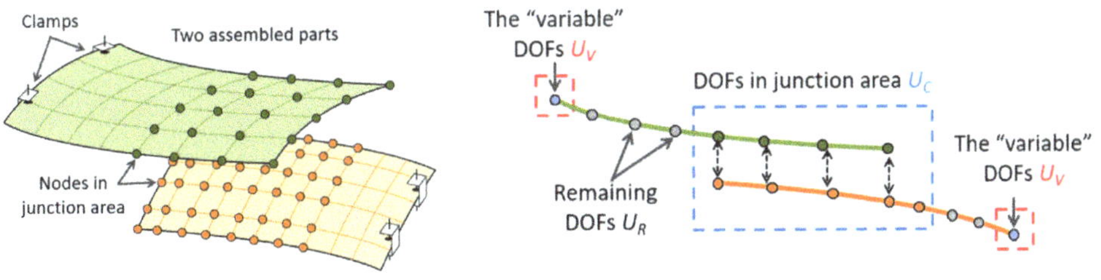

(a) (b)

Figure 1. (**a**) FE model of assembled parts; (**b**) types of FE nodes in the model.

Thus, the vector of unknown displacements in FE nodes $U = (U_i)_{i=1}^{N}$ can be constructed:

$$U = (U_C, U_V, U_R)^T,\ N = N_C + N_V + N_R \tag{1}$$

If the possibility of contact interaction is not taken into account, then the vector U is the solution of the linear system:

$$K \cdot U = F, \tag{2}$$

where $K = (k_{ij})_{i=1,j=1}^{N,N}$ is the global stiffness matrix of the FEM, $F = (f_i)_{i=1}^{N}$ is the vector of applied loads that includes both forces and moments. One can note that system (2) is obtained by discretization of the equations of elasticity theory. Since the assembled parts are mainly flexible panels, shell theory is usually used to describe the stress-strain state of the parts.

The boundary conditions correspond to the fixations of parts on the assembly stand, for example, with the help of clamps, as shown in Figure 1a:

$$U_V = U_{V0}, \tag{3}$$

where $U_{V0} \in R^l$ is the vector of fixed degrees of freedoms.

The contact interaction between the considered parts is modeled using the node-to-node contact model (see Figure 1b). According to this model, only predefined pairs of nodes can come into contact. Note that the use of such a model for this kind of problem is justified since the installed fasteners prevent significant tangential displacements of parts relative to each other.

Using the node-to-node contact model, the nonpenetration conditions can be formulated as:

$$A \cdot U \leq G, \tag{4}$$

where $A \in R^{M \times N}$ is a linear operator that defines nonpenetration between the pairs of nodes in the junction area, $G \in R^M$ is the vector of initial gaps between the panels. It can be mentioned that $A = 0$ outside the junction area, so it is possible to rewrite (4):

$$A_C \cdot U_C \leq G, \tag{5}$$

Summing up the relations (2)–(5), the problem of calculating vector U can be stated in a variational form as the minimization problem:

$$\begin{gathered} \min_{U \in S} \left(\tfrac{1}{2} U^T \cdot K \cdot U - F^T \cdot U \right) \\ S = \{ U | A_C \cdot U_C \leq G, U_V = U_{V0} \}, \end{gathered} \tag{6}$$

The statement (6) of the contact problem includes all the degrees of freedom of the structure taken into account in the FEM, which dimension N can reach 10^6–10^7 variables for industrial applications. This factor makes the solution of the contact problem in statement (6) quite resource-intensive, especially if multiple calculations are required.

On the contrary, the number of DOF in the junction area N_C is much smaller than the dimension N. The static condensation technique [31] can be applied in order to reduce the problem dimension. This approach is described in the next subsection.

2.2. Problem Reformulation

Let us divide the problem (6) into blocks according to the structure of the displacement vector U from (1):

$$K = \begin{pmatrix} K_{CC} & K_{CV} & K_{CR} \\ K_{CV}^T & K_{VV} & K_{VR} \\ K_{CR}^T & K_{VR}^T & K_{RR} \end{pmatrix}, F = \begin{pmatrix} F_C \\ F_V \\ F_R \end{pmatrix} \tag{7}$$

The loads are applied only inside the junction area, so $F_V = F_R = 0$. The system (2) providing a global minimum of (6) without accounting for S is rewritten:

$$\begin{pmatrix} K_{CC} & K_{CV} & K_{CR} \\ K_{CV}^T & K_{VV} & K_{VR} \\ K_{CR}^T & K_{VR}^T & K_{RR} \end{pmatrix} \cdot \begin{pmatrix} U_C \\ U_V \\ U_R \end{pmatrix} = \begin{pmatrix} F_C \\ 0 \\ 0 \end{pmatrix} \tag{8}$$

Opening the brackets, (8) transforms into:

$$\begin{aligned} K_{CC} \cdot U_C + K_{CV} \cdot U_V + K_{CR} \cdot U_R &= F_C, \\ K_{CV}^T \cdot U_C + K_{VV} \cdot U_V + K_{VR} \cdot U_R &= 0, \\ K_{CR}^T \cdot U_C + K_{VR}^T \cdot U_V + K_{RR} \cdot U_R &= 0 \end{aligned} \tag{9}$$

Third relation provides the formula for U_R:

$$U_R = -K_{RR}^{-1} \cdot \left(K_{CR}^T \cdot U_C + K_{VR}^T \cdot U_V \right) \tag{10}$$

Then, U_R from (10) is substituted into the first relation of system (9):

$$K_{CC} \cdot U_C - K_{CR} \cdot K_{RR}^{-1} \cdot \left(K_{CR}^T \cdot U_C + K_{VR}^T \cdot U_V \right) = F_C - K_{CV} \cdot U_V \tag{11}$$

It is recalled that $U_V = U_{V0}$ and (11) is simplified:

$$\left(K_{CC} - K_{CR} \cdot K_{RR}^{-1} \cdot K_{CR}^T \right) \cdot U_C = F_C - K_{CV} \cdot U_{V0} + K_{CR} \cdot K_{RR}^{-1} \cdot K_{VR}^T \cdot U_{V0} \tag{12}$$

From (12), it is possible to derive matrix $K_C = K_{CC} - K_{CR} \cdot K_{RR}^{-1} \cdot K_{CR}^T$ that can be interpreted as the reduced stiffness matrix and the force vector for variations of boundary conditions:

$$F_{var} = \left(-K_{CV} + K_{CR} \cdot K_{RR}^{-1} \cdot K_{VR}^T \right) \cdot U_{V0} = M_{var} \cdot U_{V0} \tag{13}$$

Adapting the notations from [6], $M_{var} = \left(-K_{CV} + K_{CR} \cdot K_{RR}^{-1} \cdot K_{VR}^T \right)$ can be called the matrix of influence coefficients.

Given all the previous transformations the reduced minimization problem equivalent to (6) can be formulated:

$$\begin{gathered} \min_{U_C \in S_C} \left(\frac{1}{2} U_C^T \cdot K_C \cdot U_C - \left(F_C^T + (M_{var} \cdot U_{V0})^T \right) \cdot U_C \right) \\ S_C = \{ U_C | A_C \cdot U_C \leq G \}, \end{gathered} \tag{14}$$

One can see that matrix K_C can be computed once, the same is applicable for matrix M_{var}. A series of assembly simulation problems can then be constructed by only using the input vector of loads F_C in the junction area, the vector of initial gaps G and the vector of boundary values U_{V0}.

In the case of matrix K, its blocks from (7) are not known; the static finite element analysis (FEA) can be exploited for calculating matrix K_C and M_{var}.

Let us suppose that $U_{V0} = 0$, and $(F_C)_j = \begin{cases} 1, & j = i \\ 0, & j \neq i' \end{cases} j \in [1, N_C]$ for i-th DOF in the junction area. The FEA analysis is performed, and the corresponding U_C is computed. Then, it can be concluded from (12) that:

$$U_C = K_C^{-1} \cdot F_C = \left\{ K_{Cj}^{-1} \right\}_{j=1}^{N_C} \tag{15}$$

After collecting all the columns of K_C^{-1} the matrix inversion is required to obtain K_C itself.

The columns of the matrix M_{var} are obtained by setting $(U_{V_0})_j = \begin{cases} 1, & j = i \\ 0, & j \neq i \end{cases}$ for the i-th DOF in the nodes to be varied ($j \in [1, N_V]$) and $F_C = 0$. U_C is computed by static FEA and revealing (12):

$$K_C \cdot U_C = \{M_{varj}\}_{j=1}^{N_C}, \tag{16}$$

So, the i-th column of M_{var} equals $K_C \cdot U_C$.

2.3. Algorithms for Quadratic Programming Problems

Obtaining matrices K_C and M_{var} for the problem (14), the variational analysis comes down to solving a set of quadratic programming problems with different input data. To achieve significant performance improvements over standard quadratic programming tools, the specific features that characterize the considered problem can be used together with:

- reformulating the quadratic programming problem in order to reduce the number of unknowns and simplify the constraints;
- using specialized optimization algorithms adapted to assembly problems;
- dividing the calculation process into a preprocessing stage (time-consuming one) and a gap calculation stage (fast one).

2.3.1. Reformulation of the Quadratic Programming Problem

In assembly problems, the formulation of the original quadratic programming problem in the form most preferable for the optimization method can help reduce computation time. More specifically, this formulation should have a low dimension with a simple constraint structure. Let us consider two additional formulations of the minimization problem (14) that meet the above requirements. The dual formulation of problem (14) is given by:

$$\max_{\Lambda_C > 0} \left(-\frac{1}{2} \Lambda_C^T \cdot Q_C \cdot \Lambda_C + P_C^T \cdot \Lambda_C + H \right), \tag{17}$$

where $\Lambda_C \in R^M$ is the vector of Lagrange multipliers, $Q_C = {A_C}^T K_C^{-1} A_C \in R^{M \times M}$ is a symmetric positive definite dense matrix, $P_C = A_C^T \cdot K_C^{-1} \cdot T_C - G \in R^M$, $H = -\frac{1}{2} T_C \cdot K_C^{-1} \cdot T_C \in R^1$, $T_C = F_C + M_{var} \cdot U_{V0}$. The physical meaning of the unknown vector Λ_C is the reaction forces arising in the contact nodes.

The relative formulation [25] of problem (14) is written as:

$$\begin{gathered} \min_{D_C \in \tilde{S}_C} \left(\tfrac{1}{2} D_C^T \cdot \tilde{K}_C \cdot D_C - \tilde{F}_C \cdot D_C \right) \\ \tilde{S}_C = \{ D_C | D_C \leq G \}, \end{gathered} \tag{18}$$

where $D_C = A_C \cdot U_C \in R^M$ is a vector of relative displacements, $\tilde{K}_C = Q_C^{-1} \in R^{M \times M}$ is a symmetric positive definite dense matrix, $\tilde{F}_C = \tilde{K}_C \cdot A_C^T \cdot K_C^{-1} T_C \in R^M$.

Solving problem (17) and (18) is equivalent to solving (14). The solution of (14) U_C is expressed from the solutions of (17) Λ_C and (18) D_C using Formula (19):

$$\begin{gathered} U_C = K_C^{-1} (T_C - A_C \cdot \Lambda_C), \\ \Lambda_C = \tilde{F}_C - \tilde{K}_C \cdot D_C \end{gathered} \tag{19}$$

The formulations (17) and (18) reduce the dimension of the quadratic programming problem (14) from N unknowns to M ($N \geq M$), the number of constraints remains unchanged, but the linear constraints S_C in general form are replaced by the bound constraints $\Lambda_C > 0$ and $\tilde{S}_C$, respectively.

2.3.2. Adaptation of Algorithms

Problem (14) itself is a convex quadratic programming problem, the solution of which exists and is unique. The optimization methods are faced with the challenge of not only finding a solution with some predetermined accuracy but also finding it quickly since it is necessary to solve a large number of similar problems. The dual Goldfarb–Idnani active set method [21,23], a Newton projection method [26], the primal-dual interior-point method [32], and Lemke's method [33] from this point of view are discussed further.

Depending on the specifics of the optimization problem, the methods are distinct because of different basic ideas. Features of assembly problems lead to the formulation of a quadratic programming problem with a block-diagonal structure of the matrix K_C, a sparse structure of the constraint matrix A_C, ill-conditioned, symmetric, and positive-definite Hessian matrices K_C, Q_C, and $\tilde{K}_C$. In addition, variation simulation provides a set of quadratic programming problems where the Hessian matrix and constraint matrix are the same, but the force and gap vectors are different. These features were used to adapt the listed methods to effectively solve assembly problems. The Newton projection method and the primal-dual interior-point method have an iterative structure, where at each iteration, the system of linear equations is solved. The adaptation (see Table 1) of these methods includes improving system solving (choice between direct and iterative methods, selection of preconditioner for iterative approaches) and reducing the number of iterations (choice of starting point, strategy for choosing a new approximation at each iteration). The dual Goldfarb–Idnani active set method and Lemke's method are based on looking over the set of active constraints. Adaptation was based on finding a strategy for modifying the working set of constraints. For all optimization methods, the structure of the Hessian and constraint matrix was taken into account to improve memory management and reduce the number of actions for matrix-vector operations. All methods presented in Table 1 are implemented in the computer code and included in the ASRP software package [30].

Table 1. Adaptation of optimization methods to assembly problems.

Newton Projection Method	Dual Goldfarb–Idnani Active Set Method	Primal-Dual Interior-Point Method	Lemke's Method
• A method for recalculating constraints to reduce the number of iterations of the method. • A method for solving a system of equations based on a combination of direct and iterative approaches. • A modification of direct methods for solving the system based on the Sherman–Morrison formula. • A method for finding step length based on the golden section method. • Using a warm start.	• Taking into account the sparse structure of the constraint matrix A_C. • Taking into account the block-diagonal structure of the Hessian matrix K_C. • Improving the procedure for adding a new constraint to the working set, based on the physical meaning of the problem. • Using a warm start.	• A method for searching for a "feasible" starting point, taking into account the specifics of assembly problems. The method makes it possible to reduce the number of iterations. • Proposing a preconditioner for solving a system of linear equations reducing iteration process of the system solution. • A modification of direct methods for solving the system based on the Sherman–Morrison formula.	• Taking into account the sparse structure of the constraint matrix A_C. • Taking into account the block-diagonal structure of the Hessian matrix K_C.

Note that the optimization problem under consideration is strictly convex (for all presented formulations) and therefore has a unique solution, which is found by all the algorithms presented here. The exit condition for all algorithms is the fulfillment of optimality conditions with an accuracy of 10^{-7}.

2.3.3. Preprocessing Stage and a Gap Calculation Stage

Preparing matrices Q_C, $\widetilde{K}_C$, and vectors P_C, $\widetilde{F}_C$ for problems (17) and (18) requires additional computational costs. However, since the matrices K_C and A_C are the same for a set of problems formed by variation analysis, the calculation of the necessary auxiliary matrices can be carried out once for each assembly model at the preprocessing stage of model preparation. Taking into account the sparse structure of the matrix A_C and the block-diagonal structure of the matrix K_C allows one to reduce the computation time of the preprocessing stage. Preparing the relative formulation (18) is the most labor-intensive. The ill-conditioning of the stiffness matrix K_C does not allow one to obtain the matrices $\widetilde{K}_C = \left(A_C{}^T \cdot K_C^{-1} \cdot A_C\right)^{-1}$ and $\widetilde{K}_C \cdot A_C^T {\cdot} K_C^{-1}$ with sufficient accuracy, when using standard inversion methods, due to the accumulation of rounding errors, and forces one to use more accurate methods of matrix inversion and multiplication [25].

Another part of data preparation is directly related to the specifics of optimization methods. Some operations, such as Hessian inversion, replacement of variables, construction of auxiliary matrices for the formation of systems of equations, and associated preconditioners necessary for the operation of the methods, can be performed in advance. The set of necessary auxiliary operations depends on the method used.

By moving all labor-intensive operations beyond the immediate solution of the quadratic programming problem, it allows one to significantly speed up the solution process at the gap calculation stage.

2.4. Methodology Application

Given the possibility of making fast multiple computations with the algorithms described in the previous section, the methodology of assembly analysis can be derived that accounts for variations from different sources.

Firstly, varying the force vector F_C means that different fastening configurations are considered. Different vectors G represent the influence of the initial gap on the assembly [34]. Varying U_{V_0} corresponds to changes in the initial positioning of the parts before the assembly.

All the variations are transferred to the preprocessing stage of the QP problem, where the most matrix computations are performed (see Figure 2). Then, the set of corresponding entities is constructed and solved. Then, the results are collected and analyzed.

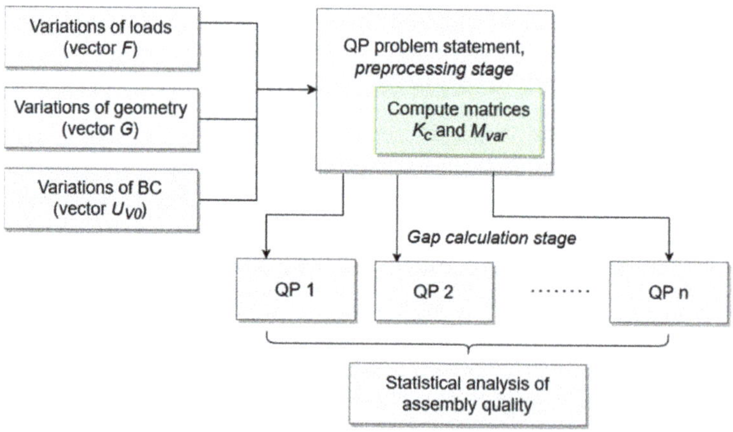

Figure 2. Variation analysis flowchart.

In order to compare simulation results, the quality criterion is to be introduced. Let us denote U_{res} the solution of problem (8), δ is the threshold gap value. If the resulting gap

$G_{res} = G - A_c \cdot U_{res}$ undergoes δ, it is assumed that the contact is achieved. The defect gap vector $d(\delta) \in R^k$ can be defined as follows for each DOF in the junction area:

$$d_j(\delta) = \begin{cases} 1, & (G_{res})_j \geq \delta \\ 0, & (G_{res})_j < \delta \end{cases} \tag{20}$$

The averaged sum of the $d(\delta)$ components shows how the gap is eliminated for the given data case G, F_C, and U_{V_0}. Further on, it is called the probability of defect:

$$P(\delta) = \frac{1}{k}\sum\nolimits_{j=1}^{k} d_j(\delta) \tag{21}$$

The lower the probability of defect $P(\delta)$, the better the quality of the assembly.

Using this methodology, it is possible to simulate what-if scenarios when it is necessary to examine the quality of the assembly under the conditions derived from the data provided by the assembly line (e.g., the information on initial gaps between parts measured for the previously manufactured aircrafts).

Another possibility is to check the given assembly against the unknown boundary conditions a priori. Typical data known for part positioning is tolerance T at each boundary point (it represents how far the actual position can vary from its nominal location): $(U_{V_0})_j \in [-T_j, T_j]$ for each boundary node to be varied $j \in [1, l]$. The statistical distribution for the variation of each point $(U_{V_0})_j$ is typically assumed to be normal (Gaussian) with a zero mean $\mu_j = 0$ and with a standard deviation $\sigma_j = T_j/3$ [35]. Thus, U_{V_0} is constructed from randomly generated $(U_{V_0})_j$ by sampling corresponding normal distributions.

It is worth investigating how the variations U_{V_0} affect the resulting displacements after the contact interaction. In [6], the same research was conducted, and the linear dependence was obtained. Since the approach presented in this paper is nonlinear, the dependence between source variations and variations in the points of observation is more complex. At the same time, using the above-described computation techniques allows for a speed of calculations that is comparable to that of linear models.

3. Results

3.1. Assembly Model of Wing-to-Fuselage Upper Joint

The described approach is applied to the model of aircraft wing-to-fuselage assembly. Figure 3a shows the scheme of the upper joint of the wing (gray) to a part of the fuselage called the cruciform (green). In the top view in Figure 3b, dots show the computational nodes that form the junction area (the area of possible contact), and gray circles show the holes where fasteners can be installed. The assembly model is full-scale. The size of the junction area is approximately 6 m by 0.3 m. There are 3612 computational nodes and 256 holes in the junction area. This model does not correspond to any specific type of aircraft, but qualitatively reproduces the characteristics of the real assembly process. The junction area is colored according to the gap between the parts.

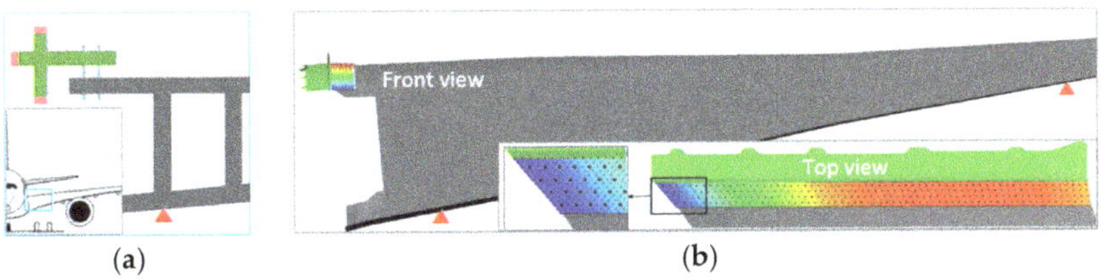

Figure 3. Scheme of the upper wing to fuselage joint (**a**) and assembly model (**b**).

The three edges of the cruciform shown in red in Figure 3a are attached to the central wing box, which is very rigid. Therefore, in the finite element model presented in Figure 4, all displacements and rotations in the nodes located on these edges are restricted. The wing is fixed at four support points depicted in Figures 3 and 4 with red triangles. At these four points, all displacements except the vertical ones are restricted, and the vertical displacements are to be varied. A finite element mesh with these boundary conditions is used to calculate the reduced stiffness matrix K_C and matrix of influence coefficients M_{var} of the assembly, as described in Section 2.2.

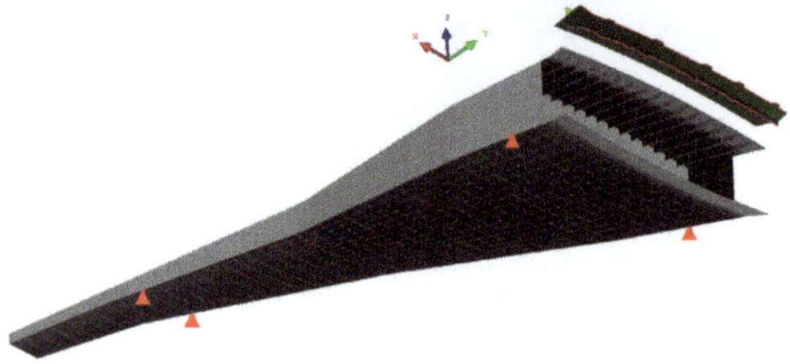

Figure 4. Finite element model of the upper wing-to-fuselage joint.

3.2. *Examples of the Influence Deviations in the Positions of Support Points on the Gap between Assembled Parts*

In all further numerical experiments in this section, it is assumed that the parts are perfectly adjacent to each other: initial gap between parts is equal to zero with no deviations of support positions. So, the source of the gap deviations in the junction area is concentrated at the variation of the vertical positions of the support points of the wing.

Note that these variations are nonlinearly reflected in the gap at the junction area. For example, Figure 5a,b show the results of moving all support points in vertical directions up and down, respectively. When all the support points deflect 3 mm downwards, the gap is opened for approximately 3 mm, and when moving support points 3 mm upwards, due to contact interaction, a gap opens up to approximately 0.5 mm at the edge of the wing.

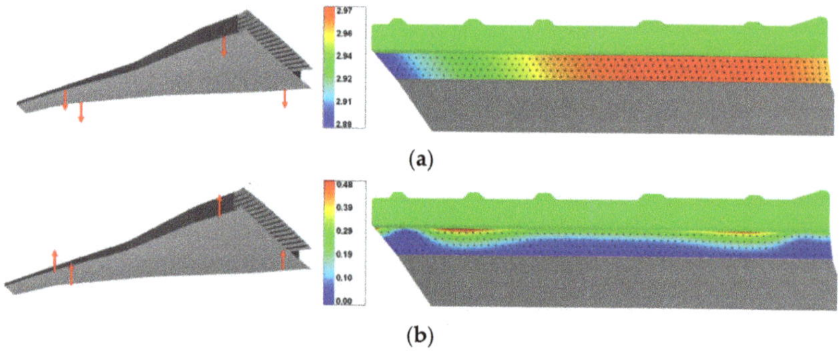

Figure 5. Gaps due to deviations in the positions of all four wing support points down 3 mm (**a**) and up 3 mm (**b**).

It is important to evaluate the impact of random deviations in the positions of the support points on the quality of the assembly. These random deviations cause various residual gaps between parts. The initial assembly stage consists of installing approximately 10% of all fasteners. As a rule, these fasteners are installed manually, which takes a relatively long time and incurs labor costs. The technological requirement is that all residual gaps with these fasteners installed are less than the specified value. The correct arrangement of fasteners at the initial assembly stage is a very important task in aircraft assembly. In this regard, the calculations of residual gaps between parts with 10% installed fasteners are considered in this work as the benchmarks. Figure 6 shows examples of such gaps without fasteners (a) and with 26 fasteners installed (that is approximately 10% of all fasteners), which corresponds to the end of the initial assembly stage (b).

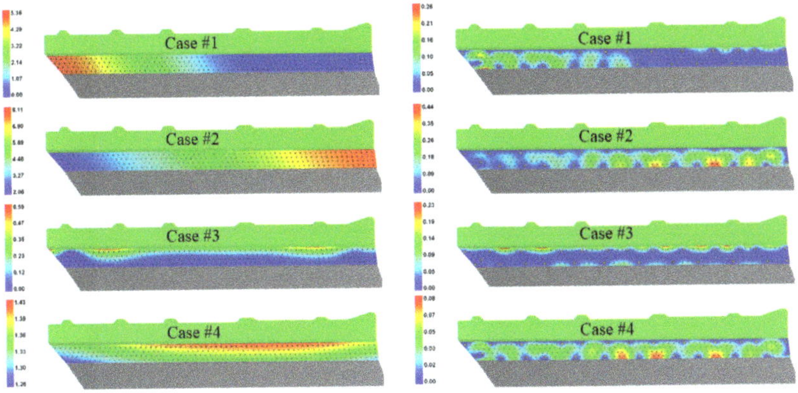

(**a**) (**b**)

Figure 6. Examples of gaps caused by random deviations in the position of wing support points (**a**) without fasteners, (**b**) with 10% fasteners installed.

3.3. Choosing a Method for Solving a Quadratic Programming Problem to Reduce Computation Time

Statistical analysis involves a lot of calculations, so choosing the fastest method is very important. Let us consider the computation time using the example of the gaps shown in Figure 6.

Figure 7 presents the computation time of the gap calculation stage spent by different optimization methods used to solve (14), (17), and (18). Similar trends are observed for different types of gaps caused by random deviations in the position of wing support. The fastest methods are the dual Goldfarb–Idnani active set method and the Newton projection method, used for the relative formulation (18). The computation time estimate of the optimization methods depends on the number of active constraints at optimum points, which is related to the number of installed fasteners (see Figure 8a). In particular, the dual Goldfarb–Idnani active set method has a proportional and Newton projection method has an inverse proportional dependence of computation time on the number of active constraints at optimum for relative problem formulation. Figure 8b helps to choose between the methods based on the number of fasteners to be installed. That is, with a small number of fixtures (less than 10%), the dual Goldfarb–Idnani active set method is preferable to use; otherwise, the Newton projection method is faster.

Note that despite the fact that the gaps for Cases 2 and 4 (see Figure 6) look qualitatively different, the dependence of the active constraints on the number of installed fasteners for these cases is almost the same (Figure 8a). This is due to the fact that both of these gaps are positive throughout the junction area, and contact occurs only after the installation of fasteners. For Cases 1 and 3, in some areas, contact is achieved before fasteners are installed. This is illustrated in Figure 9a, where the contact nodes are colored pink. From Figure 9b, it can be seen that with 10% fasteners installed, the contact nodes for Cases 2 and 4 are located only near the fasteners, while for Cases 1 and 3, they are also present in the middle of the junction area.

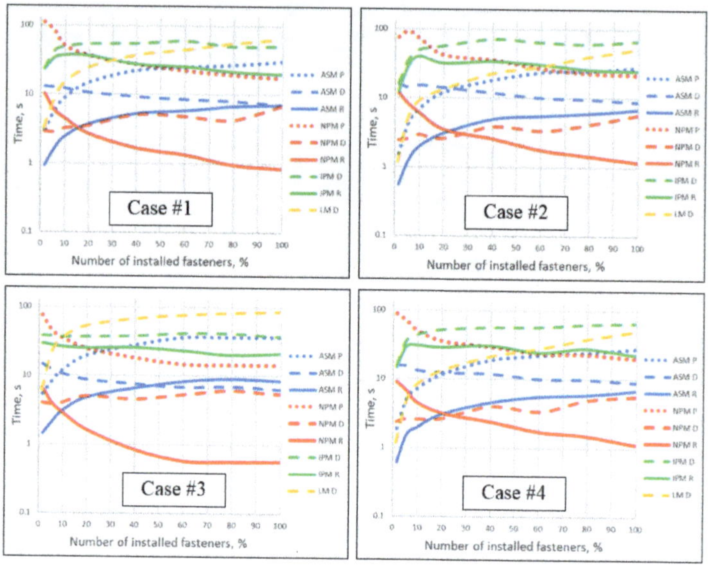

Figure 7. Calculation time depending on the number of installed fasteners.

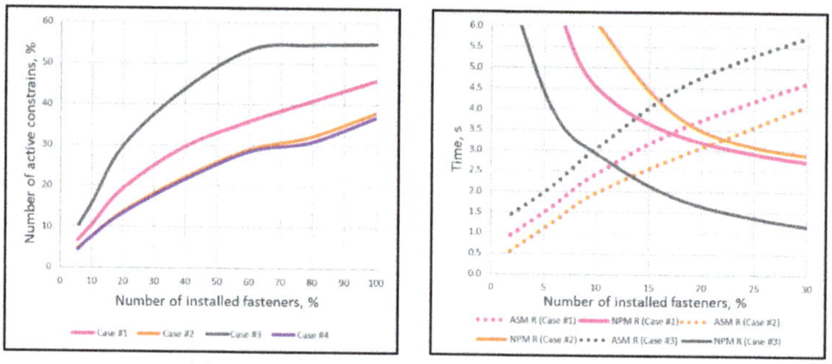

(a) (b)

Figure 8. (**a**) Dependence of active constraints on the number of installed fasteners, (**b**) dependence of calculation time on the number of installed fasteners for Cases #1–3 (relative formulation of QP problem, primal ASM, and primal NPM).

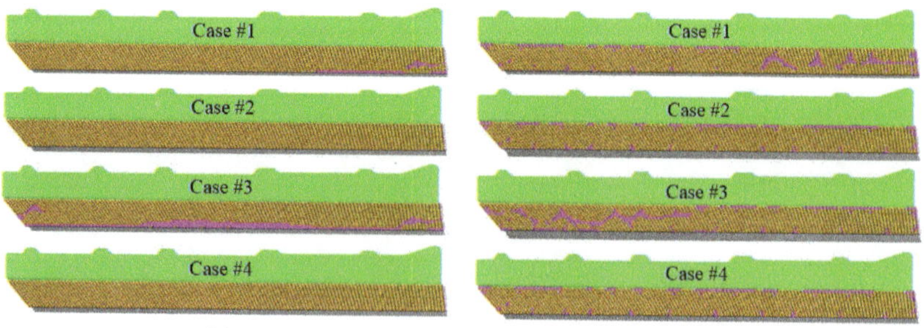

(a) (b)

Figure 9. Contact nodes: (**a**) without fasteners, (**b**) 10% fasteners installed.

Further calculations are carried out without fasteners and with 10% fasteners installed; therefore, the relative formulation of the QP problem is solved by the ASM, since for these cases it is the fastest (see Figures 7 and 8b).

3.4. The Influence of Random Deviations in the Positions of Support Points on a Assembly Process

Figure 10 shows examples of statistical distributions of support point deviations for different sample sizes. The column width in the histograms is 0.5 mm. The Gaussian distribution with a zero mean $\mu_j = 0$ and standard deviation $\sigma_j = 3$ mm is used.

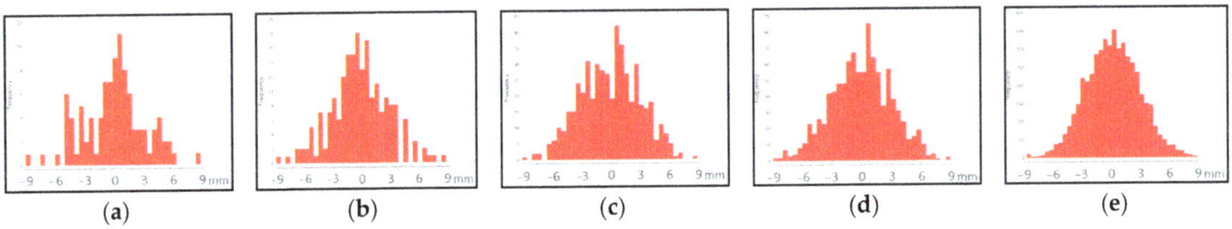

Figure 10. Examples of statistical distributions of support point deviation for different sample sizes: (**a**) N = 100, (**b**) N = 200, (**c**) N = 500, (**d**) N = 1000, and (**e**) N = 5000.

Figure 11 shows the statistical curves representing the probability of defect $P(\delta)$ for different δ. The δ values in mm are plotted on the horizontal axis, and the vertical axis shows the $P(\delta)$ as a percentage, calculated from the entire sample of the support point deviations.

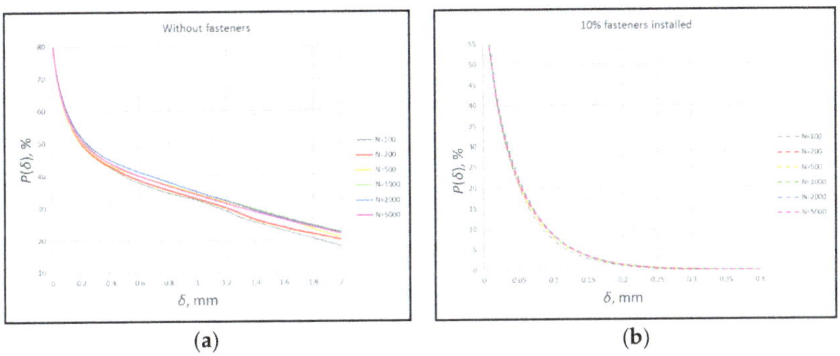

Figure 11. Statistical curves representing $P(\delta)$ for different sample size: (**a**) without fasteners, (**b**) with 10% fasteners installed.

The statistical curves without fasteners are slightly different (see Figure 11a) for different sample sizes, but after the fastener installation, this difference ceases to be noticeable starting from a sample size of 200 (Figure 11b).

Figure 12 illustrates the influence of random deviations of support points on the distribution of normal displacements of some nodes in the junction area. No fasteners are installed, but the wing and cruciform parts experience contact interaction. Figure 12a,b correspond to the case of deviation of one support point and Figure 12c to the case when two support points deviate.

As can be seen in Figure 12, despite the fact that the source variations are given by a Gaussian distribution, the deviations at the observation points have a completely different behavior. This is due to the fact that contact interactions are taken into account, which leads to the nonlinearity of the developed tolerance model. The linear model can change the parameters, but cannot change the structure of distributions (compare to Figures 8 and 9 in work [6]).

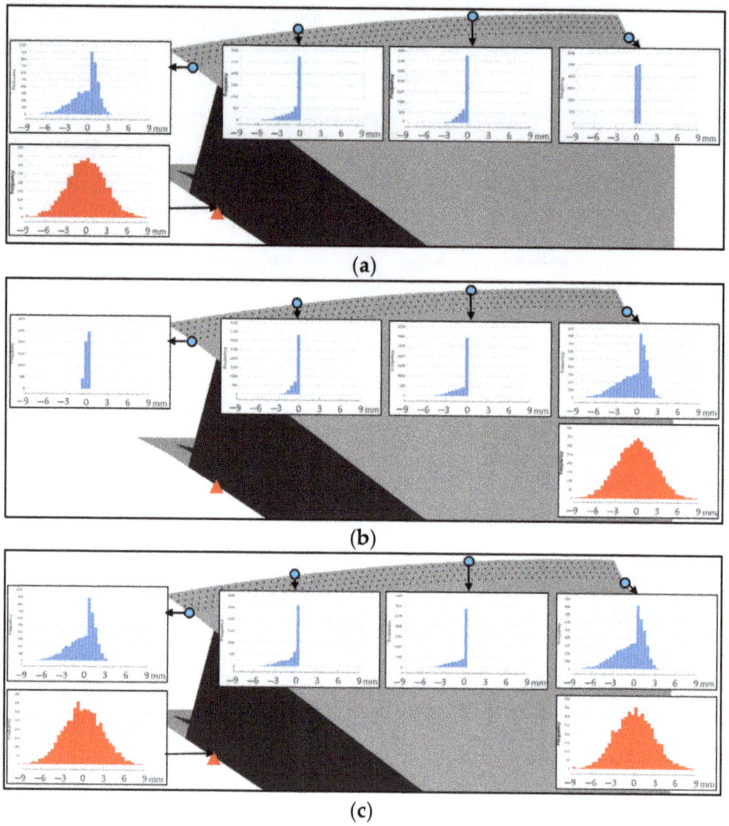

Figure 12. Normal displacements of some computational nodes: (a) in case of deviations of front support point position, (b) in case of deviations of rear support point position, (c) when two support points deviate.

3.5. Comparison of Gaps Caused by Deviations of Support Points with Generated Initial Gaps

The vector of nonuniform initial gaps between the assembled parts G can be used as an alternative source of deviations in tolerancing and variational analysis. It is of interest to compare two approaches to generating deviations during the assembly process. This comparison is the subject of the present section.

The initial gap between the assembled parts can be modeled as the homogeneous Gaussian random field $G(x, y)$ with mean μ_{RF}, standard deviation σ_{RF}, and exponential correlation function $\rho(x,\ y;\ \alpha_{RF}) = \exp(-\alpha_{RF}^2(x^2 + y^2)/2)$, where (x, y) is the local coordinate in the junction area [27]. The mean μ_{RF} represents the constant value of the initial gap in the junction area and is assumed to be zero. Thus, the modeled initial gap is defined by two parameters: the standard deviation σ_{RF} (which corresponds to the amplitude of the initial gap) and the correlation coefficient α_{RF} (which corresponds to the curvature of the initial gap) [28].

The parameters of the modeled initial gaps σ_{RF} and α were chosen in order to replicate the deviations in the vertical positions of support points. Figure 13 shows the result of statistical analysis for two groups consisting of 5000 samples of residual gaps each. The first group was obtained by the variation of vertical positions for all four support points ($\mu_j = 0$, $\sigma_j = 3$ mm) and the second group was built on the base of the generated initial gaps ($\mu_{RF} = 0$, $\sigma_{RF} = 2$ mm, $\alpha_{RF} = 0.000145$). As can be seen from Figure 13, the statistical curves for the two groups are quite close to each other, especially in the case of 10% of installed fasteners.

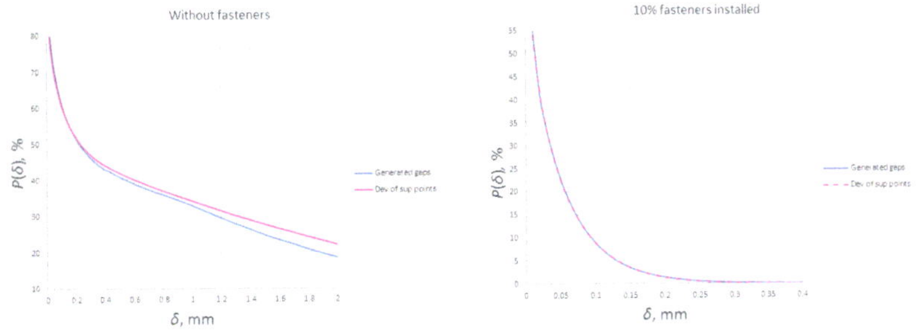

(a) (b)

Figure 13. Comparison of statistical curves for residual gaps caused by deviations in the position of wing support points and for generated initial gaps: (**a**) without fasteners, (**b**) with 10% of fasteners installed.

The same conclusion can be drawn from Figure 14, which shows four examples of the calculated residual gaps (with and without installed fasteners) based on the generated initial gaps. Comparing it with Figure 6, showing the gaps obtained by the deviation of support positions, one can see that the resulting distributions are very similar.

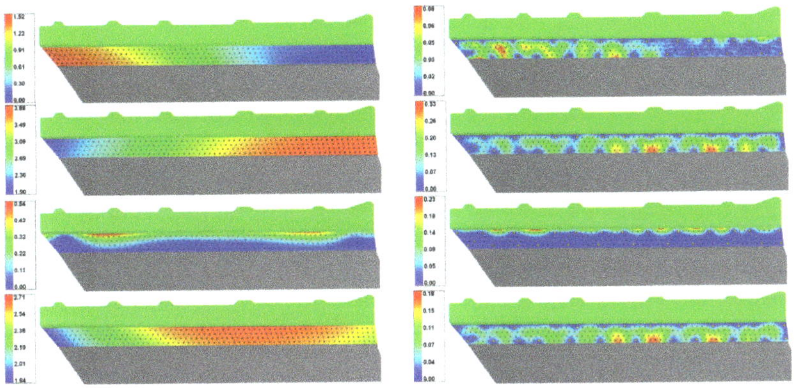

(a) (b)

Figure 14. Examples of calculated residual gaps based on the generated initial gaps: (**a**) without fasteners, (**b**) with 10% of fasteners installed.

Figure 15 presents the comparison distributions of maximal and mean values for the same groups of residual gap samples. As can be seen, the distributions of the mean and maximum gaps are quite close, both qualitatively and quantitatively. Therefore, the deviations in the shape of the assembled parts can be emulated with sufficient accuracy by setting the initial gap with proper settings.

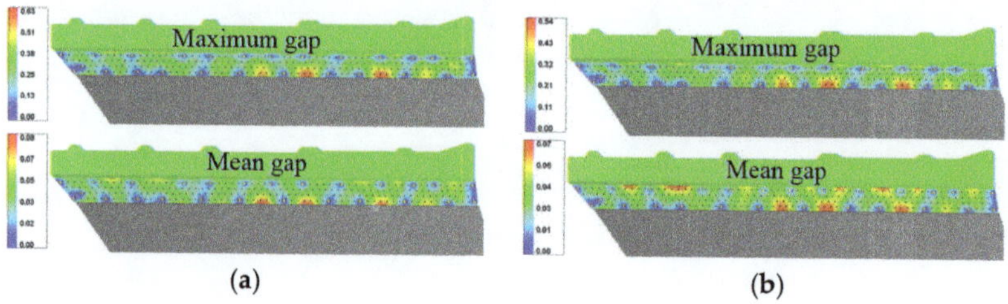

Figure 15. Maximal and mean gap with 10% fasteners installed: (**a**) in case of deviations of wing support point positions; (**b**) for generated initial gaps.

Figure 16 provides the so-called local statistics, which are the probability of obtaining a residual gap at a given point in the junction area within the prescribed range. The same two groups of gap samples are used. The distribution of local statistics provides important information for the assembly engineer, since it characterizes the quality of contact between the parts being connected at a given stage of assembly with regard to deviations of parts. In addition, local statistics make it possible to find weak points in the assembly technology under study, showing in which places the arrangement of fasteners may need to be changed.

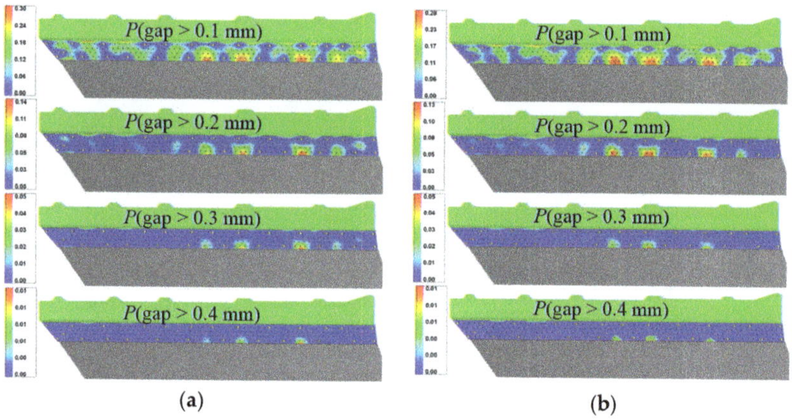

Figure 16. The probability that the gap is greater than a given number with 10% fasteners installed: (**a**) in case of deviations of wing support point positions, (**b**) for generated initial gaps.

Based on Figures 15 and 16, one can draw similar conclusions about the proximity of the final gaps obtained by different methods of generating deviations.

The analysis carried out in this section shows that positioning deviations can be adequately emulated by setting the initial gaps between assembled parts. However, selecting parameters for these initial gaps is a nontrivial task, the solution of which is impossible in the absence of information about the effect of positioning deviations.

4. Conclusions

The presented work continues the line of research started by the famous work of Liu and Hu [6], where the method of influence coefficients was proposed. The proposed generalization makes it possible to quickly and easily evaluate deviations in the resulting assembly and the quality of contact based on positioning deviations. At the same time, unlike classical MIC, the proposed method provides a mathematically correct solution to the contact problem that inevitably arises when modeling the assembly process.

Note that, along with positioning deviations, the quality of the final product is influenced by deviations in the assembled parts that appeared as a result of previous assembly operations. Such deviations can be modeled by setting a nonzero initial gap between parts, in particular according to the methods proposed in [34]. For building the most complete deviation model, these two approaches can be combined.

The main novelty of the proposed method is the full implementation of deviation modeling in the process of solving the contact problem. In this context, taking into account positioning deviations is not fundamentally different from taking into account the action of forces (for example, from fastening elements). The proposed technique allows for nonlinear tolerancing with a computation speed that is comparable to the speed of solving linear tolerancing problems of similar dimensions.

Note that the developed modeling methodology is not limited to the aerospace industry. It can be used in any industry where it is necessary to assemble structures from large, compliant parts. In particular, this takes place in the automotive industry, the assembly of cars for high-speed trains, shipbuilding, and other industries.

Author Contributions: Conceptualization, S.L.; methodology, S.L. and J.S.; software, M.P., N.Z. and M.T.; validation, M.P., J.S. and M.C.; formal analysis, M.P. and M.T.; investigation, J.S. and M.T.; resources, N.Z.; data curation, M.C.; writing—original draft preparation, S.L., M.P., M.T. and J.S.; writing—review and editing, N.Z. and M.C.; visualization, M.C.; supervision, S.L.; project administration, S.L. All authors have read and agreed to the published version of the manuscript.

Funding: The research was supported by Russian Science Foundation, project No. 22-19-00062, https://rscf.ru/en/project/22-19-00062/ (accessed on 15 January 2024).

Institutional Review Board Statement: Not applicable.

Data Availability Statement: The data presented in this study are available upon request to the authors.

Acknowledgments: The authors are grateful to Vasily Lupuleac for his valuable assistance in preparing the publication.

Conflicts of Interest: The authors declare no conflicts of interest.

References

1. Ceglarek, D.; Huang, W.; Zhou, S.; Ding, Y.; Kumar, R.; Zhou, Y. Time-Based Competition in Multistage Manufacturing: Stream-of-Variation Analysis (SOVA) Methodology—Review. *J. Flex. Manuf. Syst.* **2009**, *16*, 11–44. [CrossRef]
2. Hu, M.; Lin, Z.; Lai, X.; Ni, J. Simulation and analysis of assembly processes considering compliant, non-ideal parts and tooling variations. *Int. J. Mach. Tools Manuf.* **2001**, *41*, 2233–2243. [CrossRef]
3. Söderberg, R.; Lindkvist, L.; Wärmefjord, K.; Carlson, J.S. Virtual Geometry Assurance Process and Toolbox. *Procedia CIRP* **2016**, *43*, 3–12. [CrossRef]
4. Weber, A. Assembling the super jumbo. *Assembly* **2005**, *48*, 66–77.
5. Gao, J.; Chase, K.W.; Magleby, S.P. Comparison of Assembly Tolerance Analysis by the Direct Linearization and Modified Monte Carlo Simulation Methods. In Proceedings of the ASME Design Engineering Technical Conferences, Boston, MA, USA, 17–20 September 1995.
6. Liu, S.C.; Hu, S.J. Variation simulation for deformable sheet metal assemblies using finite element methods. *ASME J. Manuf. Sci. Eng.* **1997**, *119*, 368–374. [CrossRef]
7. Mounaud, M.; Thiebaut, F.; Bourdet, P.; Falgarone, H.; Chevassus, N. Assembly Sequence Influence on Geometric Deviations of Compliant Parts. *Int. J. Prod. Res.* **2010**, *49*, 1021–1043. [CrossRef]
8. Atik, H.; Chahbouni, M.; Amegouz, D.; Boutahari, S. Optimization tolerancing of surface in flexible parts and assembly: Influence Coefficient Method with shape defects. *Int. J. Eng. Technol.* **2017**, *7*, 90–94. [CrossRef]
9. Wang, Q.; Hou, R.; Li, J.; Ke, Y.; Maropoulos, P.G.; Zhang, X. Positioning variation modeling for aircraft panels assembly based on elastic deformation theory. *J Eng. Manuf.* **2018**, *232*, 2592–2604. [CrossRef]
10. Polini, W.; Corrado, A. Methods of influence coefficients to evaluate stress and deviation distribution of flexible assemblies—A review. *Int. J. Adv. Manuf. Technol.* **2020**, *107*, 2901–2915. [CrossRef]
11. Corrado, A.; Polini, W. Methods of influence coefficients to evaluate stress and deviation distributions of parts under operating conditions—A review. *Eng. Solid Mech.* **2021**, *9*, 41–54. [CrossRef]
12. Maltauro, M.; Passarotto, G.; Concheri, G.; Meneghello, R. Bridging the gap between design and manufacturing specifications for non-rigid parts using the influence coefficient method. *Int. J. Adv. Manuf. Technol.* **2023**, *127*, 579–597. [CrossRef]

13. Warmefjord, K.; Lindkvist, L.; Söderberg, R. Tolerance simulation of compliant sheet metal assemblies using automatic node-based contact detection. In Proceedings of the ASME International Mechanical Engineering Congress and Exposition, Boston, MA, USA, 2–6 November 2008.
14. Söderberg, R.; Wärmefjord, K.; Lindkvist, L.; Berlin, R. The influence of spot weld position variation on geometrical quality. *CIRP Ann.* **2012**, *61*, 13–16. [CrossRef]
15. Wärmefjord, K.; Söderberg, R.; Lindkvist, L. Simulation of the effect of geometrical variation on assembly and holding forces. *Int. J. Prod. Dev.* **2013**, *18*, 88–108. [CrossRef]
16. Jareteg, C.; Wärmefjord, K.; Cromvik, C.; Söderberg, R.; Lindkvist, L.; Carlson, J.; Larsson, S.; Edelvik, F. Geometry assurance integrating process variation with simulation of spring-in for composite parts and assemblies. In Proceedings of the ASME International Mechanical Engineering Congress and Exposition, Montreal, QC, Canada, 14–20 November 2014.
17. Lorin, S.; Lindkvist, L.; Söderberg, R. Variation simulation of stresses using the method of influence coefficients. *J. Comput. Inf. Sci. Eng.* **2014**, *14*, 011001. [CrossRef]
18. Söderberg, R.; Wärmefjord, K.; Lindkvist, L. Variation simulation of stress during assembly of composite parts. *CIRP Ann.* **2015**, *64*, 17–20. [CrossRef]
19. Lupuleac, S.; Kovtun, M.; Rodionova, O.; Marguet, B. Assembly simulation of riveting process. *SAE Int. J. Aerosp.* **2010**, *2*, 193–198. [CrossRef]
20. Wriggers, P. *Computational Contact Mechanics*, 2nd ed.; Springer: Berlin/Heidelberg, Germany, 2006; 518p.
21. Goldfarb, D.; Idnani, A. A numerically stable dual method for solving strictly convex quadratic programs. *Math. Program.* **1983**, *27*, 1–33. [CrossRef]
22. Burton, D.; Toint, P.L. On an instance of the inverse shortest paths problem. *Math. Program.* **1992**, *53*, 45–61. [CrossRef]
23. Powell, M. On the Quadratic Programming Algorithm of Goldfarb and Idnani. In *Mathematical Programming Essays in Honor of George B. Dantzig Part II*; Springer: Berlin/Heidelberg, Germany, 1985; pp. 46–61.
24. Gondzio, J.; Grothey, A. Direct solution of linear systems of size 109 arising in optimization with interior point methods. In *Parallel Processing and Applied Mathematics*; PPAM 2005; Lecture Notes in Computer Science; Wyrzykowski, R., Dongarra, J., Meyer, N., Waśniewski, J., Eds.; Springer: Berlin/Heidelberg, Germany, 2006; Volume 3911.
25. Stefanova, M.; Baklanov, S. The Relative Formulation of the Quadratic Programming Problem in the Aircraft Assembly Modeling. In *Optimization and Applications*; OPTIMA 2022; Lecture Notes in Computer Science; Olenev, N., Evtushenko, Y., Jaćimović, M., Khachay, M., Malkova, V., Pospelov, I., Eds.; Springer: Cham, Switzerland, 2022; Volume 13781.
26. Bertsekas, D. Projected Newton methods for optimization problems with simple constraints. *SIAM J. Control. Optim.* **1982**, *20*, 221–246. [CrossRef]
27. Lupuleac, S.; Zaitseva, N.; Stefanova, M.; Berezin, S.; Shinder, J.; Petukhova, M.; Bonhomme, E. Simulation of the wing-to-fuselage assembly process. *J. Manuf. Sci. Eng.* **2019**, *141*, 061009. [CrossRef]
28. Lupuleac, S.; Zaitseva, N.; Petukhova, M.; Shinder, Y.; Berezin, S.; Khashba, V.; Bonhomme, E. *Combination of Experimental and Computational Approaches to A320 Wing Assembly*; SAE Technical Paper 2017-01-2085; SAE International: Warrendale, PA, USA, 2017. [CrossRef]
29. Lupuleac, S.; Shinder, J.; Churilova, M.; Zaitseva, N.; Khashba, V.; Bonhomme, E.; Montero-Sanjuan, P. *Optimization of Automated Airframe Assembly Process on Example of a350 s19 Splice Joint*; SAE Technical Paper 2019-01-1882; SAE International: Warrendale, PA, USA, 2019. [CrossRef]
30. Lupuleac, S.; Petukhova, M.; Shinder, Y.; Stefanova, M.; Zaitseva, N.; Pogarskaia, T.; Bonhomme, E. *Software Complex for Simulation of Riveting Process: Concept and Applications*; SAE Technical Paper 2016-01-2090; SAE International: Warrendale, PA, USA, 2016. [CrossRef]
31. Turner, M.J.; Clough, R.W.; Martin, H.C.; Topp, L.J. Stiffness and Deflection Analysis of Complex Structures. *J. Aeronaut. Sci.* **1956**, *23*, 805–823. [CrossRef]
32. Mehrotra, S. On the implementation of a primal-dual interior point method. *SIAM J. Optim.* **1992**, *2*, 575–601. [CrossRef]
33. Lemke, C.E. On complementary pivot theory. *Math. Decis. Sci.* **1968**, *1*, 95–114.
34. Zaitseva, N.; Lupuleac, S.; Khashba, V.; Shinder, Y.; Bonhomme, E. Approaches to initial gap modeling in final aircraft assembly simulation. In Proceedings of the ASME International Mechanical Engineering Congress and Exposition, Virtual, 16–19 November 2020.
35. Scholz, F.W. Tolerance Stack Analysis Methods, a Critical Review. Available online: http://www.stat.washington.edu/fritz/DATAFILES/Rtolerancing/TOLSTACK.pdf (accessed on 15 January 2024).

axioms

Article

Calculation of Thermodynamic Quantities of 1D Ising Model with Mixed Spin-$(s,(2t-1)/2)$ by Means of Transfer Matrix

Hasan Akın [1,2]

1 Department of Mathematics, Art and Science Faculty, Harran University, Sanliurfa 63050, Turkey; akinhasan@harran.edu.tr or hakin@ictp.it
2 The Abdus Salam International Centre for Theoretical Physics (ICTP), Strada Costiera, 11, 34151 Trieste, Italy

Abstract: In this paper, we consider the one-dimensional Ising model (shortly, 1D-MSIM) having mixed spin-$(s,(2t-1)/2)$ with the nearest neighbors and the external magnetic field. We establish the partition function of the model using the transfer matrix. We compute certain thermodynamic quantities for the 1D-MSIM. We find some precise formulas to determine the model's free energy, entropy, magnetization, and susceptibility. By examining the iterative equations associated with the model, we use the cavity approach to investigate the phase transition problem. We numerically determine the model's periodicity.

Keywords: one-dimensional Ising model with mixed spin; free energy; entropy; magnetization; susceptibility; phase transition

MSC: 37A25; 37A35; 28D05; 28D20; 37A44

check for updates

Citation: Akın, H. Calculation of Thermodynamic Quantities of 1D Ising Model with Mixed Spin-$(s,(2t-1)/2)$ by Means of Transfer Matrix. *Axioms* **2023**, *12*, 880. https://doi.org/10.3390/axioms12090880

Academic Editor: Leonid Plotnikov

Received: 16 August 2023
Revised: 7 September 2023
Accepted: 11 September 2023
Published: 14 September 2023

1. Introduction

Lenz introduced the one-dimensional (1D) classical spin-1/2 Ising model [1]. Compared to the spin-1/2 model, the spin-1 Ising model, also known as the Blume–Capel model, is more appropriate; hence, it was employed in the investigation of phase transitions in systems of three states [2,3]. Then, the spin-3/2 Ising model was used to extend the spin-1 Ising model's conclusions [4,5].

The magnetic characteristics of mixed-spin systems have recently attracted both theoretical and experimental attention [6–8]. These systems are especially well suited for studying the magnetic properties of a particular class of magnetic materials, in addition to some technical applications [9]. Therefore, mixed-spin Ising models have been extensively studied in the literature lately [10]. In Ref. [11], we classified the disordered phases corresponding to the Ising model with spin-1 and spin 1/2 on the semi-finite Cayley tree (CT). Then, on the second-order CT, we examined the phase transition problem of the Ising model with spin-2 and spin 1/2. Furthermore, we discussed the chaotic behavior of the corresponding dynamical system [12]. The main methods for examining the properties of Ising models with the mixed spin on the square lattice [10], Bethe lattice [13], and CT [11,12] are Monte Carlo simulations [10], the cavity method [14–16], the Kolmogorov consistency method [11,12], the iteration method, and transfer matrix method. In [17], the magneto thermal parameters that characterize the magnetocaloric effect (MCE) behaviors, such as entropy, entropy change, and adiabatic cooling rate, are precisely calculated using the transfer matrix approach.

In the literature, there are numerous methods for determining the free energy of a given model [13,18]. The free energy of the lattice models on the Bethe lattice is calculated while boundary conditions are taken into consideration [13,18,19]. In-depth research has been performed on the thermodynamic properties of the 1D Ising model with single spin-s using the transfer matrix [20–22].

The author is not aware of any research on the thermodynamic properties of 1D Ising models with mixed spin. Due to its widespread applications, the matrix transfer approach is one of the most commonly used methods in statistical mechanics. In [21], the transfer matrix approach is considered to conduct an analytical study of the 1D Ising model with spin-s. To grasp physical issues as well as statistical mechanics, it is essential to compute thermodynamic quantities like free energy, entropy, magnetization, and magnetic susceptibility [22].

In this paper, to determine the thermodynamic quantities related to 1D-MSIM with the mixed spin-$(s,(2t-1)/2)$, we consider the transfer matrix approach. We enlarge a specific bond and establish the transfer matrix that corresponds to the bond from site j to site $j+1$. By considering the trace of several transfer matrices multiplied simultaneously, the partition function is reconstructed. Inspired by the results given for the 1D Ising model with the single spin-s [21], the partition function and the free energy of 1D-MSIM with mixed spin-$(s,\frac{(2t-1)}{2})$ having the nearest neighbor interactions and external field are calculated. To calculate the model's free energy, entropy, magnetization, and susceptibility, some exact formulas are established via the corresponding transfer matrix. By using the cavity approach to the corresponding iterative equations of the model, the phase transition issue is studied. The periodicity of the model is also estimated, numerically.

2. Preliminaries

Here, we review definitions and key findings in the construction of the partition function for the 1D-MSIM with mixed spin-$(s,\frac{(2t-1)}{2})$ on the one-dimensional lattice $\mathbb{Z}$. We denote the set of integers by $\mathbb{Z}$ and the set of strictly positive integers by $\mathbb{N}^+$. Two vertices x and y, $x,y\in\mathbb{N}^+$ are called ***nearest-neighbor (NN)*** if there exists an edge $\ell\in L$ connecting them, which is denoted by $\ell=\langle x,y\rangle$. For $x,y\in\mathbb{N}^+$, the distance $d(x,y)$ is the length (the number of edges) of the shortest path connecting x with y. For any $x\in\mathbb{N}^+$, the direct successor of the vertex x is defined by $S(x)=\{y\in\mathbb{N}^+ : d(x,y)=1\}$.

Denote the set of even natural numbers by $\mathbb{E}=\{2n : n\in\mathbb{N}\}$ and the set of odd natural numbers by $\mathbb{O}=\{2n-1 : n\in\mathbb{N}^+\}$. We consider the mixed spin-state spaces denoted by $\Phi=\{\pm s : s\in\mathbb{Z}^+\}\cup\{0\}$ and $\Psi=\{\pm\frac{(2t-1)}{2} : t\in\mathbb{Z}^+\}$. For $A\subset\mathbb{N}^+$, a spin configuration ξ_A on A is defined as a function

$$x\in A\to\xi_A(x)=\begin{cases}\xi_A(x)=\sigma(x)\in\Phi, & x\in\mathbb{E}\\ \xi_A(x)=s(x)\in\Psi, & x\in\mathbb{O}.\end{cases}$$

Let us consider finite set $\mathbb{N}_{2N+1}=\{1,2,\cdots,2N+1\}$. Let $\Omega_E=\Phi^{\mathbb{E}}$ and $\Omega_O=\Psi^{\mathbb{O}}$ be the spaces of infinite configurations and $\Omega_{+,N}=\Phi^{\mathbb{E}\cap\mathbb{N}_{2N+1}}$ and $\Omega_{-,N}=\Psi^{\mathbb{O}\cap\mathbb{N}_{2N+1}}$ be the spaces of finite configurations. $\Xi=\Omega_E\times\Omega_O$ represents the configuration space. An element of Ω_E is denoted by $\sigma(x)$, for $x\in\mathbb{E}$. Similarly, an element of Ω_O is denoted by $s(x)$, for $x\in\mathbb{O}$.

In this paper, the spins in Φ and Ψ can be placed on the odd-numbered vertices and the even-numbered vertices, respectively, when mixed spins are designed on the lattice $\mathbb{N}^+$ (see Figure 1).

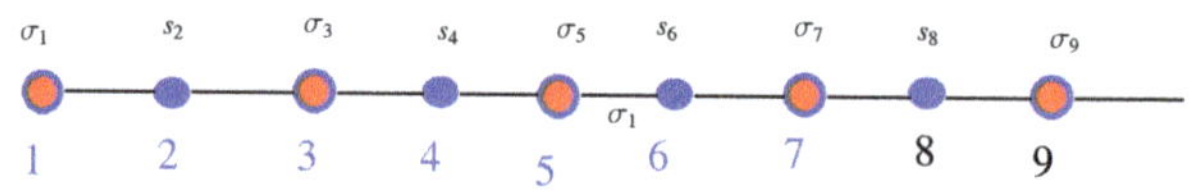

Figure 1. Configurations with a mixed spin on a one-dimensional finite block, where $\sigma_{2N-1}\in\Phi$ and $s_{2N}\in\Psi$ for $N\in\mathbb{N}^+$.

Construction of the Partition Function Associated with the Model

Let us construct the partition function corresponding to 1D-MSIM with mixed spin $(s,\frac{2t-1}{2})$ on the lattice $\mathbb{N}^+$.

Let $\xi \in \Xi$, so that one has

$$\xi(x) = \begin{cases} \sigma(x); & x \in \mathbb{E} \\ s(x); & x \in \mathbb{O}, \end{cases}$$

where $\sigma \in \Phi$ and $s \in \Psi$.

Fix a finite volume $\Lambda \subset \mathbb{N}^+$. Let $\mathcal{E} = \{\ell = \langle x,y \rangle : d(x,y) = 1, x,y \in \mathbb{N}^+\}$ be the set of nearest-neighbor edges with at least one endpoint in $\Lambda \subset \mathbb{N}^+$. We examine in detail the 1D-MSIM with mixed spin $(s, \frac{2t-1}{2})$ built on the lattice $\mathbb{N}^+$, having the Hamiltonian

$$H(\xi) = -\sum_{\langle x,y \rangle \in \mathcal{E}} J\xi(x)\xi(y) - \sum_{y \in \Lambda} h_{\xi(y)}\xi(y), \tag{1}$$

where the energy of each link tying up nearby sites is represented by the first sum, and the energy of each site is represented by the second sum.

The partial partition functions associated with the Hamiltonian (1) can be found as follows:

$$Z^{(2N+1)}_{\xi(x^{(0)})}(\beta, h_{\xi(x^{(0)})}) = \sum_{\eta_N \in \Xi^{\mathbb{N}_{2N+1}}} \exp\{-\beta H(\eta_N)\}, \tag{2}$$

where $\xi \in \sigma \times s = \Xi$, $\beta = \frac{1}{k_B T}$ and J is the coupling constant.

Given the finite set $\mathbb{N}_{2N+1}$, by considering the boundary spins, or the spins on the $(2N+1)$-th level, we can construct the summation of Equation (2). Consequently, we state that the density-free energy function is

$$F(\beta, h_{\xi(x)}) = -k_B T \lim_{N \to \infty} \frac{1}{(2N+1)} \ln Z^{(2N+1)}(\beta, h_{\xi(x)}), \tag{3}$$

where $Z^{(2N+1)}(\beta, h_{\xi(x)})$ is the total partition function.

As a result of the derivatives of the free energy function with respect to certain parameters, many authors have studied other thermodynamic properties [21,22]. If the largest eigenvalue of the transfer matrix corresponding to the given model is $\lambda_{\max}$, the thermodynamic quantities of this system are obtained as follows:

Bulk free energy

$$F(T, \beta h_{\xi(x)}) = -k_B T \ln \lambda_{\max}. \tag{4}$$

Entropy

$$S(T, \beta h_{\xi(x)}) = -\frac{\partial F(T, \beta h_{\xi(x)})}{\partial T}. \tag{5}$$

Internal energy

$$U(T, \beta h_{\xi(x)}) = -\frac{\partial \ln \lambda_{\max}}{\partial \beta}. \tag{6}$$

3. Construction of the Partition Function for the Model via Transfer Matrices

The Kramers–Wannier transfer technique may be used to construct the partition function. This method requires building a transfer matrix and determining its eigenvalues.

3.1. The Partition Function and the Boltzmann Weight

It is well known that the total partition function $Z^{(2N+1)}(\beta, h_{\xi(x)}) = Tr(-\beta H_{2N+1})$ depends on the particular realization $\{J_i, h_i\}$ (see [21,22] for details). Also, it is well known that the total partition function is equal to the Boltzmann weight's sum over all possible states. Here, we consider blocks or configurations with $(2N+1)$ length, and under zero boundary conditions, that is, the spins designed for $m > 2N+1$ are regarded as $\xi(m) = 0$.

If we want to rewrite the Hamiltonian (1) for all configurations on the configuration space Ξ^{2N+1}, we can obtain the Boltzmann weight in the form:

$$e^{-\beta H(\eta)+\sum\limits_{x\in\mathbb{N}_{2N+1}}\eta(x)h_{\eta(x)}} = e^{-\beta H(\sigma)} \tag{7}$$

From (1), for $i = 1,2,\cdots,N$, we denote the energy of the bond between sequential vertices $2i-1$ and $2i$ by

$$E(\sigma_{2i-1}, s_{2i}) = -\sum_{\langle 2i-1,2i\rangle\in\mathcal{E}} J\sigma_{2i-1}s_{2i} - \sum_{\langle 2i-1,2i\rangle\in\mathcal{E}} h_{s_{2i}}s_{2i}. \tag{8}$$

Similarly, for $i = 1,2,\cdots,N$ we denote the energy of the bond between sequential vertices $2i$ and $2i+1$ by

$$\widetilde{E}(\sigma_{2i}, s_{2i+1}) = -\sum_{\langle 2i,2i+1\rangle\in\mathcal{E}} Js_{2i}\sigma_{2i+1} - \sum_{\langle 2i,2i+1\rangle\in\mathcal{E}} \widetilde{h}_{\sigma_{2i+1}}\sigma_{2i+1}. \tag{9}$$

Note that, as can be seen in Equations (8) and (9), for $i > 1$, while the oscillating magnetic fields $h_{\sigma_{2i-1}}$ and $\widetilde{h}_{s_{2i}}$ fluctuate depending on where the vertex is located, the coupling constant J, which determines the energy between two successive vertices, remains constant.

It is well known that the partition function is equal to the sum of the Boltzmann weight $e^{-\beta H(\eta)}$ over all possible states in the lattice $\mathbb{N}$ [21,22]. Therefore, for the configuration $\eta \in \Xi^{2N+1}$, we can write the Boltzmann weight $e^{-\beta H(\sigma)}$ in the form

$$\begin{aligned} e^{-\beta H(\eta)} &= e^{-\beta H(\sigma_1,s_2,\sigma_3,\cdots,\sigma_{2N-1},s_{2N-1},\sigma_{2N+1})} \\ &= e^{-\beta E(\sigma_1,s_2)-\beta\widetilde{E}(s_2,\sigma_3)-\cdots-\beta E(\sigma_{2N-1},s_{2N})-\beta\widetilde{E}(s_{2N},\sigma_{2N+1})} \\ &= \left(e^{-\beta E(\sigma_1,s_2)}\right)\left(e^{-\beta\widetilde{E}(s_2,\sigma_3)}\right)\cdots\left(e^{-\beta E(\sigma_{2N-1},s_{2N})}\right)\left(e^{-\beta\widetilde{E}(s_{2N},\sigma_{2N+1})}\right). \end{aligned} \tag{10}$$

For this system, the canonical partition function can be written as

$$Z_{2N+1} = \sum_{\eta\in\Xi^{2N+1}} e^{-\beta H(\eta)} = \sum_{\eta\in\Xi^{2N+1}} e^{-\beta\left(\sum_{i=1}^{N} E(\sigma_{2i-1},s_{2i})+\sum_{i=1}^{N}\widetilde{E}(\sigma_{2i},s_{2i+1})\right)}. \tag{11}$$

The next step is to decompose the bonds using the bond representation and factor the Boltzmann weights into pairwise factors.

3.2. The Transfer Matrices

Here, we construct the transfer matrices corresponding to 1D-MSIM with the $(s, \frac{(2t-1)}{2})$ mixed spin on the lattice $\mathbb{N}^+$. We obtain two different transfer matrices. First, let us place the spins of the set Φ on the odd-numbered vertices of the lattice $\mathbb{N}^+$, and the spins in the set Ψ on the even-numbered vertices. So, we define the entries of the transfer matrix $\mathbb{P}$ as follows:

$$P_{\sigma_{2i-1}s_{2i}} = \exp\left(\sigma_{2i-1}s_{2i}\beta J + \beta s_{2i}\widetilde{h}_{s_{2i}}\right) = e^{-\beta E(\sigma_{2i-1},s_{2i})}, \tag{12}$$

where $\sigma_{2i-1} \in \Phi$ and $s_{2i} \in \Psi$ for $i = 1,2,\cdots$.

Similarly, we can define the entries of the transfer matrix $\mathbb{Q}$ by

$$Q_{s_{2i}\sigma_{2i+1}} = \exp\left(s_{2i}\sigma_{2i+1}\beta J + \beta\sigma_{2i+1}h_{\sigma_{2i+1}}\right) = e^{-\beta\widetilde{E}(s_{2i},\sigma_{2i+1})}, \tag{13}$$

where $s_{2i} \in \Psi$ and $\sigma_{2i+1} \in \Phi$ for $i = 1,2,\cdots$.

From (10)–(13), we obtain

$$e^{-\beta H(\eta)} = P_{\sigma_1 s_2} Q_{s_2\sigma_3} P_{\sigma_3 s_4} Q_{s_4\sigma_5} \cdots P_{\sigma_{2N-1}s_{2N}} Q_{s_{2N}\sigma_{2N+1}}. \tag{14}$$

One can easily see that the matrix $\mathbb{P}$ has Card$(\Phi)\times$Card(Ψ) dimensions while matrix $\mathbb{Q}$ has Card$(\Psi)\times$Card(Φ) dimensions. Assume that $\mathbb{P}\mathbb{Q} = \mathbb{M}$. The total partition function can be written as the trace of the product of $2N+1$ random transfer matrices.

3.3. 1D-MSIM with Mixed Spin-$(1, 1/2)$

Due to the difficulty of computation of the transfer matrix and its eigenvalues for $s > 1$ and $t > 1$, we restrict ourselves to the mixed spin $(1, 1/2)$. In this subsection, we compute the thermodynamic quantities of 1D-MSIM with mixed spin $(1, 1/2)$. So, we consider the mixed spins as $\Phi = \{-1, 0, 1\}$ and $\Psi = \{-\frac{1}{2}, \frac{1}{2}\}$.

Firstly, let us construct the transfer matrix as $\mathbb{P} = (P_{\sigma_{2i-1}, s_{2i}})$, for $\sigma_{2i-1} \in \{-1, 0, 1\}$ and $s_{2i} \in \{-\frac{1}{2}, \frac{1}{2}\}$. So, we have

$$\mathbb{P} = \begin{pmatrix} e^{\frac{\beta}{2}\left(J-h_{-\frac{1}{2}}\right)} & e^{-\frac{\beta}{2}\left(J-h_{\frac{1}{2}}\right)} \\ e^{-\frac{1}{2}\beta h_{-\frac{1}{2}}} & e^{\frac{1}{2}\beta h_{\frac{1}{2}}} \\ e^{-\frac{\beta}{2}\left(J+h_{-\frac{1}{2}}\right)} & e^{\frac{\beta}{2}\left(J+h_{\frac{1}{2}}\right)} \end{pmatrix}. \tag{15}$$

Secondly, let us place the spins of the set $\{-\frac{1}{2}, \frac{1}{2}\}$ on the even-numbered vertices of the lattice $\mathbb{N}^+$, and the spins in the set $\{-1, 0, 1\}$ on the odd-numbered vertices (see Figure 1). We can define the second transfer matrix $\mathbb{Q} = (Q_{s_{2i}, \sigma_{2i+1}})$, for $s_{2i} \in \{-\frac{1}{2}, \frac{1}{2}\}$ and $\sigma_{2i+1} \in \{-1, 0, 1\}$ as

$$\mathbb{Q} = \begin{pmatrix} e^{\frac{\beta}{2}(J-2h_{-1})} & 1 & e^{-\frac{\beta}{2}(J-2h_1)} \\ e^{-\frac{\beta}{2}(J+2h_{-1})} & 1 & e^{\frac{\beta}{2}(J+2h_1)} \end{pmatrix}. \tag{16}$$

3.4. Investigation of Thermodynamic Quantities in the Translation-Invariant Case

Assume that $h = \widetilde{h}_j = h_i$ for all $j \in \{-\frac{1}{2}, \frac{1}{2}\}$ and $i \in \{-1, 0, 1\}$. This case is called the translation-invariant property. Thus, the transfer matrices $\mathbb{P}$ and $\mathbb{Q}$ given in (15) and (16) are obtained as follows

$$\mathbb{P} = \begin{pmatrix} e^{\frac{\beta}{2}(J-h)} & e^{-\frac{\beta}{2}(J-h)} \\ e^{-\frac{1}{2}\beta h} & e^{\frac{1}{2}\beta h} \\ e^{-\frac{\beta}{2}(J+h)} & e^{\frac{\beta}{2}(J+h)} \end{pmatrix}, \tag{17}$$

$$\mathbb{Q} = \begin{pmatrix} e^{\frac{\beta}{2}(J-2h)} & 1 & e^{-\frac{\beta}{2}(J-2h)} \\ e^{-\frac{\beta}{2}(J+2h)} & 1 & e^{\frac{\beta}{2}(J+2h)} \end{pmatrix}. \tag{18}$$

Case I

When we multiply the transfer matrices $\mathbb{P}$ and $\mathbb{Q}$ given in Equations (17) and (18), we obtain the square matrix:

$$\mathbb{M} = \mathbb{P}\mathbb{Q} = \begin{pmatrix} e^{-\frac{1}{2}(3h-2J)\beta} + e^{-\frac{1}{2}(h+2J)\beta} & e^{-\frac{1}{2}(J-h)\beta} + e^{\frac{1}{2}(J-h)\beta} & e^{\frac{h\beta}{2}}\left(1+e^{h\beta}\right) \\ e^{-\frac{1}{2}(3h-J)\beta} + e^{-\frac{1}{2}(h+J)\beta} & e^{-\frac{h\beta}{2}}\left(1+e^{h\beta}\right) & e^{\frac{1}{2}(h-J)\beta} + e^{\frac{1}{2}(3h+J)\beta} \\ e^{-\frac{3h\beta}{2}}\left(1+e^{h\beta}\right) & e^{-\frac{1}{2}(h+J)\beta} + e^{\frac{1}{2}(h+J)\beta} & e^{\frac{1}{2}(h-2J)\beta} + e^{\frac{1}{2}(3h+2J)\beta} \end{pmatrix}. \tag{19}$$

After some tedious and complicated algebraic manipulations, the eigenvalues of the matrix $\mathbb{M}$ are obtained as:

$$\begin{cases} \lambda_1 = 0, \\ \lambda_2 = \frac{1}{2e^{\frac{\beta}{2}(3h+2J)}}\left(A(\beta,h,J) - \sqrt{(A(\beta,h,J))^2 - 4B(\beta,h,J)}\right), \\ \lambda_3 = \frac{1}{2e^{\frac{\beta}{2}(3h+2J)}}\left(A(\beta,h,J) + \sqrt{(A(\beta,h,J))^2 - 4B(\beta,h,J)}\right), \end{cases}$$

where

$$A(\beta,h,J) = \left(1+e^{h\beta}\right)\left(e^{h\beta} + e^{2J\beta} + e^{h\beta+J\beta} - e^{h\beta+2J\beta} + e^{2h\beta+2J\beta}\right),$$
$$B(\beta,h,J) = e^{2h\beta}\left(e^{J\beta}-1\right)^2\left(e^{J\beta}\left(1+e^{2h\beta}\right) + e^{h\beta}\left(1+e^{J\beta}\right)^2\right).$$

For the sake of simplicity, substitute the variables $\theta = e^{\frac{\beta J}{2}}$ and $v = e^{\frac{\beta h}{2}}$ to obtain

$$\begin{cases} \lambda_1 = 0, \\ \lambda_2 = \frac{(1+v^2)(\theta^4+v^2(1+\theta^2-\theta^4)+v^4\theta^4) - \sqrt{((1+v^2)(\theta^4+v^2(1+\theta^2-\theta^4)+v^4\theta^4))^2 - 4v^4(\theta^2-1)^2(\theta^2(1+v^4)+v^2(1+\theta^2)^2)}}{2v^3\theta^2}, \\ \lambda_3 = \frac{(1+v^2)(\theta^4+v^2(1+\theta^2-\theta^4)+v^4\theta^4) + \sqrt{((1+v^2)(\theta^4+v^2(1+\theta^2-\theta^4)+v^4\theta^4))^2 - 4v^4(\theta^2-1)^2(\theta^2(1+v^4)+v^2(1+\theta^2)^2)}}{2v^3\theta^2}. \end{cases} \tag{20}$$

By computing the trace of the matrix product $\mathbb{M}$, we can determine the partition function as

$$Z^{(2N+1)}(\beta, h_{\xi(x)}) = Tr[M^N] = \sum_{i=1}^{3} \lambda_i^N, \tag{21}$$

where λ_i denotes the eigenvalues of the transfer matrix $\mathbb{M}$ for $i = 1,2,3$.

From (11), we obtain

$$Z^{(2N+1)}(\beta, h_{\xi(x)}) = Tr(\mathbb{M}^N) = \sum_{j=1}^{t} \lambda_{t,j}^N = \lambda_{\max}^N\left(1 + \sum_{j=1}^{t-1}\left(\frac{\lambda_{t,j}}{\lambda_{\max}}\right)^N\right), \tag{22}$$

where $\lambda_{t,j}$ denote the eigenvalues of transfer matrix $\mathbb{M}$.

So, one has $Z^{(2N+1)}(\beta, h_{\xi(x)}) = \lambda_{\max}^N$ for $N \to \infty$. As it is well known, the model's critical behavior manifests itself in the thermodynamic limit as $N \to \infty$; therefore, the largest eigenvalue $\lambda_{\max} = \lambda_3$ of the transfer matrix $\mathbb{M}$ given in (19) stands as the only indicator of the bulk free energy:

$$\begin{aligned} F(\beta,h) &= -\frac{1}{\beta}\lim_{N\to\infty}\frac{1}{(2N+1)}\ln Z^{(2N+1)}(\beta,h) \\ &= -\frac{1}{\beta}\lim_{N\to\infty}\frac{1}{(2N+1)}\ln\lambda_3^N = -\frac{1}{2\beta}\ln\lambda_3 \\ &= -\frac{1}{2\beta}\ln\left(\frac{A(\beta,h,J) + \sqrt{(A(\beta,h,J))^2 - 4B(\beta,h,J)}}{2e^{\frac{\beta}{2}(3h+2J)}}\right). \end{aligned} \tag{23}$$

3.5. *Behavior of the Thermodynamic Quantities of 1D-MSIM with Mixed Spin-(s,(2t − 1)/2) in the Absence of a Magnetic Field*

For $h = 0$, we obtain the transfer matrix $\mathbb{M}$ given in (19) as

$$\widehat{\mathbb{M}} = \begin{pmatrix} 2\cosh(J\beta) & 2\cosh\left(\frac{J\beta}{2}\right) & 2 \\ 2\cosh\left(\frac{J\beta}{2}\right) & 2 & 2\cosh\left(\frac{J\beta}{2}\right) \\ 2 & 2\cosh\left(\frac{J\beta}{2}\right) & 2\cosh(J\beta) \end{pmatrix}. \tag{24}$$

The set of eigenvalues of the matrix $\widehat{\mathbb{M}}$ is obtained as follows:

$$\{\widehat{\lambda}_1 = 0, \widehat{\lambda}_2 = -2 + 2\cosh(J\beta), \widehat{\lambda}_3 = 4 + 2\cosh(J\beta)\}.$$

It is clear that $\widehat{\lambda}_3 = \widehat{\lambda}_{\max}$. From the formula in (23), we obtain

$$F(\beta, 0) = -\frac{1}{2\beta}\ln(4 + 2\cosh(J\beta)). \tag{25}$$

Figure 2 shows the graph of free energy $F(\beta, 0)$ given in (25) as a function of β for $J = 0.2$. From Figure 2, it can be seen that the free energy is approaching zero for $T \to 0^+$. In Figure 2, the oscillating magnetic fields $h_{s_{2i}}$ and $\widetilde{h}_{\sigma_{2i+1}}$ are assumed to be zero. If $h_{s_{2i}}$ and $\widetilde{h}_{\sigma_{2i+1}}$ are taken to be nonzero, then, by using the parametric representation, the free energy function can be plotted in the form $h \to (\beta(h), F(h))$.

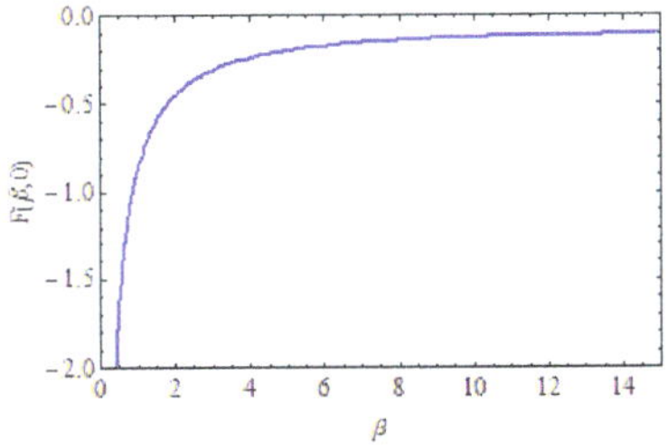

Figure 2. The graph of free energy $F(\beta, 0)$ given in (25) as a function of β for $J = 0.2$.

In the absence of an external magnetic field, from (5), the entropy of the model is obtained as

$$\begin{aligned} S(\beta, 0) &= -\frac{\partial F(\beta, 0)}{\partial T} = -\frac{\frac{\partial F(\beta,0)}{\partial \beta}}{\frac{\partial T}{\partial \beta}} \\ &= \frac{(-2\ln(4 + 2\cosh(J\beta)) - \cosh(J\beta)\ln(4 + 2\cosh(J\beta)) + J\beta\sinh(J\beta))k_B}{2(2 + \cosh(J\beta))}. \end{aligned} \tag{26}$$

Similarly, from (6), we obtain the internal energy as

$$U(\beta, 0) = \frac{-2\ln(4 + 2\cosh(J\beta)) - \cosh(J\beta)\ln(4 + 2\cosh(J\beta)]) + J\beta\sinh(J\beta])}{2\beta^2(2 + \cosh(J\beta))}. \tag{27}$$

The changes in entropy of 1D-MSIM with mixed spin $(1, 1/2)$ are given in Figure 3. Results show that the internal energy converges to zero at the higher temperature values, while becoming a constant 0 as $T \to 0$. When the temperature increases, the entropy increases until it reaches $\ln(\sqrt{6})$ and goes to zero at low temperature values.

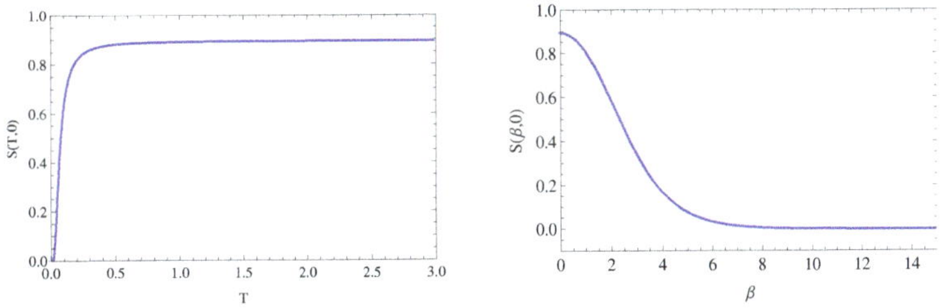

Figure 3. (**Left**) The graph of entropy $S(\beta, 0)$ given in (26) for $J = 0.2$ as a function of the temperature T in the absence of a magnetic field. (**Right**) The graph of entropy $S(\beta, 0)$ given in (26) for $J = 0.2$ as a function of β in the absence of a magnetic field.

Case II

Note that while successively placing spins on the vertices of a one-dimensional lattice, if the elements of Ψ are placed in the first vertex of the lattice $\mathbb{N}$, then a transfer matrix having 2×2 dimensions is obtained, and the eigenvalues of the transfer matrix are the same as those of the previous matrix $\mathbb{M}$.

$$\widetilde{\mathbb{M}} = \mathbb{QP} = \begin{pmatrix} e^{-\frac{h\beta}{2}} + e^{-\frac{1}{2}(3h-2J)\beta} + e^{\frac{1}{2}(h-2J)\beta} & e^{\frac{h\beta}{2}} + e^{-\frac{h\beta}{2}} + e^{\frac{3h\beta}{2}} \\ e^{-\frac{h\beta}{2}} + e^{-\frac{3h\beta}{2}} + e^{\frac{h\beta}{2}} & e^{\frac{h\beta}{2}} + e^{-\frac{1}{2}(h+2J)\beta} + e^{\frac{1}{2}(3h+2J)\beta} \end{pmatrix}. \tag{28}$$

If we substitute the variables $\theta = e^{\frac{\beta J}{2}}$ and $v = e^{\frac{\beta h}{2}}$ for simplicity, we obtain

$$\begin{aligned} \widetilde{M} = \mathbb{QP} &= \begin{pmatrix} \frac{\theta}{v^2} & 1 & \frac{v^2}{\theta} \\ \frac{1}{\theta v^2} & 1 & \theta v^2 \end{pmatrix} \cdot \begin{pmatrix} \frac{\theta}{v} & \frac{v}{\theta} \\ \frac{1}{v} & vs. \\ \frac{1}{\theta v} & \theta v \end{pmatrix} \\ &= \begin{pmatrix} \frac{(v^2 - v\theta + \theta^2)(v^2 + v\theta + \theta^2)}{v^3\theta^2} & \frac{(1 - v + v^2)(1 + v + v^2)}{v} \\ \frac{(1 - v + v^2)(1 + v + v^2)}{v^3} & \frac{(1 - v\theta + v^2\theta^2)(1 + v\theta + v^2\theta^2)}{v\theta^2} \end{pmatrix}. \end{aligned} \tag{29}$$

We obtain the set of eigenvalues of the matrix $\widetilde{\mathbb{M}}$ given in (29) as

$$\begin{cases} \widetilde{\lambda}_1 = \frac{(1+v^2)(\theta^4+v^2(1+\theta^2-\theta^4)+v^4\theta^4) - \sqrt{((1+v^2)(\theta^4+v^2(1+\theta^2-\theta^4)+v^4\theta^4))^2 - 4v^4(\theta^2-1)^2(\theta^2(1+v^4)+v^2(1+\theta^2)^2)}}{2v^3\theta^2}, \\ \widetilde{\lambda}_2 = \frac{(1+v^2)(\theta^4+v^2(1+\theta^2-\theta^4)+v^4\theta^4) + \sqrt{((1+v^2)(\theta^4+v^2(1+\theta^2-\theta^4)+v^4\theta^4))^2 - 4v^4(\theta^2-1)^2(\theta^2(1+v^4)+v^2(1+\theta^2)^2)}}{2v^3\theta^2}. \end{cases} \tag{30}$$

It should be noted here that the eigenvalues of the matrices $\mathbb{M}$ and $\widetilde{\mathbb{M}}$ are the same, except for 0. Therefore, it is obtained that $\lambda_3 = \widetilde{\lambda}_2$ (see Equations (20) and (30)). We can choose any spin on the starting vertex, while mixed spins are placed on consecutive vertices of the lattice. From (20) and (30), it is clear that $\widetilde{\lambda}_{max} = \widetilde{\lambda}_2$.

3.6. Magnetization and Magnetic Susceptibility

In this subsection, we construct magnetization and magnetic susceptibility by means of the eigenvalue $\widetilde{\lambda}_2$ in the Formula (30).

Let us consider the reduced nearest neighbor spin–spin coupling interactions $K = \frac{\beta J}{2} = \ln(\theta)$ and the reduced magnetic field $H = \frac{\beta h_{\xi(x)}}{2} = \ln(v)$, we write the magnetization as

$$m(T, H) = -\beta \frac{\partial F(T, H)}{\partial a \widetilde{\lambda}_2} \partial H, \tag{31}$$

and susceptibility

$$\chi(T, H) = -\beta^2 \frac{\partial^2 F(T, H)}{\partial H^2} = \beta \frac{\partial}{\partial H} \left(\frac{1}{\widetilde{\lambda}_2} \frac{\partial \widetilde{\lambda}_2}{\partial H} \right). \tag{32}$$

We do not give the exact expressions of the formulas for the magnetization and susceptibility here, because the operations are excessively long and complex. We numerically examine the behavior of these two quantities as functions of h and T.

The Mathematica software (Version 8.0, Wolfram Research, Inc., Champaign, IL, USA) [23] has been used to perform the calculations and to plot the figures. A three-dimensional plot of $m(T, H)$ is given in Figure 4. Taking into account the eigenvalue $\widetilde{\lambda}_2$ in the Formula (30), for $J = -2$, one can easily see that, as the temperature T increases, the smoothness of the function decreases and exhibits a behavior similar to the step function. On the other hand, for $J = 2$, the surface of the function becomes smooth. In contrast to the single-spin Ising chain's typical single step [21], in Figure 4 (left), the magnetization graph shows three distinct steps at low temperatures. This phenomenon may be explained by spins adopting distinct states within even and odd sublattices, or by all spins assuming the same state.

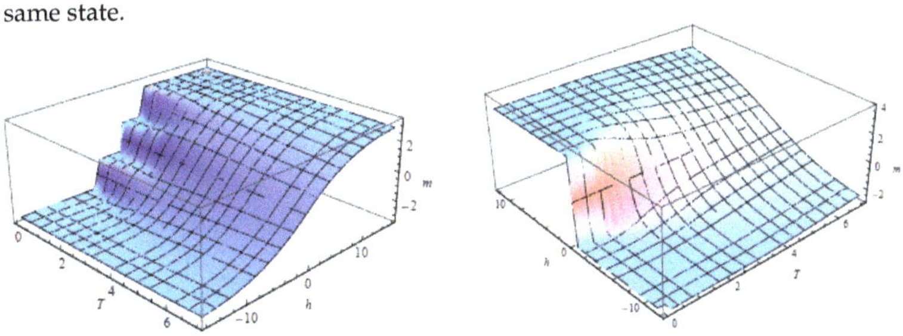

Figure 4. (**Left**) The graph of magnetization $m(T,H)$ given in (31) for $J=-2$ as a function of h and T. (**Right**) The graph of magnetization $m(T,H)$ given in (31) for $J=2$ as a function of h and T.

A three-dimensional plot of $\chi(T,H)$ is given in Figure 5. Figure 5 (Left) and (Right) show the behavior of the susceptibility function given in Equation (32). For $J=-2$, three stacks resembling a boot are observed, and a stack appears for $J=2$ in the chosen region $T\in[0.001,7]$, $h\in[-6,6]$. As seen in Figure 5 (Left), while three different susceptibility peaks appear for $J=-2$ at low temperatures, the susceptibility peaks disappear as the temperature increases. For $J=2$, only a single susceptibility peak is observed at low temperatures (see Figure 5 (Right)).

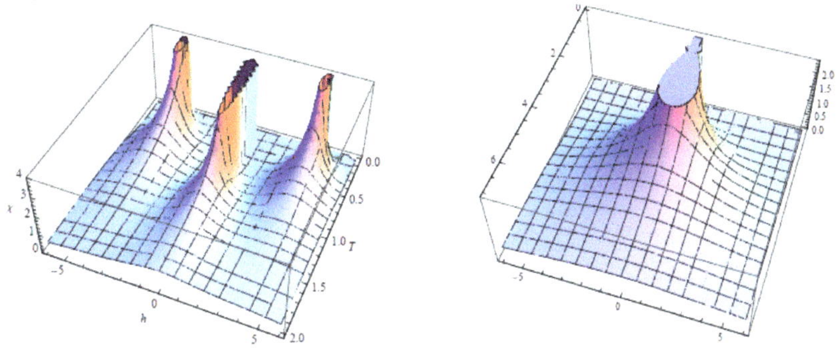

Figure 5. (**Left**) The graph of susceptibility $\chi(T,H)$ given in (32) for $J=-2$ as a function of h and T. (**Right**) The graph of susceptibility $\chi(T,H)$ given in (32) for $J=2$ as a function of h and T.

4. Nonexistence Phase Transition in the Absence of the External Magnetic Field

In this section, we study 1D-MSIM with the mixed spin-(1,1/2) employing the ERR.

Suppose that $W^{(n)}=\frac{\tilde{Z}^{(n)}_{-\frac{1}{2}}}{\tilde{Z}^{(n)}_{\frac{1}{2}}}$, $X^{(n-1)}=\frac{Z^{(n-1)}_{-1}}{Z^{(n-1)}_{0}}$ and $Y^{(n-1)}=\frac{Z^{(n-1)}_{1}}{Z^{(n-1)}_{0}}$, from Equations (43) and (44), we obtain

$$W^{(n)}=\frac{\theta^2X^{(n-1)}+\theta+Y^{(n-1)}}{X^{(n-1)}+\theta+\theta^2Y^{(n-1)}}, \tag{33}$$

$$X^{(n-1)}=\frac{\theta^2W^{(n-2)}+1}{\theta(W^{(n-2)}+1)}, \tag{34}$$

$$Y^{(n-1)}=\frac{W^{(n-2)}+\theta^2}{\theta(W^{(n-2)}+1)}, \tag{35}$$

where $\theta=e^{\frac{\beta J}{2}}$.

If we assume $x = \lim_{n\to\infty} X^{(n-1)}$, $y = \lim_{n\to\infty} Y^{(n-1)}$ and $w = \lim_{n\to\infty} W^{(n)}$, then we obtain a new three-dimensional rational dynamical system (3D-RDS) as

$$w = \frac{\theta^2 x + \theta + y}{x + \theta + \theta^2 y}, \tag{36}$$

$$x = \frac{\theta^2 w + 1}{\theta(w+1)}, \tag{37}$$

$$y = \frac{w + \theta^2}{\theta(w+1)}. \tag{38}$$

If the values of x and y in Equations (37) and (38) are substituted into Equation (36), one obtains the following rational recursive equation $f : \mathbb{R}_+ \to \mathbb{R}_+$:

$$f(w) := \frac{3\theta^2 + w(1+\theta^2+\theta^4)}{(1+\theta^2+\theta^4) + 3w\theta^2} = w. \tag{39}$$

From (39), we obtain a second-order equation

$$3(w-1)(1+w)\theta^2 = 0. \tag{40}$$

From Equation (40), it is clear that the function f given in (39) has only one positive fixed point, so there is no phase transition for the given model.

The graphs of the function f for antiferromagnetic ($\theta = 0.421$) and ferromagnetic ($\theta = 3.421$) values are plotted in Figure 6. The diagrams show that the function in both cases has just two fixed points. Additionally, there is only one positive fixed point. As a result, there is just one Gibbs measure and no phase transition in the model. As it is well known, the classical single spin 1D Ising model's phase transition issue has attracted the interest of statistical mechanics researchers for over a century, and it has been established that there is no phase transition [1]. In this present paper, we demonstrated that the mixed spin 1D Ising model has no phase transition as an example.

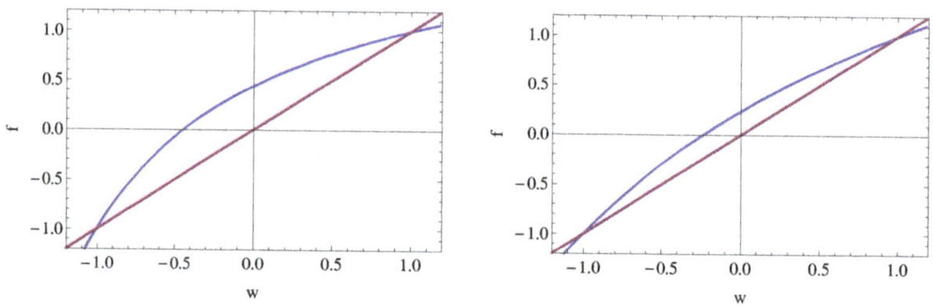

Figure 6. (**Left**) The graph of the function f given in (39) for $\theta = 0.421$. (**Right**) The graph of the function f given in (39) for $\theta = 3.421$.

Chaoticity of the Model

A dynamical system's chaotic behavior is determined by how sensitively it depends on the initial conditions [24,25]. It has long been a challenge to see whether a model's phase transition and the chaotic behavior of the corresponding dynamical system are related [26]. In this subsection, we investigate the chaoticity of the 3D-RDS given in (36)–(38). With the help of the Lyapunov exponent, we numerically study the model's chaotic behavior.

From the definition of the Lyapunov exponent, we obtain

$$\lambda = \frac{1}{n}\ln\left|\prod_{i=0}^{n-1} f'_{\theta}(w_i)\right| = \frac{1}{n}\sum_{i=0}^{n-1}\ln\left|f'_{\theta}(w_i)\right| \tag{41}$$
$$= \frac{1}{n}\sum_{i=0}^{n-1}\ln\left|\frac{(\theta^2-1)^2(1+4\theta^2+\theta^4)}{(1+(1+3w_i)\theta^2+\theta^4)^2}\right|.$$

From (41), we have

$$\lambda(\theta) := \lim_{n\to\infty}\frac{1}{n}\sum_{i=0}^{n-1}\ln\left|\frac{(\theta^2-1)^2(1+4\theta^2+\theta^4)}{(1+(1+3w_i)\theta^2+\theta^4)^2}\right|. \tag{42}$$

Figure 7 shows the Lyapunov exponent of the dynamical system corresponding to the 1D-MSIM with mixed spin-(1,1/2). It is seen that the Lyapunov exponent is always negative in the ferromagnetic region. Therefore, the rational function $f(w)$ given in (39) is periodic. The behavior of the Lyapunov exponent $\lambda(\theta)$ in the antiferromagnetic region can also be examined.

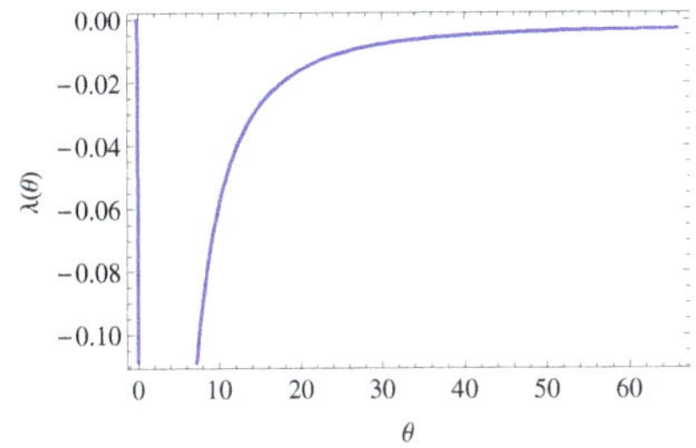

Figure 7. The graph of the function $\lambda(\theta)$ given in (42) versus $\theta = e^{\frac{J}{2T}}$ in the ferromagnetic region ($J > 0$).

5. The Average Magnetization for the Mixed Spin-(1,1/2) Ising Model

In this section, by using the exact recursion relations (ERRs) (see [9]), we obtain the partial partition functions of the 1D-MSIM with mixed spin-($s, \frac{(2t-1)}{2}$). Contrary to the previous sections, here we place the spins of Ψ in the first vertex of the lattice and the spins of Φ in the second vertex of the lattice, while placing spins at the vertices of the lattice $\mathbb{Z}_+$, so for $n = 0, 1, 2, \cdots$, we obtain $s_{2n-1} \in \Psi$ and $\sigma_{2n} \in \Phi$. With the help of the cavity method (see [14–16] for details), we obtain the partial partition functions as follows:

$$\widetilde{Z}^{(n)}_{s_1} = \sum_{\sigma_2\in\Phi}\exp(\beta(Js_1\sigma_2 + h_{\sigma_2}))Z^{(n-1)}_{\sigma_2}, \tag{43}$$
$$Z^{(n-1)}_{\sigma_2} = \sum_{s_3\in\Psi}\exp\left(\beta(J\sigma_2 s_3 + \widetilde{h}_{s_3})\right)\widetilde{Z}^{(n-2)}_{s_3}. \tag{44}$$

The Average Magnetization

In this subsection, assuming there is an external magnetic field, we obtain the magnetization equations for spins s and $\frac{(2t-1)}{2}$, respectively, as follows.

$$M^{(n)}_{1/2} = \frac{\sum_{s_1\in\Psi} s_1\exp(\beta(Js_1\sigma_2 + h_{\sigma_2}))\widetilde{Z}^{(n)}_{s_1}}{\sum_{s_1\in\Psi}\exp(\beta(Js_1\sigma_2 + h_{\sigma_2}))\widetilde{Z}^{(n)}_{s_1}}, \tag{45}$$

$$M_1^{(n-1)} = \frac{\sum\limits_{\sigma_2 \in \Phi} \sigma_2 \exp\left(\beta(J\sigma_2 s_3 + \widetilde{h}_{s_3})\right) Z_{\sigma_1}^{(n-1)}}{\sum\limits_{\sigma_2 \in \Phi} \exp\left(\beta(J\sigma_2 s_3 + \widetilde{h}_{s_3})\right) Z_{\sigma_2}^{(n-1)}}. \tag{46}$$

We are only concerned with the magnetization for mixed spin-(1,1/2) here, since the operations for obtaining partition functions are excessively long for $s > 1$ and $t > 1$. So, from (45) and (46), we obtain

$$M_{1/2}^{(n)} = \frac{e^{\frac{h\beta}{2}}\left(e^{\left(-\frac{h}{2}-\frac{J}{2}\right)\beta} - e^{\left(-\frac{h}{2}+\frac{J}{2}\right)\beta}\right)\left(e^{-h\beta} - W^{(n)}\right)}{e^{-3h\beta}\left(e^{\frac{1}{2}(4h-J)\beta} + e^{\frac{1}{2}(4h+J)\beta} + e^{2h\beta}\right) + e^{-\frac{J\beta}{2}}\left(1 + e^{\frac{J\beta}{2}} + e^{J\beta}\right)W^{(n)}}, \tag{47}$$

$$M_1^{(n-1)} = \frac{\left(e^{J\beta} - 1\right)\left(e^{2h\beta}Y^{(n-1)} - X^{(n-1)}\right)}{2\left(2e^{h\beta + \frac{J\beta}{2}} + \left(1 + e^{J\beta}\right)\left(X^{(n-1)} + e^{2h\beta}Y^{(n-1)}\right)\right)}. \tag{48}$$

The following equation results from the simple substitutions of the variables $\theta = e^{\frac{J\beta}{2}}$ and $v = e^{\frac{h\beta}{2}}$

$$M_{1/2}^{(n)} = \frac{\left(v^2 W^{(n)} - 1\right)(\theta^2 - 1)}{(1 + v^2 W^{(n)})(1 + \theta + \theta^2)}, \tag{49}$$

$$M_1^{(n-1)} = -\frac{\left(X^{(n-1)} - v^4 Y^{(n-1)}\right)(\theta^2 - 1)}{2(X^{(n-1)}(1 + \theta^2) + 2v^2\theta + v^4 Y^{(n-1)}(\theta^2 + 1))}. \tag{50}$$

Now, let us assume that the external magnetic field h ($v = 1$) is zero. In this case, after some calculations, as in Ref. [14], we can show that the fixed point for the 3D-RDS given in (36)–(38) is only $x = y = \left(\frac{1+\theta^2}{2\theta}\right)$ and $w = 1$. So, we obtain zero average magnetization around this fixed point. That is, we have

$$M_{1/2}^{(n)} = M_1^{(n-1)} = 0.$$

6. Conclusions

There are many methods for calculating the free energy of lattice models on the Bethe lattices or the d-dimensional integer lattice $\mathbb{Z}^d$ ($d \geq 1$). Some of these are the exact recursion relations [9], the cavity method [18], and vector-valued boundary conditions. One of the most widely used methods to examine thermodynamic quantities corresponding to a 1D Ising model is the transfer matrix technique [21,22]. In this present work, we constructed the partial partition functions associated with 1D-MSIM having mixed spin-$(s, \frac{(2t-1)}{2})$ via the transfer matrix. We have computed some thermodynamic quantities such as the free energy, entropy, magnetization, and susceptibility of the model. To the best of my knowledge, the thermodynamic properties of the 1D Ising model with the single spin have been investigated in several studies to date. Using the transfer matrix method, we evaluated the thermodynamic features of the 1D-MSIM for the first time. For the single spin-Isign model, Amin et al. [21] plotted the magnetization and susceptibility in two dimensions.

In our previous studies, we proved that there is a phase transition for the mixed-spin Ising model on the CT using many methods [11,12,14]. As it is known, there is no phase transition for a single spin 1D Ising model [1]. In this study, we investigated how mixed spins affect thermodynamic quantities and the existence of the phase transition. We demonstrated the uniqueness of limiting Gibbs measures associated with 1D-MSIM having mixed spin-(1,1/2). We have shown that if the external magnetic field is zero, for the aforementioned model, there is no phase transition. Our next research will deal with other issues in statistical mechanics. By considering the approach given by Albayrak [9], the isothermal entropy change of the average magnetization for 1D-MSIM will be analyzed.

The findings obtained in our present article exhibit different behaviors from the results given in previous studies [21,22]. Frankly, we cannot comment on the physical meaning of such behavior. These topics are covered in introductory courses in statistical mechanics at the undergraduate level. Therefore, we believe that the results of our present study will be of interest to a wide readership of statistical mechanics.

Funding: This research received no external funding.

Data Availability Statement: The statistical data presented in the article do not require copyright. They are freely available and are listed at the reference address in the bibliography.

Acknowledgments: The author would like to thank the Simons Foundation (10.13039/100000893) and the Institute of International Education for their support. While the author was a visiting scientist at the ICTP, a significant part of the research was completed there. The author therefore thanks ICTP for their hospitality. The author expresses gratitude to the referees for their comments and suggestions.

Conflicts of Interest: The author declare no conflict of interest.

References

1. Ising, E. Beitrag zur theorie des ferromagnetismus. *Z. Physik* **1925**, *31*, 253. [CrossRef]
2. Blume, M. Theory of the first-order magnetic phase change in UO_2. *Phys. Rev.* **1966**, *141*, 517. [CrossRef]
3. Capel, H.W. On the possibility of first-order phase transitions in Ising systems of triplet ions with zero-field splitting. *Physica* **1966**, *32*, 966. [CrossRef]
4. Lipowski, A.; Suzuki, M. On the exact solution of twodimensional spin S Ising models A. *Phys. A* **1993**, *193*, 141. [CrossRef]
5. Izmailian, N.S.; Ananikian, N.S. General spin-3/2 Ising model in a honeycomb lattice: Exactly solvable case. *Phys. Rev. B* **1994**, *50*, 6829. [CrossRef]
6. De La Espriella, N.; Arenas, A.J.; Paez Meza, M.S. Critical and compensation points of a mixed spin-2-spin-5/2 Ising ferrimagnetic system with crystal field and nearest and next-nearest neighbors interactions. *J. Magn. Magn. Mater.* **2016**, *417*, 434–441. [CrossRef]
7. De La Espriella, N.; Buendia, G.M.; Madera, J.C. Mixed spin-1 and spin-2 Ising model: Study of the ground states. *J. Phys. Commun.* **2018**, *2*, 025006. [CrossRef]
8. Kaneyoshi, T. Phase transition of the mixed spin system with a random crystal field. *Phys. A* **1988**, *153*, 556–566. [CrossRef]
9. Albayrak, E. The study of mixed spin-1 and spin-1/2: Entropy and isothermal entropy change. *Phys. A* **2020**, *559*, 125079. [CrossRef]
10. De La Espriella, N.; Buendia, G.M. Magnetic behavior of a mixed Ising 3/2 and 5/2 spin model. *J. Phys. Condens. Matter* **2011**, *23*, 176003. [CrossRef]
11. Akın, H.; Mukhamedov, F. Phase transition for the Ising model with mixed spins on a Cayley tree. *J. Stat. Mech.* **2022**, *2022*, 053204. [CrossRef]
12. Akın, H. The classification of disordered phases of mixed spin (2,1/2) Ising model and the chaoticity of the corresponding dynamical system. *Chaos Solitons Fractals* **2023**, *167*, 113086. [CrossRef]
13. Seino M., The free energy of the random Ising model on the Bethe lattice. *Phys. A* **1992**, *181*, 233–242. [CrossRef]
14. Akın, H. Quantitative behavior of (1,1/2)-MSIM on a Cayley tree. *Chin. J. Phys.* **2023**, *83*, 501–514. [CrossRef]
15. Ostilli, M. Cayley Trees and Bethe Lattices: A concise analysis for mathematicians and physicists. *Phys. A* **2012**, *391*, 3417–3423. [CrossRef]
16. Mézard, M.; Parisi, G. The Bethe lattice spin glass revisited. *Eur. Phys. J. B* **2001**, *20*, 217–233. [CrossRef]
17. Qi, Y.; Liu, J.; Yu, N.-S.; Du, A. Magnetocaloric effect in ferroelectric Ising chain magnet. *Solid State Commun.* **2016**, *233*, 1–5. [CrossRef]
18. Akın, H. Calculation of the free energy of the Ising model on a Cayley tree via the self-similarity method. *Axioms* **2022**, *11*, 703. [CrossRef]
19. Akın, H.; Ulusoy, S. A new approach to studying the thermodynamic properties of the q-state Potts model on a Cayley tree. *Chaos Solitons Fractals* **2023**, *174*, 113811. [CrossRef]
20. Salinas, S.R.A. Phase Transitions and Critical Phenomena: Classical Theories. In *Introduction to Statistical Physics. Graduate Texts in Contemporary Physics*; Springer: New York, NY, USA, 2001.
21. Amin, M.E.; Mubark, M.; Amin, Y. On the critical behavior of the spin-s ising model. *Rev. Mex. Fis.* **2023**, *69*, 021701. [CrossRef]
22. Wang, W.; Diaz-Mendez, R.; Capdevila, R. Solving the one-dimensional Ising chain via mathematical induction: An intuitive approach to the transfer matrix. *Eur. J. Phys.* **2019**, *40*, 065102. [CrossRef]
23. Wolfram Research, Inc. *Mathematica*; Version 8.0; Wolfram Research, Inc.: Champaign, IL, USA, 2010.
24. Feigenbaum, M.J. Quantitative universality for a class of nonlinear transformations. *J. Stat. Phys.* **1978**, *19*, 25–52. [CrossRef]

25. Feigenbaum, M.J. Universal behavior in nonlinear systems. *Phys. D* **1983**, *7*, 16–39. [CrossRef]
26. Hilborn, R.C. *Chaos and Nonlinear Dynamics: An Introduction for Scientists and Engineers*; Oxford University Press on Demand: Oxford, UK, 2000.

axioms

Article

An Analytic Solution for 2D Heat Conduction Problems with Space–Time-Dependent Dirichlet Boundary Conditions and Heat Sources

Heng-Pin Hsu [1], Jer-Rong Chang [1,*], Chih-Yuan Weng [2] and Chun-Jung Huang [3]

1 Department of Aircraft Engineering, Air Force Institute of Technology, 1 Associate Jyulun Road, Gang-Shan District, Kaoshiung City 820, Taiwan
2 Department of Mechanical Engineering, Air Force Institute of Technology, 1 Associate Jyulun Road, Gang-Shan District, Kaoshiung City 820, Taiwan
3 Department of Aircraft Maintenance, Far East University, 49 Zhonghua Road, Xinshi District, Tainan City 744, Taiwan
* Correspondence: jerrong.chang@gmail.com; Tel.: +886-7-6256040

Abstract: This study proposes a closed-form solution for the two-dimensional (2D) transient heat conduction in a rectangular cross-section of an infinite bar with space–time-dependent Dirichlet boundary conditions and heat sources. The main purpose of this study is to eliminate the limitations of the previous study and add heat sources to the heat conduction system. The restriction of the previous study is that the values of the boundary conditions and initial conditions at the four corners of the rectangular region should be zero. First, the boundary value problem of 2D heat conduction system is transformed into a dimensionless form. Second, the dimensionless temperature function is transformed so that the temperatures at the four endpoints of the boundary of the rectangular region become zero. Dividing the system into two one-dimensional (1D) subsystems and solving them by combining the proposed shifting function method with the eigenfunction expansion theorem, the complete solution in series form is obtained through the superposition of the subsystem solutions. Three examples are studied to illustrate the efficiency and reliability of the method. For convenience, the space–time-dependent functions used in the examples are considered separable in the space–time domain. The linear, parabolic, and sine functions are chosen as the space-dependent functions, and the sine, cosine, and exponential functions are chosen as the time-dependent functions. The solutions in the literature are used to verify the correctness of the solutions derived using the proposed method, and the results are completely consistent. The parameter influence of the time-dependent function of the boundary conditions and heat sources on the temperature variation is also investigated. The time-dependent function includes exponential type and harmonic type. For the exponential time-dependent function, a smaller decay constant of the time-dependent function leads to a greater temperature drop. For the harmonic time-dependent function, a higher frequency of the time-dependent function leads to a more frequent fluctuation of the temperature change.

Keywords: analytic solution; 2D heat conduction; space–time-dependent Dirichlet boundary conditions; space–time-dependent heat sources; rectangular cross-section

MSC: 35K05; 80M99

Citation: Hsu, H.-P.; Chang, J.-R.; Weng, C.-Y.; Huang, C.-J. An Analytic Solution for 2D Heat Conduction Problems with Space–Time-Dependent Dirichlet Boundary Conditions and Heat Sources. *Axioms* **2023**, *12*, 708. https://doi.org/10.3390/axioms12070708

Academic Editor: Leonid Plotnikov

Received: 20 June 2023
Revised: 18 July 2023
Accepted: 19 July 2023
Published: 20 July 2023

1. Introduction

Heat conduction problems commonly exist in various modern mechanical equipment and advanced instruments, and the boundary conditions and the heat sources of these heat conduction problems are often space–time-dependent. Over the years, much research has been devoted to these problems. The main methods to solve these problems include numerical methods, analytic methods, and experimental methods. The literature review below focuses on the study of heat conduction problems with time- and/or space-dependent boundary conditions and/or heat sources.

Some advanced books on heat conduction described some classical techniques for solving heat conduction problems [1–3]. These include Laplace transform, Green's function, and Duhamel's theorem. Different approximation methods have been used to study one-dimensional (1D) heat conduction problems with space- and/or time-dependent boundary conditions and heat sources, such as the iterative perturbation method used by Holy [4] and the eigenfunction expansion method used by Özisik and Murray [5]. Johansson and Lesnic [6], Chantasiriwan [7], and Young et al. [8] applied the fundamental solutions method to the time-dependent heat conduction problems. On the basis of the boundary element method, Zhu, Liu, and Lu [9] used the Laplace transform, and Bulgakov, Sarler, and Kuhn [10] used the finite difference scheme to find the solution in the time domain. Lee et al. [11] proposed an integration-free-solution method to derive the analytic closed solution to the one-dimensional heat conduction problem with time-dependent boundary conditions. Furthermore, they extended the solution method and successfully applied it to the inverse analysis of heat conduction problems [12,13].

Several numerical techniques have been proposed for two- and three-dimensional (2D and 3D) heat conduction problems with space- and/or time-dependent boundary conditions and/or heat sources. Walker [14] applied the fundamental solution of diffusion combined with time integration to solve the diffusion equation. Chen, Golberg, and Hon [15] applied the modified Helmholtz fundamental solution to the diffusion equation. Zhu [16] and Sutradhar, Paulino, and Gray [17] used the Laplace transform to deal with the time derivative term in the diffusion equation. Burgess and Mahajerin [18] used the fundamental collocation method to the problems of arbitrary shapes subjected to mixed time-dependent boundary conditions and arbitrary initial conditions. Young, Tsai, and Fan [19] solved the nonhomogeneous diffusion problems using the fundamental solution method and the dual reciprocity method. Numerical computation using the Padé scheme was proposed by Siddique [20] to solve two-dimensional diffusion problems. The finite integral transform method was used by Singh et al. [21] to solve the problem of asymmetric heat conduction in a multilayer annulus space with time-dependent boundary conditions and/or heat sources. Hematiyan et al. [22] presented a boundary element method to analyze the 2D and 3D uncoupled thermoelastic problems with space- and time-dependent heat sources. An enhanced state-space method considering laminate approximation theory was introduced by Daneshjou et al. [23] to analyze the non-Fourier heat conduction of an infinite 2D functionally graded hollow cylinder influenced by a time-dependent heat source.

Gu et al. [24] proposed the generalized finite difference method to recover the time-dependent heat sources associated with a 3D heat equation. The eigenfunction-based solutions of various heat conduction models exhibited a mismatch at the boundaries. Biswas et al. [25] homogenized the generalized time-dependent boundary conditions to remove this mismatch from the solution. Akbari et al. [26] employed Duhamel's theorem to evaluate the non-Fourier heat conduction of the 3D hollow spheres under the space- and time-dependent boundary conditions. Zhou et al. [27] proposed a polygonal boundary element method to solve the nonlinear heat conduction problems with space-dependent heat source and temperature-dependent thermal properties. The authors [28] derived a closed-form solution to the 2D heat conduction problem with the general Dirichlet boundary condition using the shifting function method with the eigenfunction expansion theorem. However, the temperature values of the boundary conditions and initial condition at the four corners of the rectangular region were restricted to zero, which limited the applicability of this study. Moreover, no heat source was considered in this document.

In this paper, a closed solution to the transient heat conduction problems in a rectangular cross-section of an infinite bar with nonhomogeneous space–time-dependent Dirichlet boundary conditions and heat sources was developed. This study addresses the limitations of the previous study [28]; that is, the values of the boundary conditions and initial conditions at the four corners of the rectangular area do not need to be limited to zero and become functions of time. The space–time-dependent heat sources are also considered in this study. The scope of application becomes much larger. This solution has a wide range of applications, such as laser heating, living tissue research, electronic devices, and microstructural applications [12,25,29,30]. The proposed method can accurately solve the heat conduction problem modeled as a rectangular cross-section of an infinite bar with Dirichlet boundary conditions and heat sources. The first step in solving the two-dimensional heat conduction problem is to change the physic system to be a dimensionless form. Afterward, the temperature function is transformed so that the temperature values at the four endpoints of the boundary conditions of the rectangular region become zero. Then, the system is divided into two separate one-dimensional subsystems. The governing differential equations with nonhomogeneous space- and time-dependent boundary conditions are transformed into differential equations with homogeneous boundary conditions by applying the shifting function method which does not require any integral transformation. Under the transformed homogeneous boundary conditions, the closed-form solution of the system can be further deduced by using the eigenfunction expansion theorem. Combining the solutions of the two individual subsystems, the analytic solution of the original two-dimensional heat conduction system can be obtained. Three examples are employed to illustrate the efficiency and reliability of the method. The solution obtained using the proposed method is in full agreement with the literature. The influence of the parameters is also investigated.

The contributions of this paper are as follows:

(1) The analytic solution to 2D heat conduction problems with the general Dirichlet boundary conditions using the shifting function method with the expansion theorem method was proposed in our previous study [28]. However, there were two restrictions, the temperature values at the four corners of the rectangular area should be zero, and the heat source was also set to zero. The greatest contribution of this work is that an analytical solution is proposed first for the 2D transient heat conduction in a rectangular cross-section of an infinite bar with the space–time-dependent Dirichlet boundary conditions and heat sources. The temperatures at the four corners of the rectangular region can be functions of time.

(2) The correctness of the solution in this study is verified by comparing the solutions of some cases using the proposed method with those of Young et al. [8], the previous work [28], and Siddique [20]. To the best of the authors' knowledge, the other cases in this paper have never been presented in past studies. Furthermore, the case studies show that the proposed method has good convergence to the solution using series expansion and can quickly reach the converged value. The parameter influence of the time-dependent function of the boundary conditions and heat sources on the temperature change is also studied.

2. Mathematical Modeling and Dimensionless Form of Physical System

Consider the transient heat conduction for a rectangular cross-section in an infinite bar with the space–time-dependent Dirichlet boundary conditions on its four sides and the heat generation inside the bar. The material of the bar is isotropic and time-independent, implying that the material properties are constants. Figure 1 shows the geometry, heat sources, boundary conditions, and initial conditions of a rectangular cross-section in an infinite bar. Therefore, the governing equation, boundary conditions, and initial conditions of the physical system are given as follows:

$$k\left[\frac{\partial^2 T(x,y,t)}{\partial x^2}+\frac{\partial^2 T(x,y,t)}{\partial y^2}\right]+g(x,y,t)=\rho c\frac{\partial T(x,y,t)}{\partial t}\ \text{in}\ 0\le x\le L_x,\ 0\le y\le L_y,\ t>0, \quad (1)$$

$$T(0,y,t)=f_1(y,t),\ \text{at}\ x=0,\ 0\le y\le L_y, \quad (2)$$

$$T(L_x, y, t) = f_2(y, t), \text{ at } x = L_x,\ 0 \le y \le L_y, \tag{3}$$

$$T(x, 0, t) = f_3(x, t), \text{ at } y = 0,\ 0 \le x \le L_x, \tag{4}$$

$$T(x, L_y, t) = f_4(x, t), \text{ at } y = L_y,\ 0 \le x \le L_x, \tag{5}$$

$$T(x, y, 0) = T_0(x, y), \text{ at } t = 0,\ 0 \le x \le L_x,\ 0 \le y \le L_y, \tag{6}$$

where $T(x, y, t)$ is the temperature function, $g(x, y, t)$ denotes the heat source inside the rectangular cross-section, x and y are the two-dimensional space variables, L_x and L_y are the thickness of the rectangular bar in the x- and y-directions, respectively, and t is the time variable. In addition, k, ρ, and c represent the thermal conductivity, mass density, and specific heat, respectively. $f_i(y, t)$ $i = 1, 2$ and $f_i(x, t)$ $i = 3, 4$ represent the general case of space–time-dependent temperatures prescribed along the surfaces at the left and right sides and at the bottom and top sides, respectively. It is worth noting that the heat source function $g(x, y, t)$ in Equation (1) is a function of space variables, which means that there are infinitely many heat sources allowed in this function. If multiple heat source functions are required to model the heat sources, the function $g(x, y, t)$ in Equation (1) can be changed to a summation term of these functions. The matching boundary conditions to initial conditions have the following properties:

$$f_1(y, 0) = T_0(0, y),\ f_2(y, 0) = T_0(L_x, y),\ f_3(x, 0) = T_0(x, 0),\ f_4(x, 0) = T_0(x, L_y). \tag{7}$$

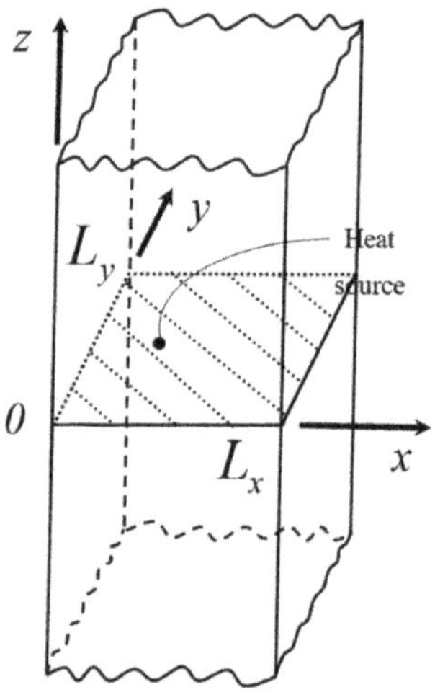

(a)

Figure 1. *Cont.*

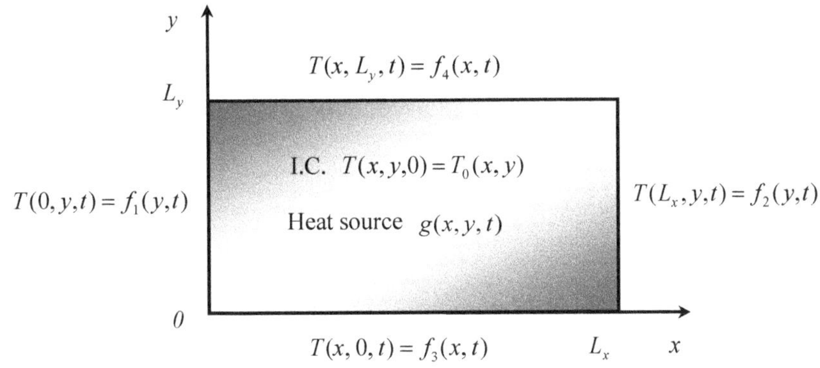

(b)

Figure 1. The 2D heat conduction in a rectangular cross-section of an infinite bar with space–time-dependent Dirichlet boundary conditions and heat sources: (a) an infinite bar with a rectangular cross-section and heat source; (b) the 2D heat conduction problem in a rectangular region (cross-section).

For ease of discussion, a dimensionless form of the 2D heat conduction system is first generated. After introducing the dimensionless parameters

$$\begin{gathered}\psi(X,Y,\tau)=\frac{T(x,y,t)}{T_r},\ \overline{G}(X,Y,\tau)=\frac{L_y^2\,g(x,y,t)}{kT_r},\tau=\frac{\alpha\,t}{L_y^2},\ X=\frac{x}{L_x},\ Y=\frac{y}{L_y},\ L_r=\frac{L_y}{L_x},\\ \overline{F}_i(Y,\tau)=\frac{f_i(y,t)}{T_r},\ i=1,2,\ \overline{F}_i(X,\tau)=\frac{f_i(x,t)}{T_r},\ i=3,4,\ \psi_0(X,Y)=\frac{T_0(x,y)}{T_r},\end{gathered} \tag{8}$$

the dimensionless form of the boundary value problem becomes

$$\left[L_r^2\frac{\partial^2\psi(X,Y,\tau)}{\partial X^2}+\frac{\partial^2\psi(X,Y,\tau)}{\partial Y^2}\right]+\overline{G}(X,Y,\tau)=\frac{\partial\psi(X,Y,\tau)}{\partial\tau},\ \text{in}\ 0<X<1,\ 0<Y<1,\tau>0, \tag{9}$$

$$\psi(0,Y,\tau)=\overline{F}_1(Y,\tau),\ \psi(1,Y,\tau)=\overline{F}_2(Y,\tau),\ \psi(X,0,\tau)=\overline{F}_3(X,\tau),\ \psi(X,1,\tau)=\overline{F}_4(X,\tau), \tag{10}$$

$$\psi(X,Y,0)=\psi_0(X,Y). \tag{11}$$

The parameter $\alpha=\frac{k}{\rho\,c}$ in Equation (8) represents the thermal diffusivity, and T_r represents the reference temperature. The matching conditions of the boundary conditions and the initial conditions also become

$$\overline{F}_1(Y,\,0)=\psi_0(0,\,Y),\ \overline{F}_2(Y,0)=\psi_0(1,Y),\ \overline{F}_3(X,0)=\psi_0(X,0),\ \overline{F}_4(X,0)=\psi_0(X,1). \tag{12}$$

3. The Solution Method

The previous study [28] presented a method to derive an analytic solution to a 2D heat conduction problem with general Dirichlet boundary conditions. However, the heat generation was not considered in the model. Moreover, the values of boundary conditions and initial conditions at each endpoint of the rectangular area were limited to zero, which limited the applicability of this method. In this study, the constraint of corner zeros of the boundary conditions was removed by first variable-transforming the temperature, and the space–time-dependent heat source was added to the heat conduction system. The system was then divided into two subsystems, each of which can be solved with a 1D heat conduction problem. By properly choosing the shifting function, the second-order governing differential equation with space–time-dependent boundary conditions was transformed into a differential equation with homogeneous boundary conditions. Due to the homogeneous boundary conditions, the eigenfunction expansion theorem could be applied to solve the closed-form solution of the subsystem. Lastly, the solutions of the two subsystems were added to obtain the analytic solution for 2D heat conduction in a rectangular region with space–time Dirichlet boundary conditions and heat sources

3.1. Temperature Variable Transformation

$a(\tau), b(\tau), c(\tau)$, and $d(\tau)$ represent the temperature at each endpoint of the rectangular section (Figure 2a) and can be expressed as follows.

$$\begin{array}{l} a(\tau) = \overline{F}_1(0,\tau) = \overline{F}_3(0,\tau),\ b(\tau) = \overline{F}_2(0,\tau) = \overline{F}_3(1,\tau), \\ c(\tau) = \overline{F}_1(1,\tau) = \overline{F}_4(0,\tau),\ d(\tau) = \overline{F}_2(1,\tau) = \overline{F}_4(1,\tau). \end{array} \quad (13)$$

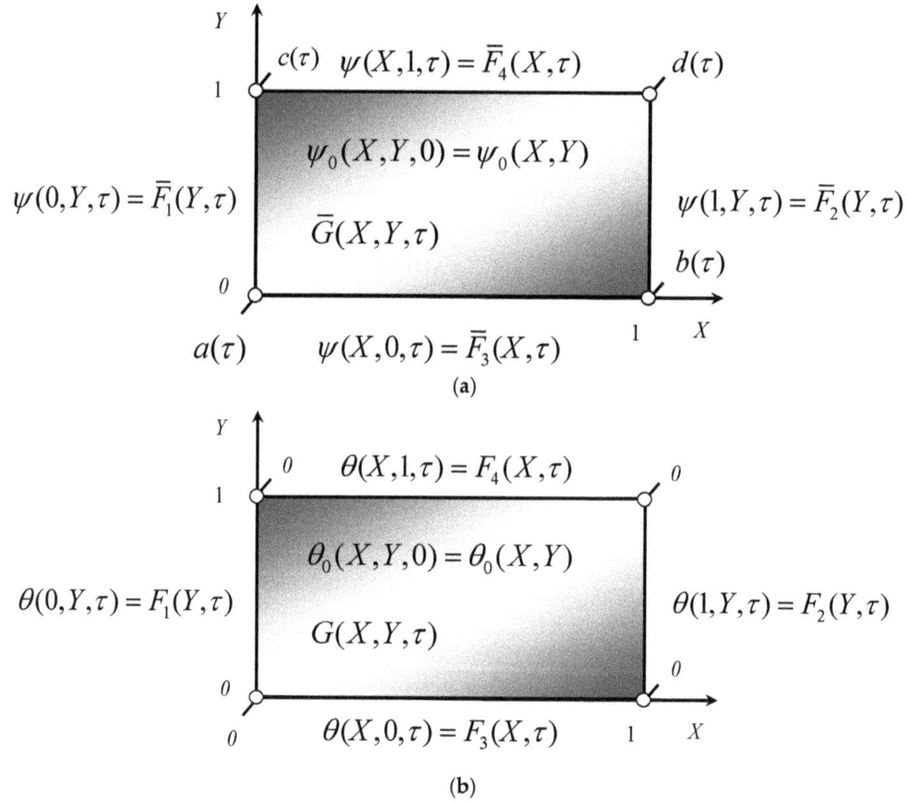

Figure 2. A 2D heat conduction system after the variable transformation of temperature: (**a**) dimensionless form of a physical system; (**b**) transformation results for dimensionless systems.

To increase the method applicability of our previous study [28], the nonzero temperatures at the four endpoints can be converted to zero first (Figure 2b). Therefore, variable transformation to achieve this goal can be expressed as follows:

$$\theta(X,Y,\tau) = \psi(X,Y,\tau) - \left\{ \begin{array}{l} a(\tau) + X\,[b(\tau) - a(\tau)] + \\ Y\,[c(\tau) - a(\tau) + X\,Y\,(d(\tau) - c(\tau) - b(\tau) + a(\tau))] \end{array} \right\}. \quad (14)$$

Substituting Equation (14) back into Equation (9) can obtain the following nonhomogeneous differential equation:

$$\left[L_r^2\frac{\partial^2\theta(X,Y,\tau)}{\partial X^2}+\frac{\partial^2\theta(X,Y,\tau)}{\partial Y^2}\right]+G(X,Y,\tau)=\frac{\partial\theta(X,Y,\tau)}{\partial\tau},\ \text{in}\ 0<X<1,\ 0<Y<1,\tau>0, \tag{15}$$

where the transformed dimensionless heat source $G(X,Y,\tau)$ is defined as

$$G(X,Y,\tau)=\overline{G}(X,Y,\tau)-\rho c\left\{\begin{array}{l}\dot{a}(\tau)+X\,[\dot{b}(\tau)-\dot{a}(\tau)]+\\ Y\,[\dot{c}(\tau)-\dot{a}(\tau)+X\,Y\,[\dot{d}(\tau)-\dot{c}(\tau)-\dot{b}(\tau)+\dot{a}(\tau)]\end{array}\right\}, \tag{16}$$

and the dot is used to indicate the differentiation with respect to the dimensionless time τ.

The dimensionless boundary conditions and initial conditions become

$$\theta(0,Y,\tau)=F_1(Y,\tau)=\overline{F}_1(Y,\tau)-\{a(\tau)+Y\,[c(\tau)-a(\tau)]\},\ \text{at}\ X=0,\ 0\le Y\le 1, \tag{17}$$

$$\theta(1,Y,\tau)=F_2(Y,\tau)=\overline{F}_2(Y,\tau)-\{b(\tau)+Y\,[d(\tau)-b(\tau)]\},\ \text{at}\ X=1,\ 0\le Y\le 1, \tag{18}$$

$$\theta(X,0,\tau)=F_3(X,\tau)=\overline{F}_3(X,\tau)-\{a(\tau)+X\,[b(\tau)-a(\tau)]\},\ \text{at}\ Y=0,\ 0\le X\le 1, \tag{19}$$

$$\theta(X,1,\tau)=F_4(X,\tau)=\overline{F}_4(X,\tau)-\{c(\tau)+X\,[d(\tau)-c(\tau)]\},\ \text{at}\ Y=1,\ 0\le X\le 1, \tag{20}$$

$$\theta(X,Y,\tau)=\theta_0(X,Y)=\psi_0(X,Y)-\left\{\begin{array}{l}a(0)+X\,[b(0)-a(0)]+\\ Y\,[c(0)-a(0)]+X\,Y\,[d(0)-c(0)-b(0)+a(0)]\end{array}\right\},\ \text{at}\tau=0,\ 0\le X\le 1,\ 0\le Y\le 1. \tag{21}$$

In addition, considering the compatibility of the boundary conditions and initial conditions, we have

$$F_1(Y,0)=\theta_0(0,Y),\ F_2(Y,0)=\theta_0(1,Y),\ F_3(X,0)=\theta_0(X,0),\ F_4(X,0)=\theta_0(X,1). \tag{22}$$

3.2. Principle of Superposition

Due to the boundary value problem with the linear characteristic, we can divide the physical system into two subsystems A and B in the X- and Y-directions through the principle of superposition, as shown in Figure 3.

$\theta(X,Y,\tau)$ is spilt into two branches as follows:

$$\theta(X,Y,\tau)=\theta_a(X,Y,\tau)+\theta_b(X,Y,\tau). \tag{23}$$

For subsystem A, the governing equation, and the boundary and initial conditions of the heat conduction problem are

$$\left[L_r^2\frac{\partial^2\theta_a(X,Y,\tau)}{\partial X^2}+\frac{\partial^2\theta_a(X,Y,\tau)}{\partial Y^2}\right]+G_a(X,Y,\tau)=\frac{\partial\theta_a(X,Y,\tau)}{\partial\tau},\ \text{in}\ 0\le X\le 1,\ 0\le Y\le 1,\tau>0, \tag{24}$$

$$\theta_a(0,Y,\tau)=F_1(Y,\tau),\ \theta_a(1,Y,\tau)=F_2(Y,\tau),\ \theta_a(X,0,\tau)=0,\ \theta_a(X,1,\tau)=0, \tag{25}$$

$$\theta_a(X,Y,0)=\theta_{a0}(X,Y),\ \text{at}\tau=0,\ 0\le X\le 1,\ 0\le Y\le 1. \tag{26}$$

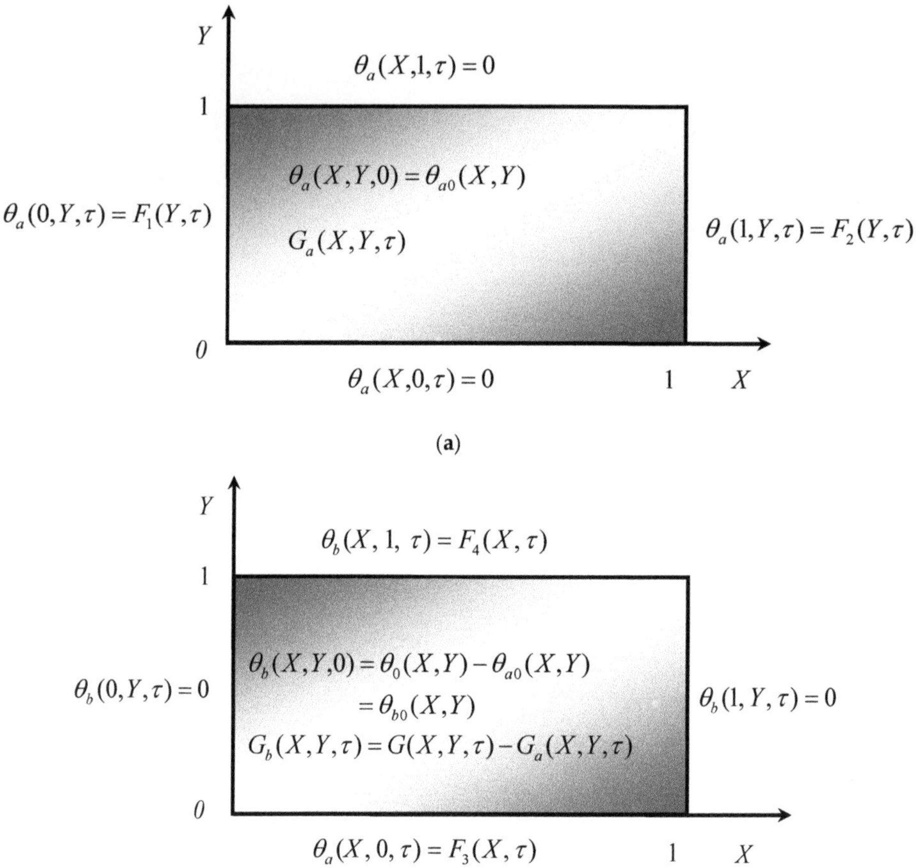

(b)

Figure 3. The two subsystems of the 2D heat conduction system with space–time-dependent Dirichlet boundary conditions and heat source in the medium: (**a**) for subsystem *A*; (**b**) for subsystem *B*.

Similarly, for subsystem *B*, the governing equation, and the boundary and initial conditions of the heat conduction problem are

$$\left[L_r^2\frac{\partial^2\theta_b(X,Y,\tau)}{\partial X^2}+\frac{\partial^2\theta_b(X,Y,\tau)}{\partial Y^2}\right]+G_b(X,Y,\tau)=\frac{\partial\theta_b(X,Y,\tau)}{\partial\tau},\ \text{in } 0\le X\le 1,\ 0\le Y\le 1, \tau>0, \tag{27}$$

$$\theta_b(0,Y,\tau)=0,\ \theta_b(1,Y,\tau)=0,\ \theta_b(X,0,\tau)=F_3(X,\tau),\ \theta_b(X,1,\tau)=F_4(X,\tau), \tag{28}$$

$$\theta_b(X,Y,0)=\theta_0(X,Y)-\theta_{a0}(X,Y)=\theta_{b0}(X,Y). \tag{29}$$

Considering the similarity of the two subsystems, for the sake of brevity, subsystem *A* is solved first, while subsystem *B* is solved in Appendix A.

3.3. Reduction to One-Dimensional Problem

Considering two homogeneous boundary conditions on opposite sides of a rectangular cross-section, namely, $Y=0$ and $Y=1$, it is reasonable to assume that the tem-

perature $\theta_a(X,Y,\tau)$, the heat source $G_a(X,Y,\tau)$, and the dimensionless boundary values $F_i(Y,\tau)$ $(i=1,2)$ defined in Equations (24) and (25) become

$$\theta_a(X,Y,\tau)=\sum_{m=1}^{\infty}[\theta_m(X,\tau)\sin(m\pi Y)], \tag{30}$$

$$G_a(X,Y,\tau)=\sum_{m=1}^{\infty}[G_{am}(X,\tau)\sin(m\pi Y)], \tag{31}$$

$$F_i(Y,\tau)=\sum_{m=1}^{\infty}\left[\overline{F}_{i,m}(\tau)\sin(m\pi Y)\right],\ i=1,2, \tag{32}$$

where $G_{am}(X,\tau)$ and $\overline{F}_{i,m}(\tau)$ $(i=1,2)$, are defined as

$$G_{am}(X,\tau)=2\int_0^1[G_a(X,Y,\tau)\sin(m\pi Y)]\,dY, \tag{33}$$

$$\overline{F}_{i,m}(\tau)=2\int_0^1 F_i(Y,\tau)\sin(m\pi Y)dY,\ i=1,2. \tag{34}$$

$\theta_m(X,\tau)$ in Equation (30) is determined by meeting the governing equation and the boundary conditions on both edges $X=0$ and $X=1$ (the first two terms in Equation (25)). Substituting Equations (30)–(32) into Equations (24)–(26) yields the following result:

$$\frac{\partial\theta_m(X,\tau)}{\partial\tau}-L_r^2\frac{\partial^2\theta_m(X,\tau)}{\partial X^2}+m^2\pi^2\theta_m(X,\tau)=G_{am}(X,\tau),\ \text{in}\ 0<X<1,\tau>0, \tag{35}$$

$$\theta_m(0,\tau)=\overline{F}_{1,m}(\tau),\ \theta_m(1,\tau)=\overline{F}_{2,m}(\tau), \tag{36}$$

$$\theta_m(X,0)=2\int_0^1\theta_{a0}(X,Y)\sin(m\pi Y)dY. \tag{37}$$

3.4. The Shifting Function Method

3.4.1. Change of Variable

To find the solution of the second-order partial differential Equation (35) with nonhomogeneous boundary conditions (Equation (36)), the following transformation equation can be employed to extend the shifting function method [11–13]:

$$\theta_m(X,\tau)=\overline{\theta}_m(X,\tau)+\sum_{i=1}^{2}[g_{i,m}(X)\overline{F}_{i,m}(\tau)], \tag{38}$$

where $\overline{\theta}_m(X,\tau)$ is the transformed function, and $g_{i,m}(X)(i=1,2)$ are the two shifting functions to be determined.

Substituting Equation (38) into Equations (35)–(37) yields

$$\begin{gathered}\dot{\overline{\theta}}_m(X,\tau)+\sum_{i=1}^{2}[g_{i,m}(X)\dot{\overline{F}}_{i,m}(\tau)]-L_r^2\left\{\overline{\theta}''{}_m(X,\tau)+\sum_{i=1}^{2}[g''{}_{i,m}(X)\overline{F}_{i,m}(\tau)]\right\}\\+m^2\pi^2\left\{\overline{\theta}_m(X,\tau)+\sum_{i=1}^{2}[g_{i,m}(X)\overline{F}_{i,m}(\tau)]\right\}=G_{am}(X,\tau)\end{gathered}, \tag{39}$$

where the double primes are used to denote twice differentiation with respect to dimensionless coordinate X. The associated boundary conditions become

$$\overline{\theta}_m(0,\tau)+\sum_{i=1}^{2}[g_{i,m}(0)\overline{F}_{i,m}(\tau)]=\overline{F}_{1,m}(\tau),\ \overline{\theta}_m(1,\tau)+\sum_{i=1}^{2}[g_{i,m}(1)\overline{F}_{i,m}(\tau)]=\overline{F}_{2,m}(\tau). \tag{40}$$

3.4.2. The Shifting Functions

For the convenience of subsequent analysis, the shifting functions can be chosen to satisfy the following differential equation and boundary conditions:

$$g''_{i,m}(X) = 0,\ i = 1, 2,\ 0 < X < 1, \tag{41}$$

$$g_{i,m}(0) = \delta_{i1},\ g_{i,m}(1) = \delta_{i2}, \tag{42}$$

where δ_{ij} is the Kronecker delta. Two shifting functions are, thus, easily specified as

$$g_{1,m}(X) = 1 - X,\ g_{2,m}(X) = X. \tag{43}$$

Substituting Equations (41)–(43) into Equations (39) and (40), the differential equation for $\overline{\theta}_m(X,\tau)$ is simplified as follows:

$$\dot{\overline{\theta}}_m(X,\tau) - L_r^2\overline{\theta}''_m(X,\tau) + m^2\pi^2\overline{\theta}_m(X,\tau) = \overline{G}_m(X,\tau), \tag{44}$$

and the homogeneous boundary conditions are

$$\overline{\theta}_m(0,\tau) = 0,\ \overline{\theta}_m(1,\tau) = 0. \tag{45}$$

$\overline{G}_m(X,\tau)$ in Equation (44) is defined as

$$\overline{G}_m(X,\tau) = -\sum_{i=1}^{2}\left\{g_{i,m}(X)\left[\dot{\overline{F}}_{i,m}(\tau) + m^2\pi^2\overline{F}_{i,m}(\tau)\right]\right\} + G_{am}(X,\tau). \tag{46}$$

In addition, the initial condition is transformed to be

$$\overline{\theta}_m(X,0) = 2\int_0^1 \theta_{a0}(X,Y)\sin(m\pi Y)dY - \sum_{i=1}^{2}\left[g_{i,m}(X)\overline{F}_{i,m}(0)\right]. \tag{47}$$

3.4.3. The Series Expansion Theorem

The solution $\overline{\theta}_m(X,\tau)$ of Equations (44) and (45) can be solved by applying the method of separation variable. In this method, the solution $\overline{\theta}_m(X,\tau)$ is expressed as

$$\overline{\theta}_m(X,\tau) = \sum_{n=1}^{\infty}\left[\overline{\theta}_{mn}(X)\ T_{mna}(\tau)\right], \tag{48}$$

where the space variable $\overline{\theta}_{mn}(X)$ is solved by the following Sturm–Liouville eigenvalue problem:

$$\overline{\theta}''_{mn}(X) + \omega_n^2\overline{\theta}_{mn}(X) = 0,\ 0 < X < 1, \tag{49}$$

$$\overline{\theta}_{mn}(0) = 0,\ \overline{\theta}_{mn}(1) = 0. \tag{50}$$

The eigenfunctions $\overline{\theta}_{mn}(X)$ $(n = 1,\ 2,\ 3,\cdots)$. and the corresponding eigenvalues are solved as

$$\overline{\theta}_{mn}(X) = \sin\omega_n X,\ \omega_n = n\pi,\ n = 1,\ 2,\ 3,\ \cdots \tag{51}$$

The eigenfunctions constitute an orthogonal set in the interval [0, 1],

$$\int_0^1 \overline{\theta}_{mi}(X)\,\overline{\theta}_{mj}(X)dX = \begin{cases} 0 & \text{for} i \neq j \\ \frac{1}{2} & \text{for} i = j \end{cases}. \tag{52}$$

Substituting Equation (51) into Equation (48), substituting Equation (48) into Equation (44), multiplying Equation (44) by $\overline{\theta}_{mn}(X)$, and integrating from 0 to 1, the following differential equation can be obtained:

$$\dot{T}_{mna}(\tau) + \lambda_{mna}^2 T_{mna}(\tau) = \gamma_{mna}(\tau), \tag{53}$$

where λ_{mna} and γ_{mna} are given as

$$\lambda_{mna} = \sqrt{m^2 + n^2 L_\gamma^2}\pi, \tag{54}$$

$$\begin{aligned} \gamma_{mna}(\tau) = 2\int_0^1 \overline{\theta}_{mn}(X)\, \overline{G}_m(X,\tau)dX = \tfrac{-2}{n\pi}\left[\dot{\overline{F}}_{1,m}(\tau) - (-1)^n \dot{\overline{F}}_{2,m}(\tau)\right] \\ -\tfrac{2m^2\pi}{n}\left[\overline{F}_{1,m}(\tau) - (-1)^n \overline{F}_{2,m}(\tau)\right] + 2\int_0^1 \overline{\theta}_{mn}(X)\, G_{am}(X,\tau)dX \end{aligned}. \tag{55}$$

$T_{mna}(0)$ is determined from the initial condition of the transformed function defined in Equation (47) as

$$T_{mna}(0) = 4\int_0^1 \sin(n\pi X)\int_0^1 \theta_{a0}(X,Y)\sin(m\pi Y)dYdX - \frac{2}{n\pi}\left[\overline{F}_{1,m}(0) - (-1)^n \overline{F}_{2,m}(0)\right]. \tag{56}$$

Equations (53) and (56) can be solved, and their general solution is as follows:

$$T_{mna}(\tau) = e^{-\lambda_{mna}^2 \tau} T_{mna}(0) + \int_0^\tau e^{-\lambda_{mna}^2(\tau-\varphi)}\gamma_{mna}(\varphi)d\varphi. \tag{57}$$

3.4.4. The Analytic Solution

After substituting the solutions for the transformation function (Equation (48)) and shifting function (Equation (43)) back into Equations (38) and (30), the closed-form solution $\theta_a(X,Y,\tau)$ for the subsystem A is derived as follows:

$$\theta_a(X,Y,\tau) = \sum_{m=1}^{\infty}\{\sum_{n=1}^{\infty}[\sin(n\pi X)T_{mna}(\tau)] + (1-X)\overline{F}_{1,m}(\tau) + X\overline{F}_{2,m}(\tau)\}\sin(m\pi Y). \tag{58}$$

Thanks to the high symmetry with the subsystem A, the solution form of the subsystem B can be easily obtained through a similar derivation process (see Appendix A for details):

$$\theta_b(X,Y,\tau) = \sum_{m=1}^{\infty}\{\sum_{n=1}^{\infty}[\sin(n\pi Y)T_{mnb}(\tau)] + (1-Y)\overline{F}_{3,m}(\tau) + Y\overline{F}_{4,m}(\tau)\}\sin(m\pi X). \tag{59}$$

The summation of the two subsystem solutions leads to an analytic solution for the heat conduction in the rectangular cross-section with space–time-dependent Dirichlet boundary conditions and heat sources as follows:

$$\begin{aligned} \theta(X,Y,\tau) = \sum_{m=1}^{\infty}\{\sum_{n=1}^{\infty}[\sin(n\pi X)T_{mna}(\tau)] + (1-X)\overline{F}_{1,m}(\tau) + X\overline{F}_{2,m}(\tau)\}\sin(m\pi Y) \\ + \sum_{m=1}^{\infty}\{\sum_{n=1}^{\infty}[\sin(n\pi Y)T_{mnb}(\tau)] + (1-Y)\overline{F}_{3,m}(\tau) + Y\overline{F}_{4,m}(\tau)\}\sin(m\pi X) \end{aligned}. \tag{60}$$

Using the relationship shown in Equation (14), the dimensionless temperature $\psi(X,Y,\tau)$ before temperature variable transformation can be calculated. In addition, using the first identity of Equation (8), the exact solution $T(x,y,t)$ with dimension can also be obtained.

3.4.5. The Extreme Case Study

If the heat source value $g(x,y,t)$ is zero, and the temperatures at four corners of the rectangular region $a(\tau)$, $b(\tau)$, $c(\tau)$, and $d(\tau)$, are all zeros, then the parameters $\overline{G}(X,Y,\tau)$, $G(X,Y,\tau)$, $G_a(X,Y,\tau)$, $G_b(X,Y,\tau)$, $G_{a\,m}(X,\tau)$, and $G_{b\,m}(Y,\tau)$ are all zeros. Therefore, the obtained exact solution $\theta(X,Y,\tau)$ will be the same as that obtained in our previous work [28]. This extreme case is studied in Example 2.

4. Examples and Verification

In order to verify the advantages of the proposed solution method, two cases involving the presence and absence of a heat source in the medium are explored in detail below. For simplicity, we take $L_x = L_y = L_r = 1$ in all examples of the 2D heat conduction problem.

4.1. With Zero Heat Source

In our previous study [28], the boundary condition was restricted such that the temperature values at the four corners of the rectangular area were zero. The method of this study is not limited to this restriction. In Example 1, we show how to solve it using the proposed method if there are nonzero temperature values at the four corners of the rectangular region.

Example 1. *Take the heat source in the bar as $g(x,y,t) = 0$, and assume the space-time-dependent Dirichlet boundary and initial conditions as follows:*

$$T(0,y,t) = f_1(y,t) = [\sin(\frac{\pi y}{2}) + \cos(\frac{\pi y}{2}) + 1]\ \eta(\alpha t),\ \text{at } x = 0,\ 0 \le y \le 1, \tag{61}$$

$$T(1,y,t) = f_2(y,t) = [\sin(\frac{\pi y}{2}) + \cos(\frac{\pi y}{2}) + 1]\ \eta(\alpha t),\ \text{at } x = 1,\ 0 \le y \le 1, \tag{62}$$

$$T(x,0,t) = f_3(y,t) = [\sin(\frac{\pi x}{2}) + \cos(\frac{\pi x}{2}) + 1]\ \eta(\alpha t),\ \text{at } y = 0,\ 0 \le x \le 1, \tag{63}$$

$$T(x,1,t) = f_4(y,t) = [\sin(\frac{\pi x}{2}) + \cos(\frac{\pi x}{2}) + 1]\ \eta(\alpha t),\ \text{at } y = 1,\ 0 \le x \le 1, \tag{64}$$

$$T(x,y,0) = \sin(\frac{\pi x}{2}) + \cos(\frac{\pi x}{2}) + \sin(\frac{\pi y}{2}) + \cos(\frac{\pi y}{2}),\ \text{at } t = 0,\ 0 \le x \le 1,\ 0 \le y \le 1. \tag{65}$$

The dimensionless form of boundary and initial conditions, and the dimensionless heat source $\overline{G}(X,Y,\tau)$ are written as

$$\psi(0,Y,\tau) = \overline{F}_1(Y,\tau) = \frac{\left[\sin(\frac{\pi Y}{2}) + \cos(\frac{\pi Y}{2}) + 1\right]\ \eta(\tau)}{T_r},\ \text{at } X = 0,\ 0 \le Y \le 1, \tag{66}$$

$$\psi(1,Y,\tau) = \overline{F}_2(Y,\tau) = \frac{[\sin(\frac{\pi Y}{2}) + \cos(\frac{\pi Y}{2}) + 1]\ \eta(\tau)}{T_r},\ \text{at } X = 1,\ 0 \le Y \le 1, \tag{67}$$

$$\psi(X,0,\tau) = \overline{F}_3(Y,\tau) = \frac{[\sin(\frac{\pi X}{2}) + \cos(\frac{\pi X}{2}) + 1]\ \eta(\tau)}{T_r},\ \text{at } Y = 0,\ 0 \le X \le 1, \tag{68}$$

$$\psi(X,1,\tau) = \overline{F}_4(Y,\tau) = \frac{[\sin(\frac{\pi X}{2}) + \cos(\frac{\pi X}{2}) + 1]\ \eta(\tau)}{T_r},\ \text{at } Y = 1,\ 0 \le X \le 1, \tag{69}$$

$$\psi(X,Y,0) = \psi_0(X,Y) = \frac{\sin(\frac{\pi X}{2}) + \cos(\frac{\pi X}{2}) + \sin(\frac{\pi Y}{2}) + \cos(\frac{\pi Y}{2})}{T_r},\ \text{at} \tau = 0,\ 0 \le X \le 1,\ 0 \le Y \le 1, \tag{70}$$

$$\overline{G}(X,Y,\tau) = 0,\ \text{at } 0 \le X \le 1,\ 0 \le Y \le 1. \tag{71}$$

Considering the matching conditions at the four corners of the rectangular section at the initial moment, we have

$$\eta(0) = 1. \tag{72}$$

In this case, $a(\tau)$, $b(\tau)$, $c(\tau)$, $d(\tau)$, and their differentiations with respect to τ are derived as

$$a(\tau) = b(\tau) = c(\tau) = d(\tau) = \frac{2\eta(\tau)}{T_r}, \quad (73)$$

$$\dot{a}(\tau) = \dot{b}(\tau) = \dot{c}(\tau) = \dot{d}(\tau) = \frac{2\dot{\eta}(\tau)}{T_r}. \quad (74)$$

Thus, the transformed temperature $\theta(X,Y,\tau)$ and heat source $G(X,Y,\tau)$ are given as

$$\theta(X,Y,\tau) = \psi(X,Y,\tau) - \frac{2\eta(\tau)}{T_r}, \quad (75)$$

$$G(X,Y,\tau) = -\frac{2\dot{\eta}(\tau)}{T_r}. \quad (76)$$

It is worth noting that, after transformation, the dimensionless heat source becomes nonzero.

Following the proposed solution procedure, the boundary and initial conditions are derived as

$$F_1(Y,\tau) = \frac{\left[\sin(\frac{\pi Y}{2}) + \cos(\frac{\pi Y}{2}) - 1\right]\eta(\tau)}{T_r}, \text{ at } X = 0,\ 0 \le Y \le 1, \quad (77)$$

$$F_2(Y,\tau) = \frac{\left[\sin(\frac{\pi Y}{2}) + \cos(\frac{\pi Y}{2}) - 1\right]\eta(\tau)}{T_r}, \text{ at } X = 1,\ 0 \le Y \le 1, \quad (78)$$

$$F_3(X,\tau) = \frac{\left[\sin(\frac{\pi X}{2}) + \cos(\frac{\pi X}{2}) - 1\right]\eta(\tau)}{T_r}, \text{ at } Y = 0,\ 0 \le X \le 1, \quad (79)$$

$$F_4(X,\tau) = \frac{\left[\sin(\frac{\pi X}{2}) + \cos(\frac{\pi X}{2}) - 1\right]\eta(\tau)}{T_r}, \text{ at } Y = 1,\ 0 \le X \le 1, \quad (80)$$

$$\theta_0(X,Y) = \frac{\sin(\frac{\pi X}{2}) + \cos(\frac{\pi X}{2}) + \sin(\frac{\pi Y}{2}) + \cos(\frac{\pi Y}{2}) - 2}{T_r}, \text{ at} \tau = 0,\ 0 \le X \le 1,\ 0 \le Y \le 1. \quad (81)$$

Next, $G(X,Y,\tau)$ and $\theta(X,Y,\tau)$ are divided into two parts as follows:

$$G_a(X,Y,\tau) = -\frac{\dot{\eta}(\tau)}{T_r},\ G_b(X,Y,\tau) = -\frac{\dot{\eta}(\tau)}{T_r}, \quad (82)$$

$$\theta_{a0}(X,Y) = \frac{\sin(\frac{\pi X}{2}) + \cos(\frac{\pi X}{2}) - 1}{T_r},\ \theta_{b0}(X,Y) = \frac{\sin(\frac{\pi Y}{2}) + \cos(\frac{\pi Y}{2}) - 1}{T_r}. \quad (83)$$

The associated dimensionless quantities $G_{am}(X,\tau)$, $G_{bm}(Y,\tau)$, and $\overline{F}_{i,m}(\tau)(i = 1,2,3,4)$ are derived as

$$G_{am}(X,\tau) = G_{bm}(X,\tau) = \frac{-2\dot{\eta}(\tau)[1-(-1)^m]}{T_r m\pi}, \quad (84)$$

$$\overline{F}_{i,m}(\tau) = \frac{2\eta(\tau)[1-(-1)^m]}{T_r m\pi(4m^2-1)},\ i = 1,2,3,4. \quad (85)$$

From Equations (54), (55), (A25), and (A26) one has

$$\lambda_{mna} = \lambda_{mnb} = \sqrt{m^2+n^2}\,\pi, \quad (86)$$

$$\gamma_{mna}(\tau) = \gamma_{mnb}(\tau) = -\frac{4m^2[1-(-1)^m][1-(-1)^n]}{T_r mn\pi^2(4m^2-1)}[4\dot{\eta}(\tau) + \pi^2\eta(\tau)]. \quad (87)$$

Likewise, $T_{mna}(0)$ and $T_{mnb}(0)$ are determined from the initial conditions of the transformed functions defined in Equations (56) and (A27) as

$$T_{mna}(0) = T_{mnb}(0) = \frac{16(m^2 - n^2)[1-(-1)^m][1-(-1)^n]}{T_r mn\pi^2(4m^2-1)(4n^2-1)}. \tag{88}$$

Therefore, $T_{mna}(\tau)$ and $T_{mnb}(\tau)$ are determined as follows:

$$\begin{array}{l} T_{mna}(\tau) = T_{mnb}(\tau) = \frac{2[1-(-1)^m][1-(-1)^n]}{T_r mn\pi^2(4m^2-1)(4n^2-1)}\{8(m^2-n^2)e^{-(m^2+n^2)\,\pi^2} \\ -4m^2(4n^2-1)\int_0^\tau e^{-(m^2+n^2)\pi^2(\tau-\phi)}[4\dot{\eta}(\phi) + \pi^2\,\eta(\phi)]\;d\phi \end{array}. \tag{89}$$

Considering the time-dependent term of exponential type,

$$\eta(\tau) = e^{-\pi^2\tau/4}, \tag{90}$$

and substituting Equation (90) back into Equations (85) and (89) yields

$$\overline{F}_{i,m}(\tau) = \frac{2e^{-\pi^2\tau/4}[1-(-1)^m]}{T_r m\pi(4m^2-1)},\; i = 1,2,3,4, \tag{91}$$

$$T_{mna}(\tau) = T_{mnb}(\tau) = \frac{16(m^2-n^2)\;[1-(-1)^m][1-(-1)^n]}{T_r mn\pi^2(4m^2-1)(4n^2-1)}e^{-(m^2+n^2)\,\pi^2}. \tag{92}$$

From Equations (60) and (14), the dimensionless solutions $\theta(X,Y,\tau)$ and $\psi(X,Y,\tau)$ become

$$\begin{array}{l} \theta(X,Y,\tau) = \sum\limits_{m=1}^{\infty}\{\sum\limits_{n=1}^{\infty}[\sin(m\pi X)\sin(n\pi Y) + \sin(m\pi Y)\sin(n\pi X)]T_{mna}(\tau)\} \\ \quad + \sum\limits_{m=1}^{\infty}[\sin(m\pi X) + \sin(m\pi Y)]\;\overline{F}_{1,m}(\tau) \end{array}, \tag{93}$$

$$\psi(X,Y,\tau) = \theta(X,Y,\tau) + \frac{2e^{-\pi^2\tau/4}}{T_r}. \tag{94}$$

$\psi(X,Y,\tau)$ is changed back to dimensional form to obtain the exact solution $T(x,y,t)$, and its series form is expressed as follows:

$$\begin{array}{l} T(x,y,t) = \sum\limits_{m=1}^{\infty}\{\sum\limits_{n=1}^{\infty}[\sin(m\pi x)\sin(n\pi y) + \sin(m\pi y)\sin(n\pi x)]\cdot \\ \frac{16(m^2-n^2)[1-(-1)^m][1-(-1)^n]}{mn\pi^2(4m^2-1)(4n^2-1)}e^{-(m^2+n^2)\pi^2}\} \\ + \sum\limits_{m=1}^{\infty}[\sin(m\pi x) + \sin(m\pi y)]\cdot\frac{2e^{-\alpha\,\pi^2 t/4}[1-(-1)^m]}{m\pi(4m^2-1)} + 2e^{-\alpha\,\pi^2 t/4} \end{array}. \tag{95}$$

Another analytic solution in compact form was derived by Young et al. [8] as follows:

$$T(x,y,t) = [\sin(\frac{\pi x}{2}) + \cos(\frac{\pi x}{2}) + \sin(\frac{\pi y}{2}) + \cos(\frac{\pi y}{2})]\;e^{-\alpha\,\pi^2\,t/4}. \tag{96}$$

Since the exact solution of Equation (95) is in the form of series summation, when the number of expansion terms m and n becomes larger, the numerical result tends to be more accurate. Table 1 shows the midpoint temperature of the rectangular area of the bar at different times. It can be seen from Table 1 that the convergence speed is very fast; moreover, when the number of terms $m = n$ is 5, the maximum error is less than 0.1%. At $0 \le \tau \le 1.2$, the temperature using $m = n \ge 10$ expansion terms almost converges to the exact solution in the literature [8].

Table 1. The temperature of the rectangular region at $x = y = 0.5$ and at different times $[\eta(t) = e^{-\pi^2 t/4}]$.

t	$T(x = 0.5, y = 0.5, t)$					
	Number of Expansion Terms $(m = n)$					
	1	3	5	10	20	Exact Solution [8]
0	2.849	2.825	2.830	2.829	2.828	2.828
0.1	2.226	2.207	2.211	2.210	2.210	2.210
0.2	1.739	1.724	1.728	1.727	1.727	1.727
0.4	1.062	1.053	1.055	1.054	1.054	1.054
0.6	0.648	0.643	0.644	0.644	0.644	0.644
0.8	0.396	0.392	0.393	0.393	0.393	0.393
1.0	0.242	0.240	0.240	0.240	0.240	0.240
1.2	0.148	0.146	0.146	0.146	0.146	0.146

4.2. With Nonzero Heat Sources

Two examples with the space–time-dependent Dirichlet boundary conditions and heat sources are used to demonstrate the proposed method.

Example 2. *We assume that the space-time-dependent heat source in the medium is given as* $g(x, y, t) = -\rho c(x + y + 1)e^{-t}$, *and that the space-time-dependent Dirichlet boundary and initial conditions are as follows:*

$$T(0, y, t) = f_1(y, t) = [\sin(\pi y)]\, e^{-\alpha\, \pi^2 t} + (y + 1)\, e^{-t}, \text{ at } x = 0,\ 0 \le y \le 1, \tag{97}$$

$$T(1, y, t) = f_2(y, t) = [\sin(\pi y)]\, e^{-\alpha\, \pi^2 t} + (y + 2)\, e^{-t}, \text{ at } x = 1,\ 0 \le y \le 1, \tag{98}$$

$$T(x, 0, t) = f_3(x, t) = [\sin(\pi x)]\, e^{-\alpha\, \pi^2 t} + (x + 1)\, e^{-t}, \text{ at } y = 0,\ 0 \le x \le 1, \tag{99}$$

$$T(x, 1, t) = f_4(x, t) = [\sin(\pi x)]\, e^{-\alpha\, \pi^2 t} + (x + 2)\, e^{-t}, \text{ at } y = 1,\ 0 \le x \le 1, \tag{100}$$

$$T(x, y, 0) = T_0(x, y) = \sin(\pi x) + \sin(\pi y) + x + y + 1, \text{ at } t = 0,\ 0 \le x \le 1,\ 0 \le y \le 1. \tag{101}$$

Therefore, the dimensionless form of boundary and initial conditions can be expressed as

$$\psi(0, Y, \tau) = \overline{F}_1(Y, \tau) = \frac{\sin(\pi Y)\, e^{-\pi^2 \tau} + (Y + 1)\, e^{-\tau/\alpha}}{T_r}, \text{ at } X = 0,\ 0 \le Y \le 1, \tag{102}$$

$$\psi(1, Y, \tau) = \overline{F}_2(Y, \tau) = \frac{\sin(\pi Y)\, e^{-\pi^2 \tau} + (Y + 2)\, e^{-\tau/\alpha}}{T_r}, \text{ at } X = 1,\ 0 \le Y \le 1, \tag{103}$$

$$\psi(X, 0, \tau) = \overline{F}_3(Y, \tau) = \frac{\sin(\pi X)\, e^{-\pi^2 \tau} + (X + 1)\, e^{-\tau/\alpha}}{T_r}, \text{ at } Y = 0,\ 0 \le X \le 1, \tag{104}$$

$$\psi(X, 1, \tau) = \overline{F}_4(Y, \tau) = \frac{\sin(\pi X)\, e^{-\pi^2 \tau} + (X + 2)\, e^{-\tau/\alpha}}{T_r}, \text{ at } Y = 1,\ 0 \le X \le 1, \tag{105}$$

$$\psi(X, Y, 0) = \psi_0(Y, \tau) = \frac{\sin(\pi X) + \sin(\pi Y) + X + Y + 1}{T_r}, \text{ at} \tau = 0,\ 0 \le X \le 1,\ 0 \le Y \le 1, \tag{106}$$

$$\overline{G}(X,Y,\tau) = -\frac{(X+Y+1)\ e^{-\tau/\alpha}}{\alpha\ T_r},\ \text{at}\ 0 \leq X \leq 1,\ 0 \leq Y \leq 1. \quad (107)$$

In this case, the time-dependent functions at four corners of the rectangular cross-section and their differentiation with respect to τ are given as follows:

$$a(\tau) = \frac{e^{-\tau/\alpha}}{T_r},\ b(\tau) = \frac{2e^{-\tau/\alpha}}{T_r},\ c(\tau) = \frac{2e^{-\tau/\alpha}}{T_r},\ d(\tau) = \frac{3e^{-\tau/\alpha}}{T_r}, \quad (108)$$

$$\dot{a}(\tau) = \frac{-e^{-\tau/\alpha}}{\alpha T_r},\ \dot{b}(\tau) = \frac{-2e^{-\tau/\alpha}}{\alpha T_r},\ \dot{c}(\tau) = \frac{-2e^{-\tau/\alpha}}{\alpha T_r},\ \dot{d}(\tau) = \frac{-3e^{-\tau/\alpha}}{\alpha T_r}. \quad (109)$$

The transformed variables $\theta(X,Y,\tau)$ and $G(X,Y,\tau)$ using Equations (14) and (16) become

$$\theta(X,Y,\tau) = \psi(X,Y,\tau) - \frac{(X+Y+1)\ e^{-\tau/\alpha}}{T_r},\ G(X,Y,\tau) = 0. \quad (110)$$

In addition, the transformed boundary and initial conditions are given as follows:

$$F_1(Y,\tau) = F_2(Y,\tau) = \frac{\sin(\pi Y)\ e^{-\pi^2\tau}}{T_r},\ \text{at}\ 0 \leq Y \leq 1, \quad (111)$$

$$F_3(X,\tau) = F_4(X,\tau) = \frac{\sin(\pi X)\ e^{-\pi^2\tau}}{T_r},\ \text{at}\ 0 \leq X \leq 1, \quad (112)$$

$$\theta_0(X,Y) = \sin(\pi X) + \sin(\pi Y),\ \text{at}\tau = 0,\ 0 \leq X \leq 1,\ 0 \leq Y \leq 1. \quad (113)$$

$G(X,Y,\tau)$ and $\theta_0(X,Y,\tau)$ can be separated into two parts:

$$G_a(X,Y,\tau) = G_b(X,Y,\tau) = 0, \quad (114)$$

$$\theta_{a0}(X,Y) = \frac{\sin\ (\pi Y)}{T_r},\ \theta_{b0}(X,Y) = \frac{\sin\ (\pi X)}{T_r}. \quad (115)$$

In the case, using one-term expansion ($m = n = 1$) in Equation (60), the analytic solution can be obtained as

$$\begin{aligned}\theta(X,Y,\tau) = & [\sin(\pi X)T_{11a}(\tau) + (1-X)\overline{F}_{1,1}(\tau) + X\ \overline{F}_{2,1}(\tau)]\sin(\pi Y) \\ & +[\sin(\pi Y)T_{11b}(\tau) + (1-Y)\overline{F}_{3,1}(\tau) + Y\ \overline{F}_{4,1}(\tau)]\sin(\pi X)\end{aligned}, \quad (116)$$

where the quantities $\overline{F}_{i,1}(\tau)(i = 1,2,3,4)$ are given as

$$\overline{F}_{i,1}(\tau) = \frac{e^{-\pi^2\tau}}{T_r},\ i = 1,2,3,4. \quad (117)$$

From Equations (54), (55), (A25), and (A26), one has

$$\lambda_{11\ a} = \lambda_{11\ b} = \sqrt{2}\pi,\ \gamma_{11\ a}(\tau) = \gamma_{11\ b}(\tau) = 0. \quad (118)$$

Likewise, $T_{11\ a}(0)$ and $T_{11\ b}(0)$ are determined from the initial conditions of the transformed functions defined in Equations (56) and (A27) as

$$T_{11\ a}(0) = T_{11\ b}(0) = 0. \quad (119)$$

Therefore, one obtains

$$T_{11\ a}(\tau) = T_{11\ b}(\tau) = 0. \quad (120)$$

From Equation (116), the solution in dimensionless form is as follows:

$$\theta(X,Y,\tau) = \frac{\sin(\pi X) + \sin(\pi Y)}{T_r}\ e^{-\pi^2\tau}. \quad (121)$$

It is worth noting that the solution $\theta(X, Y, \tau)$ of the boundary value problem is the same as the analytical solution for case 1 of Example 1 [28]. After the temperature $\theta(X, Y, \tau)$ is transformed from the dimensionless temperature, the heat source and the associated boundary and initial conditions of the heat conduction system become an extreme case (case 1 of Example 1 in our previous work [28]). This can verify the correctness of the proposed method.

Furthermore, the dimensionless temperature before the transformation of this example can be derived as

$$\psi(X, Y, \tau) = \frac{[\sin(\pi X) + \sin(\pi Y)]\, e^{-\pi^2 \tau} + (X + Y + 1)\, e^{-\tau/\alpha}}{T_r}. \tag{122}$$

Finally, the analytic solution $T(x, y, t)$ of this example becomes

$$T(x, y, t) = [\sin(\pi x) + \sin(\pi y)]\, e^{-\alpha\, \pi^2\, t} + (x + y + 1)\, e^{-\, t}. \tag{123}$$

Example 3. *The heat source in the medium is given as* $g(x, y, t) = -\rho\, c\, (x^2 + y^2 + 4)\ \eta_0(\alpha\, t)$, *where* $\eta_0(\alpha\ t)$ *is a time-dependent function. Assume that the space-time-dependent Dirichlet boundary and initial conditions are as follows:*

$$T(0, y, t) = f_1(y, t) = 1 + y^2\, \eta(\alpha\, t),\ \text{at}\ x = 0,\ 0 \le y \le 1, \tag{124}$$

$$T(1, y, t) = f_2(y, t) = 1 + (1 + y^2)\ \eta(\alpha\, t),\ \text{at}\ x = 1,\ 0 \le y \le 1, \tag{125}$$

$$T(x, 0, t) = f_3(x, t) = 1 + x^2 \eta\ (\alpha\, t),\ \text{at}\ y = 0,\ 0 \le x \le 1, \tag{126}$$

$$T(x, 1, t) = f_4(x, t) = 1 + (1 + x^2)\ \eta(\alpha\, t),\ \text{at}\ y = 1,\ 0 \le x \le 1, \tag{127}$$

$$T(x, y, 0) = T_0(x, y) = 1 + x^2 + y^2,\ \text{at}\ t = 0,\ 0 \le x \le 1,\ 0 \le y \le 1. \tag{128}$$

The dimensionless form of boundary and initial conditions becomes

$$\psi(0, Y, \tau) = \overline{F}_1(Y, \tau) = \frac{1 + Y^2\ \eta(\tau)}{T_r},\ \text{at}\ X = 0,\ 0 \le Y \le 1, \tag{129}$$

$$\psi(1, Y, \tau) = \overline{F}_2(Y, \tau) = \frac{1 + (1 + Y^2)\ \eta(\tau)}{T_r},\ \text{at}\ X = 1,\ 0 \le Y \le 1, \tag{130}$$

$$\psi(X, 0, \tau) = \overline{F}_3(Y, \tau) = \frac{1 + X^2\ \eta(\tau)}{T_r},\ \text{at}\ Y = 0,\ 0 \le X \le 1, \tag{131}$$

$$\psi(X, 1, \tau) = \overline{F}_4(X, \tau) = \frac{1 + (1 + X^2)\ \eta(\tau)}{T_r},\ \text{at}\ Y = 1,\ 0 \le X \le 1, \tag{132}$$

$$\psi(X, Y, 0) = \psi_0(X, Y) = \frac{1 + X^2 + Y^2}{T_r},\ \text{at}\tau = 0,\ 0 \le X \le 1,\ 0 \le Y \le 1, \tag{133}$$

$$\overline{G}(X, Y, \tau) = -\frac{(X^2 + Y^2 + 4)\ \eta_0(\tau)}{\alpha T_r},\ \eta(0) = 1,\ \text{at}\ 0 \le X \le 1,\ 0 \le Y \le 1. \tag{134}$$

In this case, $a(\tau)$, $b(\tau)$, $c(\tau)$, $d(\tau)$, and their differentiation with respect to τ are

$$a(\tau) = \frac{1}{T_r},\ b(\tau) = \frac{1 + \eta(\tau)}{T_r},\ c(\tau) = \frac{1 + \eta(\tau)}{T_r},\ d(\tau) = \frac{1 + 2\eta(\tau)}{T_r}, \tag{135}$$

$$\dot{a}(\tau)=0,\ \dot{b}(\tau)=\frac{\dot{\eta}(\tau)}{T_r},\ \dot{c}(\tau)=\frac{\dot{\eta}(\tau)}{T_r},\ \dot{d}(\tau)=\frac{2\dot{\eta}(\tau)}{T_r}. \tag{136}$$

The transformed $\theta(X,Y,\tau)$ and $G(X,Y,\tau)$ are given as

$$\theta(X,Y,\tau)=\psi(X,Y,\tau)-\frac{1+(X+Y)\,\eta(\tau)}{T_r}, \tag{137}$$

$$G(X,Y,\tau)=-\frac{(X^2+Y^2+4)\,\eta_0(\tau)+\alpha(X+Y)\,\dot{\eta}(\tau)}{\alpha T_r}. \tag{138}$$

Following the proposed procedure, one has the space–time-dependent boundary and initial conditions as follows:

$$F_1(Y,\tau)=\frac{(Y^2-Y)\,\eta(\tau)}{T_r},\ \text{at } X=0,\ 0\le Y\le 1, \tag{139}$$

$$F_2(Y,\tau)=\frac{(Y^2-Y)\,\eta(\tau)}{T_r},\ \text{at } X=1,\ 0\le Y\le 1, \tag{140}$$

$$F_3(X,\tau)=\frac{(X^2-X)\,\eta(\tau)}{T_r},\ \text{at } Y=0,\ 0\le X\le 1, \tag{141}$$

$$F_4(X,\tau)=\frac{(X^2-X)\,\eta(\tau)}{T_r},\ \text{at } Y=1,\ 0\le X\le 1, \tag{142}$$

$$\theta_0(X,Y)=\frac{X^2+Y^2-(X+Y)}{T_r},\ \text{at}\tau=0,\ 0\le X\le 1,\ 0\le Y\le 1. \tag{143}$$

One separates $G(X,Y,\tau)$ and $\theta(X,Y,\tau)$ into two parts as follows:

$$G_a(X,Y,\tau)=-\frac{(X^2+2)\,\eta_0(\tau)+\alpha\,X\,\dot{\eta}(\tau)}{\alpha\,T_r},\ G_b(X,Y,\tau)=-\frac{(Y^2+2)\,\eta_0(\tau)+\alpha\,Y\,\dot{\eta}(\tau)}{\alpha\,T_r}, \tag{144}$$

$$\theta_{a0}(X,Y)=\frac{X^2-X}{T_r},\ \theta_{b0}(X,Y)=\frac{Y^2-Y}{T_r}. \tag{145}$$

The associated dimensionless quantities $G_{am}(X,\tau)$, $G_{bm}(Y,\tau)$, and $\overline{F}_{i,m}(\tau)(i=1,2)$ are

$$\begin{aligned}G_{am}(X,\tau)&=\frac{-2[(X^2+2)\eta_0(\tau)+\alpha X\dot{\eta}(\tau)][1-(-1)^m]}{\alpha T_r m\pi},\\ G_{bm}(Y,\tau)&=\frac{-2[(Y^2+2)\eta_0(\tau)+\alpha Y\dot{\eta}(\tau)][1-(-1)^m]}{\alpha T_r m\pi},\end{aligned} \tag{146}$$

$$\overline{F}_{i,m}(\tau)=\frac{-4\eta(\tau)}{T_r m^3\pi^3}[1-(-1)^m],\ i=1,2,3,4. \tag{147}$$

In addition, one can determine $T_{mna}(0)$ and $T_{mnb}(0)$ from the initial conditions of the transformed functions defined in Equations (56) and (A27) as

$$T_{mna}(0)=T_{mnb}(0)=\frac{-8(m^2-n^2)[1-(-1)^m][1-(-1)^n]}{T_r m^3 n^3\pi^4}. \tag{148}$$

Likewise, from Equations (54), (55), (A25), and (A26), one has

$$\lambda_{mna}=\lambda_{mnb}=\sqrt{m^2+n^2}\,\pi, \tag{149}$$

$$\begin{aligned}\gamma_{mna}(\tau)=\gamma_{mnb}(\tau)=&\ \frac{4[1-(-1)^m][2-(2-m^2\,\pi^2)(-1)^n]}{T_r\,m^3\,n\pi^4}\dot{\eta}(\tau)\\ &+\frac{8[1-(-1)^m][1-(-1)^n]}{T_r mn\pi^2}\eta(\tau)-\frac{4[1-(-1)^m][2(n^2\,\pi^2-1)\,-(3n^2\,\pi^2-2)(-1)^n]}{\alpha T_r mn^3\pi^4}\eta_0(\tau)\end{aligned}. \tag{150}$$

Therefore, one can determine $T_{mna}(\tau)$ and $T_{mnb}(\tau)$ as follows:

$$\begin{aligned} T_{mna}(\tau) = T_{mnb}(\tau) = & \frac{-8(m^2-n^2)[1-(-1)^m][1-(-1)^n]}{T_r\ m^3\ n^3\pi^4}\ e^{-(m^2+n^2)\ \pi^2} \\ & + \int_0^\tau e^{-(m^2+n^2)\ \pi^2(\tau-\phi)}\{\frac{4[1-(-1)^m][2-(2-m^2\ \pi^2)(-1)^n]}{T_r\ m^3 n\pi^4}\dot{\eta}(\phi) \\ & + \frac{8[1-(-1)^m][1-(-1)^n]}{T_r\ m\ n\pi^2}\eta(\phi) - \frac{4[1-(-1)^m][2(n^2\pi^2-1)-(3n^2\pi^2-2)(-1)^n]}{\alpha T_r\ m\ n^3\pi^4}\eta_0(\phi)\}\ d\phi \end{aligned} \quad (151)$$

(Case 1) Consider the time-dependent functions to be of exponential type as follows:

$$\eta_0(\tau) = e^{-D_0\tau},\ \eta(\tau) = e^{-D\tau}, \quad (152)$$

where D_0 and D represent the decay constants for the heat source and boundary conditions, respectively. From Equations (147) and (151), one obtains

$$\overline{F}_{i,m}(\tau) = \frac{-4e^{-D\tau}}{T_r\ m^3\ \pi^3}[1-(-1)^m],\ i = 1,2,3,4. \quad (153)$$

$$\begin{aligned} T_{mna}(\tau) = T_{mnb}(\tau) = & \frac{-8(m^2-n^2)[1-(-1)^m][1-(-1)^n]}{T_r\ m^3\ n^3\pi^4}\ e^{-(m^2+n^2)\ \pi^2} \\ & - \frac{4D\ [1-(-1)^m][2-(2-m^2\ \pi^2)(-1)^n]}{T_r\ m^3\ n\pi^4[(m^2+n^2)\ \pi^2-D]}[e^{-D\tau} - e^{-(m^2+n^2)\pi^2\tau}] \\ & + \frac{8\ [1-(-1)^m][1-(-1)^n]}{T_r\ m\ n\pi^2\ [(m^2+n^2)\ \pi^2-D]}[e^{-D\tau} - e^{-(m^2+n^2)\pi^2\tau}] \\ & - \frac{4[1-(-1)^m][2(n^2\pi^2-1)-(3n^2\pi^2-2)(-1)^n]}{\alpha T_r\ m\ n^3\pi^4\ [(m^2+n^2)\ \pi^2-D_0]}[e^{-D_0\tau} - e^{-(m^2+n^2)\pi^2\tau}] \end{aligned} \quad (154)$$

Furthermore, from Equations (60) and (14), the solutions $\theta(X,Y,\tau)$ and $\psi(X,Y,\tau)$ in dimensionless form are obtained as

$$\begin{aligned} \theta(X,Y,\tau) = & \sum_{m=1}^{\infty}\{\sum_{n=1}^{\infty}[\sin(n\pi X)\sin(m\pi Y) + \sin(n\pi Y)\sin(m\pi X)]\ T_{mna}(\tau)\} \\ & + \sum_{m=1}^{\infty}[\sin(m\pi Y) + \sin(m\pi X)]\ \overline{F}_{1,m}(\tau) \end{aligned} \quad (155)$$

$$\psi(X,Y,\tau) = \theta(X,Y,\tau) + \frac{1+(X+Y)e^{-D\tau}}{T_r}. \quad (156)$$

$\psi(X,Y,\tau)$ is returned to the dimensional form as

$$T(x,y,t) = T_r\,\theta + 1 + (x+y)\ e^{-\alpha D\ t}. \quad (157)$$

Considering the case $D_0 = D = \alpha = 1$, it can be verified that the above series solution is the same as that given by Siddique [20] (shown below).

$$T(x,y,t) = 1 + (x^2+y^2)\ e^{-\ t}. \quad (158)$$

$T_{mna}(\tau)$ and $T_{mnb}(\tau)$ can be re-expressed as

$$\begin{aligned} T_{mna}(\tau) = T_{mnb}(\tau) = & \frac{-8(m^2-n^2)[1-(-1)^m][1-(-1)^n]}{T_r\ m^3\ n^3\pi^4}\ e^{-(m^2+n^2)\ \pi^2} \\ & + \frac{8(m^2-n^2)\ [1-(-1)^m][1-(-1)^n]}{T_r m^3 n^3\pi^4[(m^2+n^2)\pi^2-1]}[e^{-\tau} - e^{-(m^2+n^2)\pi^2\tau}] \end{aligned} \quad (159)$$

The numerical result for the temperature of the rectangular region of the bar at the middle point in different times is shown in Table 2. From this table, one can find that the largest error was less than 0.1% when the number of terms $m = n$ was 5. The calculated temperature converged when 10-term expansion is applied. The temperature at $0 \le \tau \le 1.2$ was also verified with the results of the literature [20]. It turns out that the two results were identical when applying $m = n \ge 10$ expansion terms to obtain the proposed solution.

Table 2. The temperature of the rectangular region at $x = y = 0.5$ at different times $[\eta(t) = \eta_0(t) = e^{-t}]$.

t	$T(x = 0.5, y = 0.5, t)$					
	Number of Expansion Terms ($m = n$)					
	1	**3**	**5**	**10**	**20**	**Exact Solution [20]**
0	1.484	1.503	1.499	1.500	1.500	1.500
0.1	1.438	1.455	1.452	1.452	1.452	1.452
0.2	1.396	1.412	1.409	1.409	1.409	1.409
0.4	1.324	1.337	1.335	1.335	1.335	1.335
0.6	1.266	1.276	1.274	1.274	1.274	1.274
0.8	1.218	1.226	1.224	1.225	1.225	1.225
1.0	1.178	1.185	1.184	1.184	1.184	1.184
1.2	1.146	1.152	1.150	1.151	1.151	1.151

To gain insight into the effect of the decay constant for the boundary conditions and the heat source on the temperature in the middle of the rectangular region, two figures are established. First, Figure 4a plots the three temperature curves for the three decay constants of the boundary conditions ($D = 1, 1.5, 2$ and $D_0 = 1$). All three curves decayed exponentially and converge to 1. It can be found that, for the smallest decay constant $D = 1$ of the boundary conditions, the decrease rate of the temperature curve was largest as expected. This could mean that a smaller decay constant for the boundary conditions causes slower heat dissipation at the boundary; thus, the heat dissipation in the middle is faster and its temperature drops more quickly. Second, the temperature curves for the three cases of the different decay constants of the heat source are shown in Figure 4b ($D_0 = 1, 1.5, 2$ and $D = 1$). It can be seen that, for the smallest decay constant $D_0 = 1$ of the boundary conditions, the decrease rate of the temperature curve was largest. This can be explained by the fact that a smaller decay constant of the heat source leads to faster heat dissipation and, therefore, a faster temperature drop.

(Case 2) Consider the time-dependent functions to be of periodic type as follows:

$$\eta_0(\tau) = \cos(\omega_0 \tau),\ \eta(\tau) = \cos(\omega\tau), \tag{160}$$

where ω_0 and ω represent the frequency for the heat source and boundary conditions, respectively. Then, one has

$$\dot{\eta}(\tau) = -\omega \sin(\omega\tau), \tag{161}$$

and one can obtain $\overline{F}_{i,m}(\tau)$, $\gamma_{mna}(\tau) = \gamma_{mnb}(\tau)$, and $T_{mna}(\tau) = T_{mnb}(\tau)$ as follows:

$$\overline{F}_{i,m}(\tau) = \frac{-4\cos(\omega\tau)}{T_r m^3 \pi^3}[1 - (-1)^m],\ i = 1, 2, 3, 4, \tag{162}$$

$$\begin{aligned} \gamma_{mna}(\tau) = \gamma_{mnb}(\tau) = \tfrac{4[1-(-1)^m][2-(2-m^2\pi^2)(-1)^n]}{T_r m^3 n \pi^4}[-\omega\ \sin(\omega\tau)] \\ + \tfrac{8[1-(-1)^m][1-(-1)^n]}{T_r m n \pi^2}\cos(\omega\tau) - \tfrac{4[1-(-1)^m][2(n^2\pi^2-1)-(3n^2\pi^2-2)(-1)^n]}{\alpha T_r m n^3 \pi^4}\cos(\omega_0\tau) \end{aligned}, \tag{163}$$

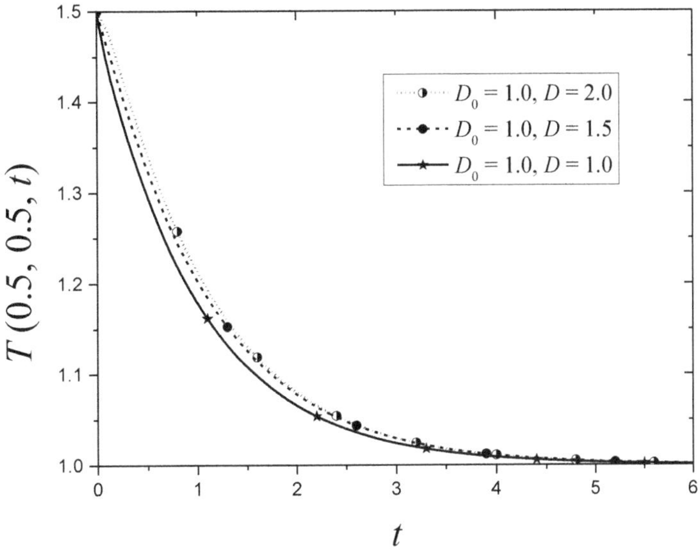

(**a**)

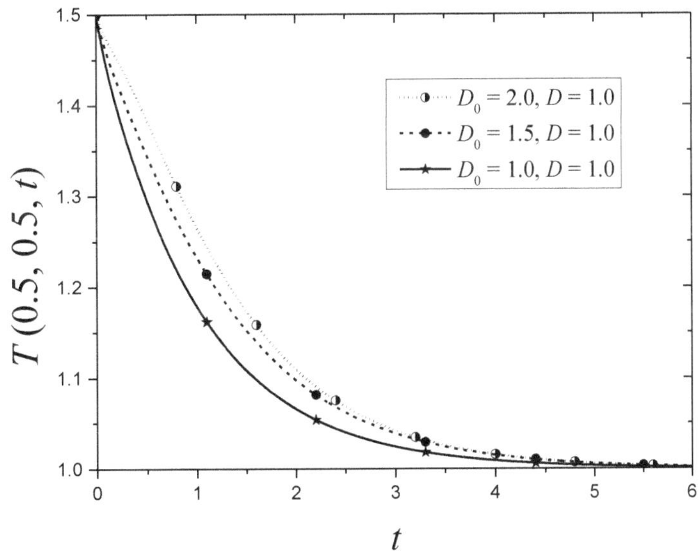

(**b**)

Figure 4. Temperature variation in the middle of the rectangular region with different decay constants of exponential-type time-dependent functions (case 1 of Example 3): (**a**) for different decay constants of the boundary conditions; (**b**) for different decay constants of the heat source.

$$
\begin{aligned}
T_{mna}(\tau) = T_{mnb}(\tau) = & \frac{-8(m^2-n^2)[1-(-1)^m][1-(-1)^n]}{T_r\ m^3\ n^3\pi^4}\ e^{-(m^2+n^2)\ \pi^2} \\
& + \frac{4[1-(-1)^m][2-(2-m^2\ \pi^2)(-1)^n]}{T_r\ m^3\ n\pi^4} \cdot \frac{\omega^2\ [\cos(\omega\tau)-e^{-(m^2+n^2)\ \pi^2\tau}]-\omega\ (m^2+n^2)\pi^2 \sin(\omega\tau)}{(m^2+n^2)^2\pi^4+\omega^2} \\
& + \frac{8[1-(-1)^m][1-(-1)^n]}{T_r\ m\ n\pi^2} \cdot \frac{(m^2+n^2)\pi^2\ [\cos(\omega\tau)-e^{-(m^2+n^2)\ \pi^2\tau}]+\omega\ \ \sin(\omega\tau)}{(m^2+n^2)^2\pi^4+\omega^2} \\
& - \frac{4[1-(-1)^m][2(n^2\pi^2-1)-(3n^2\pi^2-2)(-1)^n]}{\alpha T_r\ m\ \ n^3\pi^4} \cdot \frac{(m^2+n^2)\pi^2\ [\cos(\omega_0\tau)-e^{-(m^2+n^2)\ \pi^2\tau}]+\omega_0\ \ \sin(\omega_0\tau)}{(m^2+n^2)^{\ 2}\ \pi^4+\omega_0{}^2}
\end{aligned}
\quad . \tag{164}
$$

Therefore, from Equation (60), the solution in dimensional form is derived as

$$
\begin{aligned}
\theta(X,Y,\tau) = & \sum_{m=1}^{\infty} \{ \sum_{n=1}^{\infty} [\sin(n\pi X)\sin(m\pi Y) + \sin(n\pi Y)\sin(m\pi X)]\ T_{mna}(\tau) \} \\
& + \sum_{m=1}^{\infty} [\sin(m\pi Y) + \sin(m\pi X)]\ \overline{F}_{1,m}(\tau)
\end{aligned}
\quad , \tag{165}
$$

$$
\psi(X,Y,\tau) = \theta(X,Y,\tau) + \frac{1+(X+Y)\cos(\omega\tau)}{T_r}. \tag{166}
$$

Recovering it to the dimensional form yields

$$
T(x,y,t) = T_r\,\theta + 1 + (x+y)\ \cos(\alpha\ \omega\ t). \tag{167}
$$

Temperature variation in the middle of the rectangular region with different decay constants of harmonic type for the boundary conditions or the heat source was studied numerically. Figure 5a shows the temperature variation in the middle of the rectangular region for the three cases, including the frequencies $\omega = \pi, 5, 7$ and $\omega_0 = \pi$. Likewise, the results for the frequencies $\omega_0 = \pi, 5, 7$ and $\omega = \pi$ are shown in Figure 5b. These plots show that all the temperature curves oscillated up and down with the horizontal line $T = 1$. When the frequency of the heat source or boundary conditions was higher, the fluctuation of temperature change was more frequent.

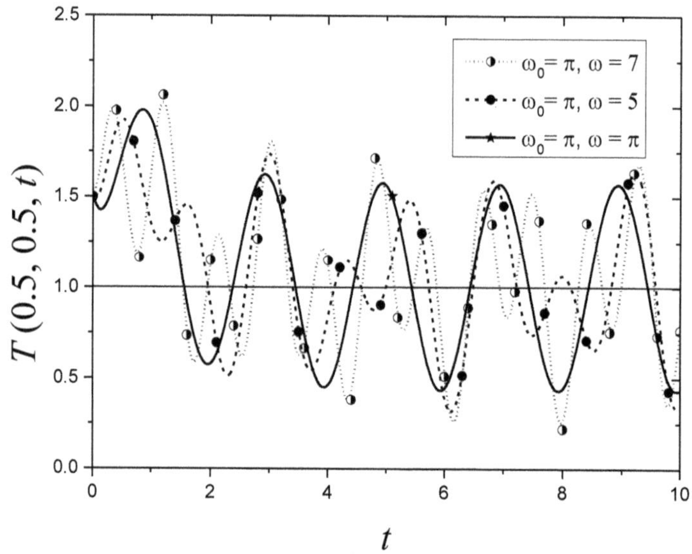

(a)

Figure 5. *Cont.*

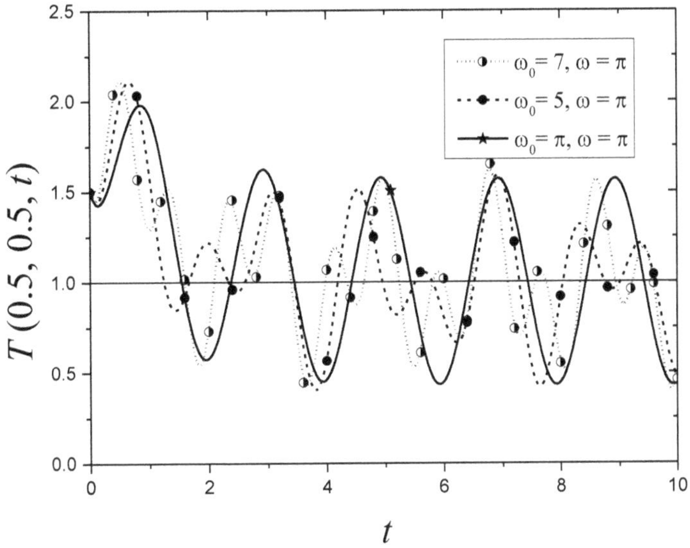

(**b**)

Figure 5. Temperature variation in the middle of the rectangular region with different decay constants of harmonic-type time-dependent functions (case 2 of Example 3): (**a**) for different decay constants of the boundary conditions; (**b**) for different decay constants of the heat source.

5. Conclusions

In this paper, the analytical solution of 2D transient heat conduction in a rectangular cross-section of an infinitely long bar with the space–time-dependent Dirichlet boundary conditions and heat sources was solved using a new analytical solution method that does not require any kind of integral transformation. Three examples were studied to illustrate the efficiency and reliability of the proposed method. Some results were verified to be the same as those in the literature [8,20,28].

The new findings of the present study are as follows:

(1) The purpose of this study was to complete the future work of our previous study [28], i.e., to remove the limitations of the previous study and add a heat source to the heat conduction system. The restriction of the temperatures of the boundary conditions and initial conditions at the four corners of the rectangular region to zero in the previous study was successfully eliminated. The zero temperature could be replaced by a function of time.

(2) From the examples, it was found that the convergence speed was very fast, and the maximum error was less than 0.1% when only five terms were used in the series expansion of the solution. Compared with the literature, the temperature could converge to the exact solution.

(3) The space–time-dependent functions used for the boundary conditions and heat sources in this study were considered separable in the space–time domain. The influence of the time-dependent function of the boundary conditions and heat sources on the temperature variation was investigated. For the exponential time-dependent function, a smaller decay constant (D_0 and D) of the time-dependent function ($e^{-D_0\tau}$ and $e^{-D\tau}$) led to a greater temperature drop. The temperatures with different decay constants converged to the same value. For the harmonic time-dependent function, a higher frequency (ω_0, and ω) of the time-dependent function ($\cos(\omega_0\tau)$ and $\cos(\omega\tau)$)

led to a more frequent fluctuation of the temperature change. All temperature curves oscillated above and below a horizontal line.

Author Contributions: Conceptualization, H.-P.H. and J.-R.C.; formal analysis, H.-P.H., J.-R.C., C.-Y.W. and C.-J.H.; methodology, H.-P.H., J.-R.C., C.-Y.W. and C.-J.H.; software, H.-P.H., C.-Y.W. and C.-J.H.; supervision, J.-R.C.; validation, C.-Y.W.; writing—original draft, H.-P.H. and J.-R.C.; writing—review and editing, J.-R.C. and C.-J.H. All authors have read and agreed to the published version of the manuscript.

Funding: This research received no external funding.

Data Availability Statement: Not applicable.

Acknowledgments: The authors specially thank Te-Wen Tu (Department of Mechanical Engineering, Air Force Institute of Technology, Taiwan (R.O.C.)) for his valuable advice, especially regarding the solution of Fourier series expansion.

Conflicts of Interest: The authors declare no conflict of interest.

Glossary

A, B	two subsystems
$a(\tau)$	dimensionless time-dependent function at the lower left corner of the cross-section
$b(\tau)$	dimensionless time-dependent function at the lower right corner of the cross-section
c	specific heat (W·S/kg·°C)
$c(\tau)$	dimensionless time-dependent function at the upper left corner of the cross-section
$d(\tau)$	dimensionless time-dependent function at the upper right corner of the cross-section
D_0, D	the decay constants for the heat source and boundary conditions, respectively
$f_i(y,t), i = 1, 2$	temperatures along the surface at the left end and the right end of the rectangular region
$f_i(x,t), i = 3, 4$	temperatures along the surface at the bottom end and the top end of the rectangular region
$\overline{F}_i(Y,\tau), i = 1, 2$	dimensionless quantity defined in Equation (8)
$\overline{F}_i(X,\tau), i = 3, 4$	dimensionless quantity defined in Equation (8)
$F_i(x,t), i = 3, 4$	transformed temperatures along the surface at the bottom and top end of the rectangular region
$F_i(y,t), i = 1, 2$	transformed temperatures along the surface at the left and right end of the rectangular region
$\overline{F}_{i,m}(\tau), i = 1, 2$	dimensionless quantity defined in Equation (32)
$\overline{F}_{i,m}(\tau), i = 3, 4$	dimensionless quantity defined in Equation (A7)
$g(x,y,t)$	the heat source inside the rectangular cross-section
$g_{i,m}(X), i = 1,2$	shifting functions
$g_{i,m}(Y), i = 3,4$	shifting functions
$\overline{G}(X,Y,\tau), G(X,Y,\tau)$	dimensionless heat sources
$G_a(X,Y,\tau), G_b(X,Y,\tau)$	dimensionless heat sources for subsystems A and B
$\overline{G}_m(X,\tau), \overline{G}_m(Y,\tau)$	nonhomogeneous terms in differential eqaquations of the transformed subsystems A and B
$G_{am}(X,\tau), G_{bm}(Y,\tau)$	series expansion of $G_a(X,Y,\tau), G_b(X,Y,\tau)$
k	thermal conductivity (W/m °C)
L_r	aspect ratio L_y/L_x defined in Equation (8)
L_x, L_y	thickness of the two-dimensional rectangular region in x- and y-directions (m)
$T(x,y,t)$	temperature function (°C)

$T_{mna}(\tau), T_{mnb}(\tau)$	dimensionless time variable of the transformed function defined in Equations (48) and (A22)
T_r	reference temperature (°C)
$T_0(x, y)$	initial temperature (°C)
t	time variable (s)
x	space variable in x-direction of a rectangular region (m)
X	dimensionless space variable in x-direction of a rectangular region
y	space variable in y-direction of a rectangular region (m)
Y	dimensionless space variable in y-direction of a rectangular region
α	thermal diffusivity (m^2/s)
ϕ	auxiliary integration variable
$\gamma_{mna}(\tau), \gamma_{mnb}(\tau)$	dimensionless quantity defined in Equations (55) and (A26)
$\eta_0(\tau), \eta(\tau)$	dimensionless time-dependent boundary conditions
$\lambda_{mna}, \lambda_{mnb}$	n-th eigenvalues dependent on ω_n defined in Equations (54) and (A25)
θ	dimensionless temperature
θ_0	dimensionless initial temperature
$\theta_a,\ \theta_b$	dimensionless temperature for subsystems A and B
$\theta_m(X, \tau)$	generalized Fourier coefficient defined in Equation (30)
$\overline{\theta}_m(X, \tau)$	transformed function defined in Equation (38)
$\overline{\theta}_{mn}(X, \tau)$	n-th eigenfunctions of the transformed function defined in Equation (48)
ρ	density (kg/m^3)
τ	dimensionless time
ψ	dimensionless temperature function
ω_0, ω	frequencies for the heat source and boundary conditions, respectively
ω_n	n-th eigenvalues for Sturm–Liouville problem defined in Equation (51)
Subscripts	
$0, 1, 2, 3, 4, a, b, i, m, n, r$	

Appendix A. An Analytic Solution of Subsystem B

For subsystem B, the boundary value problem is as follows:

$$\left[L_r^2\frac{\partial^2\theta_b(X,Y,\tau)}{\partial X^2}+\frac{\partial^2\theta_b(X,Y,\tau)}{\partial Y^2}\right]+G_b(X,Y,\tau)=\frac{\partial\theta_b(X,Y,\tau)}{\partial\tau},\ \text{in } 0<X<1,\ 0<Y<1, \tau>0, \tag{A1}$$

$$\theta_b(0,Y,\tau)=0,\ \theta_b(1,Y,\tau)=0, \tag{A2}$$

$$\theta_b(X,0,\tau)=F_3(X,\tau),\ \theta_b(X,1,\tau)=F_4(X,\tau), \tag{A3}$$

$$\theta_b(X,Y,0)=\theta_{b0}(X,Y). \tag{A4}$$

Because the boundary conditions of the bar with a rectangular cross-section at two opposite edges $X = 0$ and $X = 1$ are homogeneous, the temperature $\theta_b(X,Y,\tau)$ and the dimensionless quantities $F_3(X,\tau)$, $F_4(X,\tau)$ defined in Equation (A3) can be written as

$$\theta_b(X,Y,\tau)=\sum_{m=1}^{\infty}[\theta_m(Y,\tau)\sin(m\pi X)], \tag{A5}$$

$$G_b(X,Y,\tau)=\sum_{m=1}^{\infty}[G_{bm}(Y,\tau)\sin(m\pi X)], \tag{A6}$$

$$F_i(X,\tau)=\sum_{m=1}^{\infty}\left[\overline{F}_{i,m}(\tau)\sin(m\pi X)\right],\ i=3,4, \tag{A7}$$

where m denotes a positive integer, and $G_{b\,m}(Y,\tau)$ and $\overline{F}_{i,m}(\tau)(i=3,\,4)$ are expressed as

$$G_{bm}(Y,\tau)=2\int_0^1[G_b(X,Y,\tau)\sin(m\pi X)]dX, \tag{A8}$$

$$\overline{F}_{i,m}(\tau)=2\int_0^1 F_i(X,\tau)\sin(m\pi X)dX,\ i=3,4. \tag{A9}$$

Substituting Equations (A5)–(A7) into Equations (A1)–(A4), one has

$$\frac{\partial\theta_m(Y,\tau)}{\partial\tau}-\frac{\partial^2\theta_m(Y,\tau)}{\partial Y^2}+m^2\pi^2\theta_m(Y,\tau)=G_{bm}(Y,\tau), \tag{A10}$$

$$\theta_m(0,\tau)=\overline{F}_{3,m}(\tau),\ \theta_m(1,\tau)=\overline{F}_{4,m}(\tau), \tag{A11}$$

$$\theta_m(Y,0)=2\int_0^1\theta_{b0}(X,Y)\sin(m\pi X)dX. \tag{A12}$$

To find the solution for the second-order differential Equation (A10) with nonhomogeneous boundary conditions (A11), one uses the shifting function method by taking

$$\theta_m(Y,\tau)=\overline{\theta}_m(Y,\tau)+\sum_{i=3}^{4}g_{i,m}(Y)\overline{F}_{i,m}(\tau), \tag{A13}$$

where $\overline{\theta}_m(Y,\tau)$ is the transformed function, while $g_{i,m}(Y)(i=3,4)$ indicates the shifting functions to be specified.

Substituting Equation (A13) back into Equations (A10)–(A12), one obtains

$$\begin{aligned}&\dot{\overline{\theta}}_m(Y,\tau)+\sum_{i=3}^{4}[g_{i,m}(Y)\dot{\overline{F}}_{i,m}(\tau)]-\left[\overline{\theta}''{}_m(Y,\tau)+\sum_{i=3}^{4}g''{}_{i,m}(Y)\overline{F}_{i,m}(\tau)\right]\\&+m^2\pi^2L_y^2\left\{\overline{\theta}_m(Y,\tau)+\sum_{i=3}^{4}[g_{i,m}(Y)\overline{F}_{i,m}(\tau)]\right\}=G_{bm}(Y,\tau)\end{aligned}. \tag{A14}$$

Then, the associated boundary conditions become

$$\overline{\theta}_m(0,\tau)+g_{3,m}(0)\overline{F}_{3,m}(\tau)+g_{4,m}(0)\overline{F}_{4,m}(\tau)=\overline{F}_{3,m}(\tau), \tag{A15}$$

$$\overline{\theta}_m(1,\tau)+g_{3,m}(1)\overline{F}_{3,m}(\tau)+g_{4,m}(1)\overline{F}_{4,m}(\tau)=\overline{F}_{4,m}(\tau). \tag{A16}$$

Like the derivation procedure, these two shifting functions are determined as

$$g_{3,m}(Y)=1-Y,\ g_{4,m}(Y)=Y. \tag{A17}$$

After substituting Equation (A17) into Equations (A14)–(A16), one has the differential equation for $\overline{\theta}_m(Y,\tau)$,

$$\dot{\overline{\theta}}_m(Y,\tau)-\overline{\theta}''{}_m(Y,\tau)+m^2\pi^2L_y^2\,\overline{\theta}_m(Y,\tau)=\overline{G}_m(Y,\tau), \tag{A18}$$

and the associated homogeneous boundary conditions

$$\overline{\theta}_m(0,\tau)=0,\ \overline{\theta}_m(1,\tau)=0, \tag{A19}$$

where $\overline{G}_m(Y,\tau)$ is defined as

$$\overline{G}_m(Y,\tau)=-\sum_{i=3}^{4}\left[m^2\pi^2L_y^2\,g_{i,m}(Y)\,\overline{F}_{i,m}(\tau)+g_{i,m}(Y)\,\dot{\overline{F}}_{i,m}(\tau)\right]+G_{b\,m}(Y,\tau). \tag{A20}$$

Furthermore, the initial condition is transformed to be

$$\overline{\theta}_m(Y,0) = 2\int_0^1 \theta_{b0}(X,Y)\sin(m\pi X)dX - \sum_{i=3}^{4}[g_{i,m}(Y)\overline{F}_{i,m}(0)]. \tag{A21}$$

The solution $\overline{\theta}_m(Y,\tau)$ specified by Equations (A18)–(A21) can be expressed in the form of eigenfunctions as

$$\overline{\theta}_m(Y,\tau) = \sum_{n=1}^{\infty} \overline{\theta}_{mn}(Y)T_{mnb}(\tau), \tag{A22}$$

where $\overline{\theta}_{mn}(Y)$ are

$$\overline{\theta}_{mn}(Y) = \sin(n\,\pi\,Y),\ n = 1,\ 2,\ 3,\ \cdots \tag{A23}$$

Substituting Equation (A22) into Equation (A18), multiplying it by $\overline{\theta}_{mn}(Y)$, and integrating from 0 to 1, one has

$$\dot{T}_{mnb}(\tau) + \lambda_{mnb}^2 T_{mnb}(\tau) = \gamma_{mnb}(\tau), \tag{A24}$$

where λ_{mnb} and $\gamma_{mnb}(\tau)$ are

$$\lambda_{mnb} = \sqrt{m^2L_r^2 + n^2\pi}, \tag{A25}$$

$$\begin{aligned}\gamma_{mnb}(\tau) = 2\textstyle\int_0^1 \overline{\theta}_{mn}(Y)\,\overline{G}_m(Y,\tau)dY = \frac{-2}{n\pi}\left[\dot{\overline{F}}_{3,m}(\tau) - (-1)^n\dot{\overline{F}}_{4,m}(\tau)\right] \\ -\frac{2m^2\pi\,L_r^2}{n}\left[\overline{F}_{3,m}(\tau) - (-1)^n\overline{F}_{4,m}(\tau)\right] + 2\textstyle\int_0^1 \overline{\theta}_{mn}(Y)\,G_{bm}(Y,\tau)dY\end{aligned}. \tag{A26}$$

$T_{mnb}(0)$ can be determined from the initial condition of the transformed function defined in Equation (A21) as

$$T_{mnb}(0) = 4\int_0^1 \sin(n\pi Y)\int_0^1 \theta_{b0}(X,Y)\sin(m\pi X)dXdY - \frac{2L_r^2}{n\pi}[\overline{F}_{3,m}(0) - (-1)^n\overline{F}_{4,m}(0)]. \tag{A27}$$

The general solution of Equation (A24) with the initial condition is obtained as

$$T_{mnb}(\tau) = e^{-\lambda_{mnb}^2\tau}T_{mnb}(0) + \int_0^\tau e^{-\lambda_{mnb}^2(\tau-\phi)}\gamma_{mnb}(\phi)d\phi. \tag{A28}$$

References

1. Carslaw, H.; Jaeger, J. *Heat in Solids*, 2nd ed.; Clarendon Press: Oxford, UK, 1959.
2. Özisik, M.N. *Heat Conduction*; John Wiley & Sons: New York, NY, USA, 1993.
3. Cole, K.D.; Beck, J.V.; Haji-Sheikh, A.; Litkouhi, B. *Heat Conduction Using Green's Functions*; Taylor & Francis: Boca Raton, FL, USA, 2010.
4. Holy, Z. Temperature and stresses in reactor fuel elements due to time-and space-dependent heat-transfer coefficients. *Nucl. Eng. Des.* **1972**, *18*, 145–197. [CrossRef]
5. Özişik, M.N.; Murray, R. On the solution of linear diffusion problems with variable boundary condition parameters. *J. Heat Transf.* **1974**, *96*, 48–51. [CrossRef]
6. Johansson, B.T.; Lesnic, D. A method of fundamental solutions for transient heat conduction. *Eng. Anal. Bound. Elem.* **2008**, *32*, 697–703. [CrossRef]
7. Chantasiriwan, S. Methods of fundamental solutions for time-dependent heat conduction problems. *Int. J. Numer. Methods Eng.* **2006**, *66*, 147–165. [CrossRef]
8. Young, D.; Tsai, C.; Murugesan, K.; Fan, C.; Chen, C. Time-dependent fundamental solutions for homogeneous diffusion problems. *Eng. Anal. Bound. Elem.* **2004**, *28*, 1463–1473. [CrossRef]
9. Zhu, S.P.; Liu, H.W.; Lu, X.P. A combination of LTDRM and ATPS in solving diffusion problems. *Eng. Anal. Bound. Elem.* **1998**, *21*, 285–289. [CrossRef]
10. Bulgakov, V.; Šarler, B.; Kuhn, G. Iterative solution of systems of equations in the dual reciprocity boundary element method for the diffusion equation. *Int. J. Numer. Methods Eng.* **1998**, *43*, 713–732. [CrossRef]
11. Chen, H.T.; Sun, S.L.; Huang, H.C.; Lee, S.Y. Analytic closed solution for the heat conduction with time dependent heat convection coefficient at one boundary. *Comput. Model. Eng. Sci.* **2010**, *59*, 107–126.

12. Lee, S.Y.; Huang, T.W. A method for inverse analysis of laser surface heating with experimental data. *Int. J. Heat Mass Transf.* **2014**, *72*, 299–307. [CrossRef]
13. Lee, S.Y.; Yan, Q.Z. Inverse analysis of heat conduction problems with relatively long heat treatment. *Int. J. Heat Mass Transf.* **2017**, *105*, 401–410. [CrossRef]
14. Walker, S. Diffusion problems using transient discrete source superposition. *Int. J. Numer. Methods Eng.* **1992**, *35*, 165–178. [CrossRef]
15. Chen, C.; Golberg, M.; Hon, Y. The method of fundamental solutions and quasi-Monte-Carlo method for diffusion equations. *Int. J. Numer. Methods Eng.* **1998**, *43*, 1421–1435. [CrossRef]
16. Zhu, S.P. Solving transient diffusion problems: Time-dependent fundamental solution approaches versus LTDRM approaches. *Eng. Anal. Bound. Elem.* **1998**, *21*, 87–90. [CrossRef]
17. Sutradhar, A.; Paulino, G.H.; Gray, L. Transient heat conduction in homogeneous and non-homogeneous materials by the Laplace transform Galerkin boundary element method. *Eng. Anal. Bound. Elem.* **2002**, *26*, 119–132. [CrossRef]
18. Burgess, G.; Mahajerin, E. Transient heat flow analysis using the fundamental collocation method. *Appl. Therm. Eng.* **2003**, *23*, 893–904. [CrossRef]
19. Young, D.; Tsai, C.; Fan, C. Direct approach to solve nonhomogeneous diffusion problems using fundamental solutions and dual reciprocity methods. *J. Chin. Inst. Eng.* **2004**, *27*, 597–609. [CrossRef]
20. Siddique, M. Numerical computation of two-dimensional diffusion equation with nonlocal boundary conditions. *Int. J. Appl. Math.* **2010**, *40*, 26–31.
21. Singh, S.; Jain, P.K.; Uddin, R. Finite integral transform method to solve asymmetric heat conduction in a multilayer annulus with time-dependent boundary conditions. *Nucl. Eng. Des.* **2011**, *241*, 144–154. [CrossRef]
22. Hematiyan, M.R.; Mohammadi, M.; Marin, L.; Khosravifard, A. Boundary element analysis of uncoupled transient thermo-elastic problems with time- and space-dependent heat sources. *Appl. Math. Comput.* **2011**, *218*, 1862–1882. [CrossRef]
23. Daneshjou, K.; Bakhtiari, M.; Parsania, H.; Fakoor, M. Non-Fourier heat conduction analysis of infinite 2D orthotropic FG hollow cylinders subjected to time-dependent heat source. *Appl. Therm. Eng.* **2016**, *98*, 582–590. [CrossRef]
24. Gu, Y.; Lei, J.; Fan, C.M.; He, X.Q. The generalized finite difference method for an inverse time-dependent source problem associated with three-dimensional heat equation. *Eng. Anal. Bound. Elem.* **2018**, *91*, 73–81. [CrossRef]
25. Biswas, P.; Singh, S.; Srivastava, A. A unique technique for analytic solution of 2-D dual phase lag bio-heat transfer problem with generalized time-dependent boundary conditions. *Int. J. Therm. Sci.* **2022**, *147*, 106139. [CrossRef]
26. Akbari, S.; Faghiri, S.; Poureslami, P.; Hosseinzadeh, K.; Shafii, M.B. Analytical solution of non-Fourier heat conduction in a 3-D hollow sphere under time-space varying boundary conditions. *Heliyon* **2022**, *8*, e12496. [CrossRef] [PubMed]
27. Zhou, L.; Sun, C.; Xu, B.; Peng, H.; Cui, M.; Gao, X. A new general analytical PBEM for solving three-dimensional transient nonlinear heat conduction problems with spatially-varying heat generation. *Eng. Anal. Bound. Elem.* **2023**, *152*, 334–346. [CrossRef]
28. Hsu, H.P.; Tu, T.W.; Chang, J.R. An analytic solution for 2D heat conduction problems with general Dirichlet boundary conditions. *Axioms* **2023**, *12*, 416. [CrossRef]
29. Zhang, L.; Zhao, H.; Liang, S.; Liu, C. Heat transfer in phase change materials for integrated batteries and power electronics systems. *Appl. Therm. Eng.* **2023**, *232*, 120997. [CrossRef]
30. Zhang, H.; Gu, D.; Ma, C.; Guo, M.; Yang, J.; Wang, R. Effect of post heat treatment on microstructure and mechanical properties of Ni-based composites by selective laser melting. *Mat. Sci. Eng. A-Struct.* **2019**, *765*, 138294. [CrossRef]

axioms

Article

Fluid Dynamics Calculation in SF6 Circuit Breaker during Breaking as a Prerequisite for the Digital Twin Creation

Vladislav V. Popovtsev, Alexandra I. Khalyasmaa * and Yurii V. Patrakov

Ural Power Engineering Institute, Ural Federal University Named after the First President of Russia B.N. Yeltsin, 620002 Ekaterinburg, Russia; vladislav.popovtsev@urfu.ru (V.V.P.); iuvpatrakov@urfu.ru (Y.V.P.)
* Correspondence: a.i.khaliasmaa@urfu.ru

Abstract: The requirements to switching the capacities of SF6 circuit breakers submitted by Russian Grid companies are difficult to satisfy. The first limitation is related to material and financial costs in order to create a new requirement-satisfying switching device. The second limitation is dictated by the necessity of calculating complex physical processes in a circuit braker interrupter during fault–current making or breaking before creating a prototype. The latter task is reduced to the problem of simulating the processes of interaction between the switching arc and the SF6 gas flow. This paper deals with the solution of the problem both analytically by a special method and numerically by a numerical software package through the creation of a mathematical model of the interaction process. The switching arc is taken into account as a form of a temperature source, based on experimental data on measuring the temperature of the arc column. The key feature of the research is to use the finite element method based on a moving mesh—the Arbitrary Lagrangian Eulerian (ALE) method. Such a problem statement allows us to take the contact separation curve of the circuit breaker into account as the input data of the model. The calculations were carried out during fault-current breaking by a 110 kV SF6 dead-tank circuit breaker. The calculations of pressure and mass flow in the under-piston volume change, gas flow speed, and temperature depending on the contact separation are given. The proposed model of the switching arc was used to simulate the process of 25 kA symmetrical fault–current breaking and was compared with an experiment.

Keywords: computational fluid dynamics; SF6 circuit breaker; switching arc; moving mesh; ALE; arc quenching

MSC: 68T20

Citation: Popovtsev, V.V.; Khalyasmaa, A.I.; Patrakov, Y.V. Fluid Dynamics Calculation in SF6 Circuit Breaker during Breaking as a Prerequisite for the Digital Twin Creation. *Axioms* **2023**, *12*, 623. https://doi.org/10.3390/axioms12070623

Academic Editor: Leonid Plotnikov

Received: 2 May 2023
Revised: 12 June 2023
Accepted: 19 June 2023
Published: 22 June 2023

1. Introduction

The increase in power consumption in the Unified National Power Grid (UNPG) of Russia, along with the expansion of the technical and regulatory framework, quantitatively increases the used equipment or their replacements. One of the most important elements of the Electric Power System (EPS) ensuring its reliability is a such electrical device as a power circuit breaker (CB). Its main tasks are to interrupt short-circuit currents and isolate faulty parts of an EPS. However, conflicting requirements are simultaneously imposed on the CB. On the one hand, during contingencies—fault–current breaking or making—the CB must turn them off and provide an infinitely large resistance between the arcing contacts. On the other hand, under normal conditions, the operating currents through its contact system and its resistance must be infinitely small in order to avoid unnecessary losses of the power to be transmitted [1].

Thus, the following requirements are imposed on the CB:

- Low resistance in normal conditions (in the normally closed contact);
- High-voltage proof of external and internal insulation, which makes it possible to withstand lightning and switching overvoltage, as well as transient recovering voltage (TRV) after the arc is extinguished;

- The ability of both making and breaking the short-circuit currents —the CB must reliably extinguish the arc without its re-ignition;
- Ensuring fast transition from the closed to open position and vice versa, especially in automatic reclosing cycles.

In addition to the short-circuit currents, the CB must also provide the switching capacity of capacitive currents of unloaded overhead transmission lines, capacitor banks, and inductive currents of shunt reactors in accordance with Russian [2] and foreign [3–5] standards.

The principle of operation in most CBs as mechanical switching devices is based on actuating the operating mechanism—the drive. When the protective relay sends a command to the opening solenoid, the CB must operate within a very short period of time (the total opening time of gas CBs with a rated voltage of $U_{nom} = 110$ kV is 55 ± 5 ms).

When the contacts open, an electric arc occurs between them. At its core, an electric arc is an independent arc discharge, which is a low-temperature plasma channel characterized by a high current density and a low cathode voltage drop [6,7].

In CBs with a rated voltage of 110 kV and above, the electric arc is a burning high-pressure arc in oil. It can also be compressed air or another gaseous dielectric with good arc-extinguishing properties, such as SF6 gas. The interruption of the current is carried out by cooling the arc plasma in such a way that the resulting electric arc disappears at the first occurring current zero after the contact separation. This process of cooling or extinguishing can be carried out in various ways, due to which the CBs are classified according to the type of arc quenching medium and the interrupter type. Currently, most CBs use compressed SF6 gas as an arc extinguishing medium—it has a high voltage proof (2.5 times higher than that of air) and a high heat transfer coefficient [8]. The arc extinguishing process itself lies in the fact that the arc under the high speed is blown by the cold high-pressure gas—puffer and self-blast technologies [9]. A modern interrupter is designed in such a way that the gas flow, cooling the arc, is supersonic (Mach numbers > 1) in order to level out the re-ignition of the arc in the next power frequency half-cycle when the current zero approaches [10].

Internationally, special scientific interest has been shown to SF6 as an arc-quenching medium in CBs. The main reason is that it is quite difficult to find a worthy alternative with the same arc-quenching characteristics so that the CB can provide high switching capacity [11]. Despite the active search for alternatives to SF6, due to its high global warming potential, an environmentally friendly and dielectrically equivalent medium has not been found. Thus, the leading manufacturers of high voltage CBs continue to use SF6.

In Russia, the interest in SF6 CBs is supported by the regulatory documents from the operator of the Russian power grids—the Federal Grid Company (FGC)—Rosseti (Regulation "On the Unified Technical Policy in the Integrated Power Grids"). It regulates the preferential use of SF6 CBs for the rated voltages of 110 kV and higher. In addition, according to the digital policy of the FGC–Rosseti (the concept of "Digital Transformation 2030"), one of the best methods to accurately control the operation of electrical equipment is to use an effective monitoring system.

Given the growth of short-circuit currents in power grids of 110 kV and above [12,13], there is a problem in increasing the CBs' switching capacities. One of the solutions is to optimize the design of the CBs or create new devices [14]. However, the development and design of high voltage switching devices is an expensive undertaking, as it requires numerous experimental tests, both on physical models and prototypes. In addition, there is a need to take into account the conflicting requirements for the CB interrupter in terms of its switching capacity, mechanical characteristics, electrical insulation level, etc., which also requires time and financial costs.

Thus, the calculation and modeling of internal processes occurring in the interrupter of SF6 CB is of particular scientific interest, both in the field of the operation of switching equipment and in the field of its design. Therefore, a course should be taken towards the development of approaches to modeling complex physical (gas-, thermo-, electrodynamic, and electrophysical) processes that occur in SF6 CB interrupters during fault–current making or breaking, with the possibility of verifying the resulting model—creating its digital twin.

2. Switching of SF6 Circuit Breakers

2.1. Interrupter Types in SF6 Circuit Breakers

Arc extinguishing in modern SF6 CBs occurs in the interrupter or the interruption chamber. These are special chambers in which the process of cooling the arc should be intensified, removing heat from it. Such a process is called blast, which is possible on exposure to the SF6 gas, flowing at the speed of sound, relative to the arc column. It is possible to organize effective blasting in SF6 CBs of 110 kV and higher in several ways, according to which the following designs of interrupters are distinguished [10,11,15–17]:

(1) Puffer type, single pressure: the blast is created by the means of a built-in compression device that creates excess pressure due to the drive's energy;

(2) Double pressure: equipped with a longitudinal blowing system, in which pre-compressed gas is supplied from a reservoir with a relatively high pressure of SF6 gas, which, after the arc is extinguished, undergoes a recompression process;

(3) Self-blast/Auto-puffer: blast occurs during the thermal expansion of SF6 gas using the energy of the arc itself to (partly) produce the pressure necessary to blast the arc;

(4) Magnetic arc rotation: interrupters with electromagnetic blast;

(4.1) The arc is extinguished by its rapid movement in a stationary SF6 gas under the influence of a radial magnetic field, created by arc itself;

(4.2) A longitudinal blowing system, in which the increase in pressure in the SF6 gas occurs when it is heated by an arc, rotating in a special chamber under the influence of a magnetic field.

Basically, in SF6 CBs with a rated voltage of 110 kV and above, the first two types of interrupter designs are used, along with the thermal expansion of SF6 gas under the action of a switching arc. For higher rated voltages (220 kV and above), or for higher switching capacity requirements for CBs with a voltage of 110 kV (rated breaking current of 50 kA and above), the operation of the interrupter can be further optimized. For example, by using the so-called double-motion principle, which consists of moving two arcing contacts in opposite directions. Another option is using a double-speed mechanism, which consists of dividing the mass of the moving contacts into two parts (upper and lower) and temporarily transferring part of the kinetic energy from the lower mass to the upper mass.

Single pressure SF6 circuit breaker operating.

In such designs, the overpressure that provides the gas flow is formed during the switching. Interrupter types are divided into a puffer type and a self-blast type.

Figure 1 shows a diagram of a double-blast, puffer-type interrupter. Inside the sealed insulating chamber filled with SF6 gas, the two contacts 1 and 2 are rigidly connected to each other. They are connected to the power drive mechanism through an insulating rod (is not shown in Figure 1). First, during the fault–current breaking, practically without discharges, the main contacts 1 and 3 open, and then the current passes into the arcing contact zone between contacts 2 and 5, where arc 4 burns. The insulating rod moves the entire moving system relative to the fixed piston 8, whereas, as it moves, SF6 gas is compressed in the working capacity of the cylinder cavity *B*—the under-piston volume. Thus, the principle of self-blast is implemented. Arc 4, which occurs between the arcing contacts 2 and 5, is drawn into the nozzles 6 and 7 by the flow of the compressed SF6 gas. A double-blast principle is provided, which intensively affects the arc shaft, which goes out in one of the current zeroes.

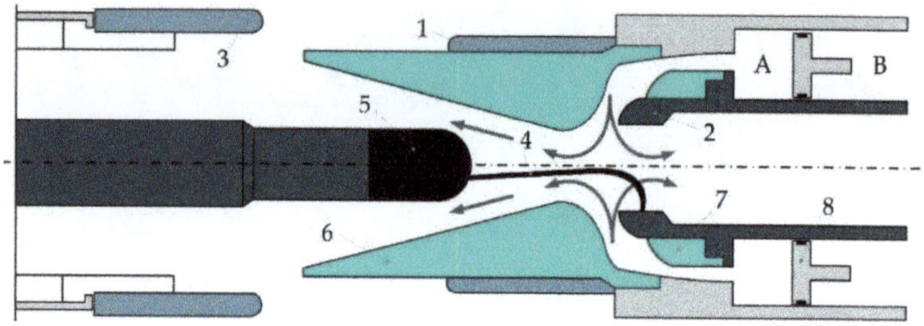

Figure 1. Puffer-type interrupter: 1—moving main contact; 2—moving arcing contact; 3—fixed main contact; 4—arc; 5—fixed arcing contact; 6—PTFE main nozzle; 7—PTFE auxiliary nozzle; 8—piston; A—above-piston volume; B—under-piston volume.

The improvement of SF6 CBs due to the increased requirements for switching capacity, on the one hand, is associated with an increase in the rated breaking current per one pole of the CB [16,17]. On the other hand, the goal is to reduce the power of the drive mechanism. However, these two methods contradict each other since a decrease in drive power causes the pressure drop at the moment of arc quenching, which means a decrease in breaking capacity. A fundamentally different way to increase the efficiency of the arc extinguishing of SF6 CBs during electric arc burning in the nozzle channel is possible when the nozzle is made of an insulating material—polytetra-fluoro-ethylene (PTFE). The impact of radiation energy on the inner surface of the insulating nozzle initiates an additional gas blast due to the ablation of the insulating walls, accompanied by the release of C2 and CF4. This leads to an increase in pressure in the contact gap and a consumption effect that limits the access of the arc-quenching medium to the contact gap at the maximum of the current to be broken. According to Figure 2a, the ablation of the insulating walls of the PTFE nozzle 2 occurs between the arcing contacts 3 and 4. This effect makes it possible to increase the gas pressure in chamber *K* not only due to the high temperature but also due to the additional mass flow from the gas-generating walls of the chamber [9,10].

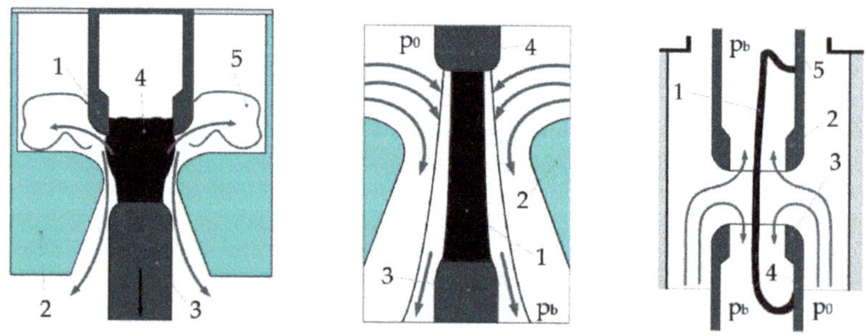

(a) (b) (c)

Figure 2. Interrupter types: (**a**) Self-blast; (**b**) Single-blast; (**c**) Full double-blast.

Double-pressure SF6 circuit breakers.

In such designs, a pre-created pressure drop is formed during the shutdown process, which provides the gas flow. The arc-quenching devices of this type are divided into single-blast (Figure 2b) and double-blast (Figure 2c).

In the case of a longitudinal blast (Figure 2a), it is possible to obtain a better SF6 flow and the absence of the so-called "dead zone". In this case, the direction of gas movement coincides with the longitudinal axis of the interrupter and the axis of the switching arc. The arc that occurs between contacts 3 and 4 interacts with the longitudinal gas flow formed by the nozzle 2. The gas flow is provided by the pressure difference in the upper and lower parts of the flow—p_0 and p_b, respectively. During the arc quenching, this difference is not constant. However, in the optimal case of arc quenching, it provides the supercritical gas outflow case ($p_0 > p_b$)—the critical gas mass flow rate for the longest possible period of time [9]. The main parameters of the interrupter that affect the formation of the gas flow are given in Section 3, according to [15,18].

In case of double blast (Figure 2b), the direction of the gas flow's movement, blowing over the arc, was the opposite. The arc 1 burned between the contacts 3 and 4 in the gas flow, formed by the two nozzles 2. These flows were formed by the channel made up of the tips of the contacts. The flow was first directed perpendicular to the arc and then rotated 90°. In this case, on the axis of the nozzle channel, where two radially directed jets met (at the point where the flow turns), a stagnation area, or a "dead zone", was formed, where the effect of the SF6 gas on the arc column was minimal. In this "dead zone", the residual arc column had an increased diameter, the decay processes of the residual arc shaft were slowed down, and a cloud of hot, conductive gas remained there for a long time, which negatively affected the extinguishing process [15].

CBs with double-pressure interrupter types are structurally complex. Firstly, they require an automatic compressor to recompress the gas from the low-pressure reservoir to the high-pressure one; secondly, the heating of the SF6 gas in the high-pressure reservoir [15,16] is required. The latter is related to the fact that SF6 liquefies at about 10 °C and at a pressure of about 1.6 MPa. Therefore, it is necessary to install heaters in the high-pressure reservoir to avoid SF6 liquefaction [8,16].

2.2. Models of Switching Arc Interaction with SF6 Gas Flow

All calculations of gas (fluid) dynamics processes are based on three well-known equations of fluid and gas mechanics:

- Continuity equation (law of conservation of mass);
- Equation of the second law (law of conservation of momentum);
- Energy equation (law of conservation of energy).

In order to describe the processes occurring in the arc within the framework of interaction with the SF6 gas flow, it is necessary to set a qualitative mathematical model.

Analytical models.

One of the first models was proposed by Cassie in 1939 [19]. It describes the process of arcing at high currents and is based on the assumption that the voltage of the arc column is constant. The model calculates an arc channel with a constant temperature, current density, and electric field strength. Changes in the conductivity of the arc are caused by the changes in the cross-section of the arc. The tension on the arc shaft is constant and does not depend on the arc current.

Mayer's model, proposed in 1943 [20], describes the process of arcing at currents, close to zero, and is based on the assumption that the power removed from the arc is unchanged:

$$\frac{1}{g}+\frac{dg}{dt}=\frac{d\ln g}{dt}=\frac{1}{\tau}\left(\frac{ui}{P}-1\right) \tag{1}$$

where g is arc conductivity; t is arcing time; u is arc voltage; and i is arc current.

The arc is represented as a non-linear resistance in an equivalent circuit. The main idea of the model is that only convection causes power losses, i.e., the temperature in the arc is not constant. This means that the cross-sectional area of the arc is proportional to the current and that the arc voltage is constant. Furthermore, it is believed that power losses are caused by thermal conduction at low currents. This means that the conductivity strongly depends on temperature but does not depend on the cross-sectional area of the arc.

The Cassie and Mayer models have modifications—hybrid models that allow a more accurate description of the breaking arc, for example, Brown's model [21], which he subsequently applied to analyze the breaking process in the post-zero arcing period, characterized by energy balance [22]. Undoubtedly, the above models are useful but have limited applications, as they are based on ordinary differential equations. In other words, they cannot be used to study in detail the physical processes during arcing, i.e., evaluate the Interaction with the SF6 gas flow since it is described by the equations of gas dynamics, which are differential equations in partial derivatives [23].

Modified arc models.

More promising arc models [24–29] have limitations in their uses, described in [23]. They are based on the continuity, momentum, and energy equations of gas dynamics and Ohm's Law.

Continuity equation:

$$\frac{\partial \rho}{\partial t} + \frac{\partial(\rho \vartheta_{\mathbf{z}})}{\partial \mathbf{z}} + \frac{1}{\mathbf{r}} \cdot \frac{\partial(\mathbf{r}\rho\vartheta_{\mathbf{r}})}{\partial \mathbf{r}} = 0 \tag{2}$$

Axial momentum equation:

$$\rho\frac{\partial \vartheta_{\mathbf{z}}}{\partial t} + \rho\vartheta_{\mathbf{z}}\frac{\partial \vartheta_{\mathbf{z}}}{\partial \mathbf{z}} + \rho\vartheta_{\mathbf{r}}\frac{\partial \vartheta_{\mathbf{z}}}{\partial \mathbf{r}} = -\frac{\partial p}{\partial \mathbf{z}} + \frac{1}{\mathbf{r}} \cdot \frac{\partial}{\partial \mathbf{r}}\left[(\eta + \eta_t) \cdot \mathbf{r}\frac{\partial \vartheta_{\mathbf{z}}}{\partial \mathbf{r}}\right] \tag{3}$$

Energy equation:

$$\rho\frac{\partial h_0}{\partial t} + \rho\vartheta_{\mathbf{z}}\frac{\partial h_0}{\partial \mathbf{z}} + \rho\vartheta_{\mathbf{r}}\frac{\partial h_0}{\partial \mathbf{r}} = \sigma\mathbf{E}^2 - U + \frac{1}{\mathbf{r}} \cdot \frac{\partial}{\partial \mathbf{r}}\left[(k + k_t) \cdot \mathbf{r}\frac{\partial T}{\partial \mathbf{r}}\right] \tag{4}$$

Ohm's law:

$$I = \mathbf{E}\int_0^{r_1} 2\pi \mathbf{r}\sigma \cdot d\mathbf{r} \tag{5}$$

In the equations above, ρ is gas density; $\vartheta_{\mathbf{z}}$ is axial velocity; $\vartheta_{\mathbf{r}}$ is radial velocity; p is gas pressure; η is molecular viscosity; η_t is turbulent viscosity; σ is electrical conductivity; $\mathbf{E}$ is the voltage gradient; U is the net emission coefficient; k is thermal conductivity; k_t is turbulent thermal conductivity; and h_0 is total enthalpy.

Generally, the axial symmetric problem statement means that Equations (2)–(5) are integrated over the radius within limits from a to b, and equations from [23] are obtained.

Based on the above integral equations of the arc, it is possible to adequately describe the interaction of the arc with the blown SF6 gas flow. In other words, in contrast to the problems of magnetohydrodynamics, in which the movement of a conducting gas in an electromagnetic field manifests itself in two effects—the Lorentz force and the Joule heat release—the calculation of processes in arc extinguishing devices is reduced to a gas-dynamic problem in the presence of only one electromagnetic effect—Joule heat release. This principle of arc modeling has been described in studies [14,30–33].

To calculate gas dynamics, you can also use the technique from [15,18], which is based on the analytical determination of the parameters of the piston device and its speed, in order to obtain a pressure drop that provides a supercritical gas flow mode. To determine the expiration mode, the pressure ratio under the piston p_0/p_i is considered, where i is the calculation step, and p_0 is the initial pressure in the under-piston volume. In addition to the serious assumptions used, the main disadvantage of this technique is that it does not take into account the interaction of the gas flow with the arc.

Experimental KEMA model.

The model is based on 79 short-circuit tests of CBs with a rated voltage range of 123–550 kV in test center "KEMA" (now "CESI") [34–37]. The model accuracy of the interruption prediction (prediction of post-current-zero events from pre-current-zero information). The arc parameters for each arc model are extracted by an optimization approach in a defined time interval to minimize the difference between the measured and simulated voltage waveforms.

At its core, this experimental model is a modified Mayer model, assembled from three submodels:

$$\frac{dg_i}{dt} = \frac{1}{P_i\tau_i} \cdot g_i^{\lambda_i} \cdot u_i^2 - \frac{1}{\tau_i} g_i, \quad \text{for } i = 1,2,3 \tag{6}$$

By solving each equation:

$$\frac{1}{g} = \frac{1}{g_1} + \frac{1}{g_2} + \frac{1}{g_3} \tag{7}$$

The model has six arc parameters: three time constants (τ_i) and three cooling power constants (Pi) [36], $\lambda_1 = 1.4, \lambda_2 = 1.9$, and $\lambda_3 = 2.0$. The model consists of three differential equations of $g = i/u$, the same ones as the earlier model, each representing different time intervals of the interruption processes (Figure 3).

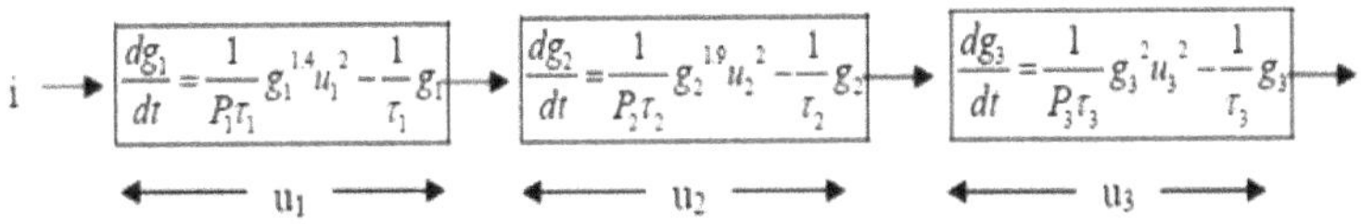

Figure 3. The tree−part arc model of three arcs in series.

Parameters of the KEMA model are represented in Table 1.

Table 1. KEMA model parameters.

Parameter Description	Parameter	
	Symbol	Formula
Arc parameters (varying from test to test)		
Time constant	τ_1	$\frac{k_t}{l_a - l_T}$
Cooling power constants	B_1	–
	B_2	–
Parameters related to CB design		
Distance between arcing contacts	l_a	–
Empirical constant (depends on tested CB and for conditions of the short-line fault)	l_t, k_t	According to [3]
Time constants	τ_2	$\frac{\tau_1}{k_1}$
	τ_3	$\frac{\tau_2}{k_2}$
Constants representing the breaker design	k_1	According to [37]
	k_2	According to [37]
	k_3	According to [37]
Cooling power	P_1	$B_1 \cdot g_1^{0.6}$
	P_2	$B_2 \cdot g_2^{0.1}$
	P_3	$\frac{B_2}{k_3}$

Magnetohydrodynamic model.

Fluid models describe plasma in terms of smoothed quantities, such as density and average velocity around each position. The fluid model in the magnetohydrodynamic approach considers plasma as a single fluid, as described by the systems of the Maxwell and Navier–Stokes equations [7]. A more general description is a two-fluid plasma, where ions and electrons are described separately. Fluid models are often accurate when the collision probability is high enough to keep the plasma velocity distribution close to the Maxwell–Boltzmann distribution [38]. Fluid models typically describe plasma in terms of a single flow at a certain temperature in each spatial location. They cannot capture high-velocity spatial structures, such as beams or double layers, nor resolve particle wave effects [39,40].

The gas dynamics and electrodynamics of the electric arc, in addition to the equations of continuity (2), motion (3), energy (4), and Ohm's law (5), must be supplemented with Maxwell's equations and material relations:

$$\frac{\partial \mathbf{E_r}}{\partial \mathbf{z}} - \frac{\partial \mathbf{E_z}}{\partial \mathbf{r}} = 0 \tag{8}$$

$$\frac{1}{\mathbf{r}}\frac{\partial \mathbf{r}\mathbf{H_\varphi}}{\partial \mathbf{r}} = j_z \tag{9}$$

$$-\frac{\partial \mathbf{H_\varphi}}{\partial \mathbf{z}} = j_r \tag{10}$$

where according to Ohm's law:

$$j_z = \sigma \mathbf{E_z};\ j_r = \sigma \mathbf{E_r}; \tag{11}$$

$$\mathbf{B} = \mu_0 \mathbf{H}; \mathbf{D} = \varepsilon_0 \mathbf{E}; \tag{12}$$

where μ_0 is magnetic permeability, and ε_0 is dielectric permittivity.

In general, μ_0 and ε_0 are tensors.

The above equations are supplemented by dependencies:

$$\begin{gathered} \rho = \rho(T,p);\ \sigma = \sigma(T,p);\ \lambda = \lambda(T,p);\ \eta = \eta(T,p);\ c_p = c_p(T,p); \\ h = h(T,p);\ \psi = \psi(T,p) \end{gathered} \tag{13}$$

Boundary conditions in this case (under axisymmetric conditions—cylindrical coordinates):

$$\mathbf{r} = 0, z > 0;\ \vartheta_\mathbf{r} = 0;\ \mathbf{H_\varphi} = 0;\ \frac{\partial \vartheta_\mathbf{z}}{\partial \mathbf{r}} = 0;\ \frac{\partial T}{\partial \mathbf{r}} = 0;\ \frac{\partial \mathbf{E_z}}{\partial \mathbf{r}} = 0;\ \omega = 0; \tag{14}$$

where $\mathbf{E_r}$ and $\mathbf{E_z}$ are radial and axial voltage gradients, respectively; $\mathbf{H_\varphi}$ is azimuthal magnetic field strength; j_r and j_z are radial and axial electric current density, respectively; ρ is the gas density; p is pressure; λ is thermal conductivity; c_p is specific heat capacity at constant pressure; h is specific enthalpy; and ψ is emissivity.

The equations written above can be used to describe an arc discharge in a gas. However, from the side of low currents, the boundary of the region under study is determined, as a rule, by the fulfillment of the condition of local thermodynamic equilibrium. When high currents are turned off, the limitation occurs due to the influence of reabsorption from radiation.

Hydrokinetic model.

The uniqueness of this model lies in the fact that it has a huge advantage over the magnetohydrodynamic approach. The reason is that it is possible to calculate the processes of restoration of the dielectric strength after the extinction of the arc within the framework of hydrokinetic modeling. The model is characterized by four calculation stages [41,42]:

(1) The arc plasma properties (in insulation media as SF6) for basic input data are needed, including thermodynamic properties, transport coefficients, and radiation coefficients, which are being calculated according to the Chapman–Enskog theory under local thermodynamic equilibrium (LTE) [40];

(2) A 1D hydrokinetic model is being used for the arc decaying process description (assuming cylindrical symmetry);

(3) Based on the 1D modeling results, the average radial temperature, the arc conductance, and the average critical electric field strength are calculated the three recovery stages: thermal recovery rate, the predielectric recovery rate, and the postdielectric;

(4) The postdielectric recovery stage is being calculated relying on the Boltzmann equation [43], which describes the electron transport behaviors during the last phase of extinguishing arcs.

In this model, the arc was assumed to be wall stabilized in a cylinder with 1D geometry. The governing equations describing this 1D arc were written as follows:

Continuity equation:

$$\frac{\partial \rho}{\partial t} + \frac{1}{\mathbf{r}} \cdot \frac{\partial (\mathbf{r}\rho\vartheta_{\mathbf{r}})}{\partial \mathbf{r}} = 0 \tag{15}$$

Energy equation:

$$\rho c_p \left(\frac{\partial T}{\partial t} + \vartheta_{\mathbf{r}} \frac{\partial T}{\partial \mathbf{r}} \right) = \sigma \frac{i^2}{g^2} - E_{rad} + \frac{1}{\mathbf{r}} \cdot \frac{\partial}{\partial \mathbf{r}} \left(k\mathbf{r} \frac{\partial T}{\partial \mathbf{r}} \right) \tag{16}$$

where $\mathbf{r}$ is radial distance; q is mass density; $\vartheta_{\mathbf{r}}$ is radial component of the velocity; c_p is specific heat at constant pressure; T is temperature; σ is electrical conductivity; j the thermal conductivity; i is current; g is arc conductance; and E_{rad} is radiation energy loss (net emission coefficient U in (4)).

One of the significant disadvantages of this model is the huge computing power. However, if the problem statement is made in 1D, then the calculation time will be significantly reduced.

Kinetic model.

Kinetic models describe the particle velocity distribution function at each point in the plasma and, therefore, should not assume a Maxwell–Boltzmann distribution. A kinetic description is often necessary for a collision-less plasma [44,45]. Such a plasma can be considered as the one in which the density is low enough and/or in which the temperature is high enough so that collisions can be neglected, due to the fact that the characteristic times are shorter than the particle collision time [39,44].

There are two general approaches to the kinetic description of plasma. One is based on the representation of the smoothed distribution function on the grid in terms of velocity and its position (distribution function of particles in coordinates and momenta):

$$f = f(t, \mathbf{r}, \mathbf{p}) \tag{17}$$

where $\mathbf{r}$ is particle center of mass coordinates, and $\mathbf{p}$ is the impulses of the center of mass of particles.

Function (17) in the LTE state has the form of a Maxwellian distribution and is generally found from the Boltzmann equation:

$$\frac{\partial f}{\partial t} + \vartheta \frac{\partial f}{\partial \mathbf{r}} + F \frac{\partial f}{\partial \rho} = C(f) \tag{18}$$

wherein:

$$\mathbf{F} = e\mathbf{E} + \left(\frac{e}{c} \right) [\vartheta \mathbf{B}] \tag{19}$$

where $\mathbf{F}$—an external force acting on a charged particle; and $C(f)$—taking into account mutual collisions of particles.

When considering fast motions of particles, collisions can often be neglected, assuming $C(f) \approx 0$. Then, the kinetic equation is called the collision-less Vlasov equation with self-consistent fields $\mathbf{E}$ and $\mathbf{B}$ (they are themselves determined by the motion of charged particles).

The Vlasov equation can be used to describe the dynamics of a system of charged particles interacting with an electromagnetic field. In magnetized plasma, the hydrokinetic approach can significantly reduce the computational cost of a fully kinetic simulation [40,45,46].

Another method, known as the particle-in-cell method, incorporates kinetic information by following the trajectories of a large number of individual particles. In fact, this method is used to solve nonstationary problems of magnetohydrodynamics [44]. Kinetic models tend to be more computationally expensive than fluid models.

Analyzing the works devoted to the modeling of gas-dynamic processes in SF6 CBs during their switching, the main focus was on solving the following scientific and practical problems:

1. An exploration of the electrophysical impact of the arc energy on the nozzle ablation process in order to both create advanced systems for monitoring the residual switching life and to study the effect of an auto-puffer to increase the switching capacity [47–49];

The optimization of the design of elements of the arc quenching device, the nozzle section in particular, to increase the flow rate of SF6 and the switching capacity of the CBs [50,51];

An evaluation of the residual conductivity of the arc stem to study its effect on the power proof of the contact gap after its extinction [52,53];

The main difficulty in the vast majority of studies has been related to the implementation of a suitable mathematical model of heat and mass transfer of a cold SF6 flow and a non-isothermal plasma channel. Some generalizations of works devoted to the construction of mathematical models of the interaction of an SF6 flow with an opening arc in SF6 CBs are given in Table 2.

Table 2. Studies devoted to interaction of SF6 gas flow with the arc.

№	Ref.	Problem under Study	The Model of Arc Interaction with SF6 Flow	Computational Numerical Model
1	[32]	Predicting arc extinction by simulating outgassing with nozzle ablation	Conservation equations, Joule heating, and radiation transfer	A two-dimensional axisymmetric
2	[33]	Exploration of the arc extinguishing process, when the capacitive current is turned off by a self-generating switch	Conservation equations, Joule heating, radiation transfer	A two-dimensional axisymmetric
3	[47]	Exploration of the nozzle ablation process for breaking capacity	Conservation equations, radiation transfer	A two-dimensional planar
4	[54]	Elimination of an impulse wave in front of a stationary arcing contact inside the nozzle, causing a decrease in the flow rate of SF6 gas in the nozzle	Conservation equations	A two-dimensional planar
5	[55]	Arc re-ignition prediction	Conservation equations, Joule heating and radiation transfer	A two-dimensional axisymmetric
6	[56]	Influence of impurities, arising in the process of nozzle ablation on the process of arc quenching	Conservation equations	A two-dimensional planar
7	[57]	The reconstruction of a digital model of an arc in cylindrical nozzles	Conservation equations	A two-dimensional planar
8	[58]	Exploration of the influence of the aperiodic component of the tripping current on the process of arcing	Magnetohydrodynamic: conservation equations, Maxwell's equations	A two-dimensional axisymmetric
9	[59]	Creation of a software package for modeling arc extinguishing processes	Conservation equations, Joule heating	A two-dimensional planar
10	[60]	Exploration of the process of arc extinguishing by a self-blast CB, taking into account the ablation of the nozzle	Conservation equations	A two-dimensional axisymmetric
11	[61]	Investigation of the process of arc extinguishing by a self-generated switch, taking into account the ablation of the nozzle	Conservation equations, radiation transfer	A two-dimensional axisymmetric
12	[62]	Exploration of the arc extinguishing process in a supersonic nozzle	Conservation equations	A two-dimensional axisymmetric
13	[63]	Improved accuracy at low breaking currents (wire arc).	Magnetohydrodynamic: conservation equations, Maxwell's equations	3D

2.3. Methods for Calculating the Processes of Interaction between Arc and the SF6 Flow

Analytical methods.

Considering that the flow of SF6 gas is described by the system of Navier–Stokes equations, it was worth noting that analytical integration was possible only in a limited number of cases. Furthermore, most methods for calculating the equations of fluid dynamics today are reduced to numerical ones. In magnetohydrodynamics, when the system of Maxwell equations is added to the Navier–Stokes equations, the situation only worsens in the framework of precisely finding the analytical solution. The fact is that the significant multidimensionality of the processes under study, especially in the plasma channel of the shutdown arc blown by a nonisothermal flow, as well as the presence of various types of weak and strong discontinuities, made it difficult to use numerical analysis methods [64]. Therefore, analytical studies aimed at describing the features associated with the nonlinear and multidimensional nature of plasma motions based on exact solutions are topical. A universal method that allows for analytically solving nonlinear equations that describe magnetohydrodynamics is the method of group analysis of differential equations [65]. Group theory methods were later applied to problems in fluid mechanics [66].

Numerical methods.

Numerical modeling of the processes occurring in the tripping arc mainly prevailed over other methods due to the growth of computing power. First of all, this referred to the calculations of gas-dynamic fields, current density fields, and electromagnetic forces. This, in turn, led to the so-called pinch effect and plasma acceleration and also to a more correct description of the radiation transfer in the arc column. The main numerical methods of calculation, as applied to the problems of fluid mechanics, including magnetohydrodynamics, were the finite element method and the finite volume method [67].

Non-Numerical methods.

For the calculation of physical processes that have uncertainties in the formulation of the problem or its solution (in particular, only some particular solution is found by numerically solving the system of Navier–Stokes equations), polynomial chaos methods are gaining popularity. The essence of the method is to represent random processes on a stochastic polynomial space in the form of Hermite polynomials [68].

In this study, the solution of the problem of gas dynamics—the determination of the pressure and flow rate of SF6 gas—was implemented analytically (according to the method [15,18]) and numerically (by the finite element method, in the numerical simulation software COMSOL Multiphysics 6.0).

3. Analytical Calculation of SF6 Circuit Breaker Breaking

Analytical calculation was carried out according to the method specified in [15,18]. Based on the methodology, the following main assumptions were made:

(1) There was no supply and removal of heat during the outflow of gas (adiabatic process);
(2) The process of gas outflow had a steady character;
(3) There were no friction losses;
(4) The gas was considered ideal;

All the main ratios necessary for the calculation were taken from [15]. The block diagram of the analytical calculation could be represented as follows—Figure 4:

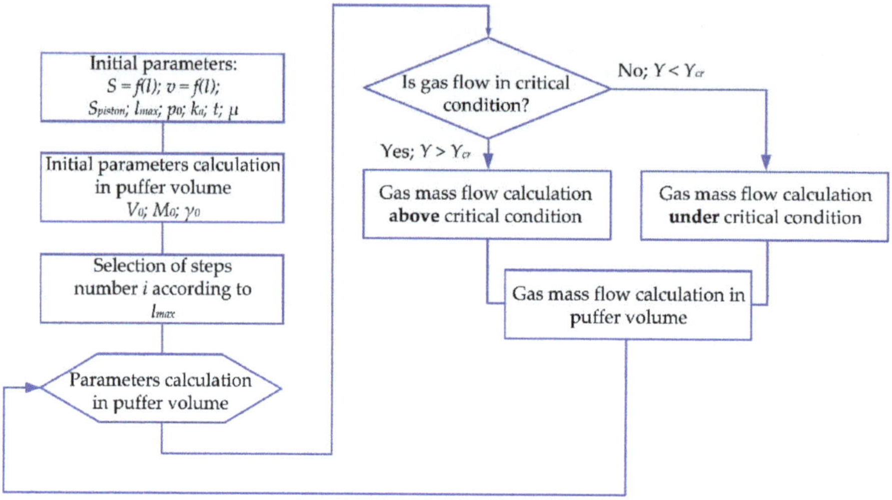

Figure 4. Block diagram of the analytical method: S_p—piston cross section; l—full contact separation; p_0—pressure in the under-piston volume during the calculation; k_a—adiabatic index for SF6 gas; V_0—initial volume of gas under the piston; M0—initial mass of SF6 gas under the piston; γ_0—initial density of SF6 gas in the under-piston volume; γ_i—SF6 density in the under-piston volume during the calculation; $Y = p_0/p_i$—relative backpressure.

The main purpose of the calculation was to determine the parameters of the piston device and its speed in order to obtain a pressure drop that provided a supercritical gas outflow regime.

3.1. SF6 Circuit Breaker under Study

Dead-tank SF6 CB for rated voltage 110 kV(RU)/126 kV(EU)—Figure 5 was chosen as the research object for the calculation of gas-dynamic processes.

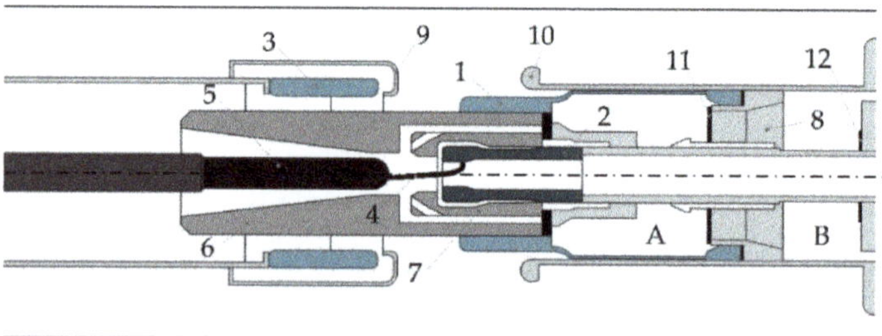

Figure 5. A 110 kV dead-tank SF6 CB: 1—moving main contact; 2—moving arcing contact; 3—fixed main contact; 4—arc; 5—fixed arcing contact; 6—PTFE main nozzle; 7—PTFE auxiliary nozzle; 8—piston; 9—tube of fixed main contact; 10—tube of moving main contact; 11—valve in the piston; 12—valve closed when contacts are opened; A—above-piston volume; B—under-piston volume.

The current flow path in the arc extinguisher of the selected CB is conceptually shown in Figure 1: in the closed state, most of the current flows from the live part of the bushing through the tulip contact (is not shown) to the tubes of both the fixed and moving main contacts 9 and 10, passing to the main contacts 1 and 3. When the main contacts open, the current also flows through the tubes 9 and 10 but passes through the conductive parts into the arcing contacts 5 and 7.

Arc extinguishing in the interrupter of this CB is based on combining self-blast and puffer-type principles. The interrupter had PTFE nozzles, providing an additional gas blast when the arc is burning.

3.2. Computational Model of the Circuit Breaker under Study

To calculate the gas dynamics, the following data were required: the full contact separation, the cross section of the piston, as well as the characteristics of the dependence of the cross section of the SF6 gas outlet on the contact separation $S = f(l)$. All these parameters were taken from the approximate geometric dimensions of the tank CB under study:

- Full contact separation $L_{max} = 120$ mm;
- The contact separation before blast start is $L_{ext} = 18$ mm;
- Piston cross section $S_p = 8.953$ mm^2;
- Ambient medium temperature $\vartheta = 40$ °C $= 313$ K;
- Pressure inside CB $p_0 = 0.42$ Mpa;
- The flow coefficient μ at all stages of the outflow was assumed to be 0.9 (the outflow coefficient, which took into account the decrease in the actual cross section of the hole due to the compression of the jet in it);
- Adiabatic exponent for SF6 gas $k_a = 1.086$;
- We set the discretization step of the calculation; for this, we divided the entire piston stroke into 20 identical sections: $n = 20$. The contact separation in each section would be:

$$\Delta l = \frac{L_{max}}{n} = \frac{120}{20} = 6 \text{ mm} \quad (20)$$

Figure 6a,b shows the dependence of the SF6 outlet cross section on the piston stroke $S = f(l)$ and the dependence of the piston speed on the contact separation $V = f(l)$ for the selected research object—110 kV SF6 dead-tank CB. The data were taken from the instruction manual of the manufacturer of this CB. Figure 7 shows the contact travel curve, which was taken from [14]. This curve was necessary in numerical implementation when solving a gas-dynamic problem with a moving grid. The total shutdown time was taken as equal to 55 ms.

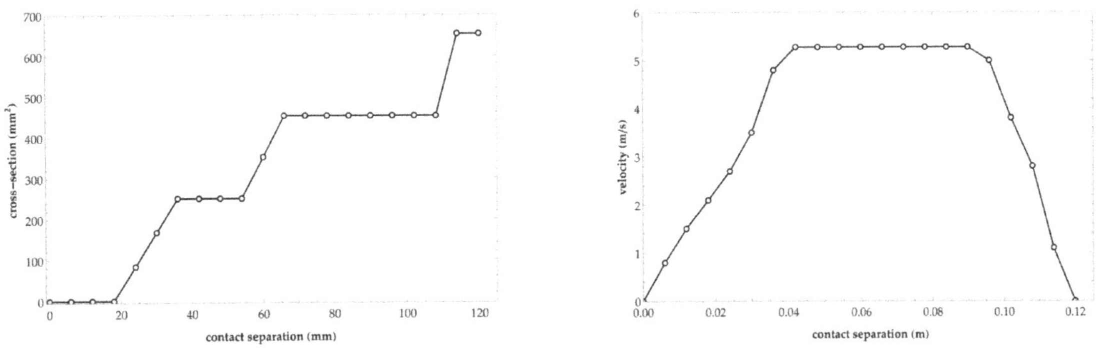

(a) (b)

Figure 6. (**a**) Dependence of the SF6 outlet cross section on the contact separation; (**b**) Dependence of the piston velocity on the contact separation.

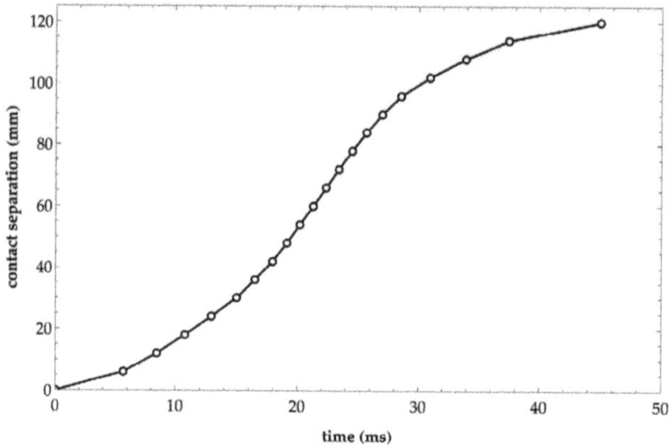

Figure 7. Dependence of moving contact separation on time.

3.3. Calculation Results

The results of the analytical calculation were the pressure changes in the under-piston volume and the mass flow rate per second depending on the contact separation. Table 3 presents the results of calculating each step by the analytical method according to the method [15,18]: pressure in the under-piston volume pi; gas mass flow G_i; mass loss of SF6 during blast; ΔM_i; mass of SF6 gas in the under-piston volume M_i; and relative backpressure Y_i. The piston speed vav was taken from Figure 6b. Parameter $\Psi(Y) = \sqrt{\frac{2k_a}{k_a-1}\left(Y^{\frac{2}{k_a}} - Y^{\frac{k_a+1}{k_a}}\right)}$ characterized the outflow of gas from the volume above the piston for each i calculation step.

Table 3. Gas dynamics parameters obtained analytically.

l, mm	V_i, mm^3	p_i, MPa	Y_i	Ψ_i	$v_{avg,i}$, m/s	G_i, kg/s	ΔM_i, kg·10^3	M_i, kg·10^3
6	1.209	0.420	1.000	0	0	0	0	29.674
12	1.155	0.441	0.975	0.218	0.40	0	0	29.674
18	1.101	0.465	0.938	0.336	1.15	0	0	29.674
24	1.048	0.491	0.895	0.424	1.80	0.114	0.380	29.294
30	0.994	0.512	0.856	0.482	2.40	0.271	0.676	28.617
36	0.940	0.530	0.823	0.520	3.10	0.455	0.881	27.736
42	0.886	0.547	0.795	0.546	4.15	0.494	0.715	27.022
48	0.833	0.569	0.766	0.569	5.04	0.533	0.635	26.387
54	0.779	0.596	0.734	0.588	5.28	0.574	0.652	25.735
60	0.725	0.627	0.701	0.604	5.28	0.863	0.981	24.755
66	0.671	0.653	0.671	0.614	5.28	1.177	1.337	23.417
72	0.618	0.673	0.646	0.620	5.28	1.231	1.399	22.019
78	0.564	0.695	0.625	0.623	5.28	1.279	1.453	20.565
84	0.510	0.719	0.604	0.625	5.28	1.326	1.507	19.058
90	0.457	0.747	0.582	0.625	5.28	1.374	1.561	17.497
96	0.403	0.780	0.559	0.625	5.28	1.427	1.622	15.875
102	0.349	0.820	0.535	0.625	5.14	1.491	1.740	14.135
108	0.295	0.867	0.508	0.625	4.40	1.564	2.133	12.002
114	0.242	0.902	0.486	0.625	3.30	2.360	4.291	7.711
120	0.188	0.733	0.526	0.625	1.95	2.188	6.732	0.979

Pressure changes in the under-piston volume and mass flow rates depending on the contact separation, plotted according to Table 3 data, are shown in the calculation results in comparison with the numerical calculations.

4. Numerical Calculation of SF6 Gas Circuit Breaker Switching

The analysis was carried out in COMSOL Multiphysics 6.0 software (Academic version) using the finite element method with a moving mesh (ALE).

The Navier–Stokes equations describe nonstationary flows and express momentum balances and mass conservations for a viscid and compressible fluid in the following form [69,70]:

$$\frac{\partial \rho}{\partial t} + \nabla \cdot (\rho \mathbf{u}) = 0 \tag{21}$$

where ρ—density; $\mathbf{u}$—flow velocity m/s; and t—time, s.

$$\rho \frac{\partial \mathbf{u}}{\partial t} + \rho \mathbf{u} \cdot \nabla \mathbf{u} = -\nabla \mathrm{p} + \nabla \cdot \left(\mu \left(\nabla u + (\nabla \mathbf{u})^T \right) - \frac{2}{3} \mu (\nabla \cdot \mathbf{u}) \mathbf{I} \right) + \mathbf{F} \tag{22}$$

where p—pressure, Pa; $\mathbf{F}$—volume force vector N/m^3; and $\mathbf{I}$—turbulence intensity.

The Reynolds number is one of the important criteria in the analysis. The Reynolds number (Re) is non-dimensional and describes the ratio between inertial forces and those of viscous friction in viscid fluids and gases.

When the Reynolds number exceeded a boundary value, the precise analytical solution for the dimensional flow or flat plate flow became chaotic, which marked the emergence of the turbulent flow. The Navier–Stokes equations were highly sensitive to changes in coefficient values in the turbulent flow conditions [69,70].

Values of the Reynolds number varied within $(0.4 - 10) \cdot 10^6$ depending on SF6 gas pressure in interruption chamber and geometrical properties of SF6 gas concurrent flow, which blew round the arc [71,72].

There were plenty of various models for turbulent flow analysis. The main idea behind such a model boiled down to the assumption about the existence of average flow velocity and the average deviation from flow velocity. The models that are listed below were used in various engineering applications with different accuracy requirements. Almost all of them have been implemented in modern software for fluid dynamics analysis.

The main models are listed in the order of increasing complexity [67,70]:

- The Boussinesq model;
- The Spallart–Allmaras model;
- The $k - \varepsilon$ model;
- The $k - \omega$ model;
- The Reynolds stress model;
- The direct numerical simulation (DNS);
- The large eddy simulation.

The $k - \varepsilon$ model was chosen to be used in numerical analysis, with the turbulent flow of gas being taken into account. This model is mainly used in nozzle arc simulations [73,74]. Furthermore, the $k - \omega$ model was used to simulate CB interruption without arc (no-load mode) due to existence of a boundary layer, which is described below. This model could be efficiently applied in the wall turbulence analysis without additional special functions.

4.1. The $k - \varepsilon$ Model of Turbulent Flow

The turbulence viscosity ratio μ_T for the $k - \varepsilon$ model is:

$$\mu_T = \rho C_\mu \frac{k^2}{\varepsilon} \tag{23}$$

where ε is velocity of turbulent dissipation, m^2/s^3; C_μ is coefficient of the $k - \varepsilon$ model; and k is turbulent kinetic energy, m^2/s^2.

The resultant equations of the $k-\varepsilon$ model are the following [75]:
The convection–diffusion equation for turbulent kinetic energy k:

$$\rho\frac{\partial k}{\partial t}+\rho(\mathbf{u}\cdot\nabla)k=\nabla\cdot\left[\left(\mu+\frac{\mu_T}{\sigma_k}\right)\nabla \mathrm{k}\right]+\mathrm{P}_k-\rho\varepsilon \tag{24}$$

The convection–diffusion equation for the dissipation ε:

$$\rho\frac{\partial \varepsilon}{\partial t}+\rho(\mathbf{u}\cdot\nabla)\varepsilon=\nabla\cdot\left[\left(\mu+\frac{\mu_T}{\sigma_\varepsilon}\right)\nabla \varepsilon\right]+C_{\varepsilon 1}\frac{\varepsilon}{k}P_k-C_{\varepsilon 2}\rho\frac{\varepsilon^2}{k} \tag{25}$$

where ε, $C_{\varepsilon 1}$, and $C_{\varepsilon 2}$ are model parameters, and their values are listed in Table 4.

Table 4. Computational model parameters for numerical calculation.

Description	Parameter	
	Designation	**Value**
Pressure inside the interrupter	p	0.42 MPa
Initial gas flow velocity	u	0 m/s
Ambient temperature	T	313 K
Von Karman constant	k_v	0.41
Parameters of k-e turbulence model		
–	$C_{\varepsilon 1}$	1.44
–	$C_{\varepsilon 2}$	1.92
–	C_μ	0.09
Turbulent kinetic energy	σ_k	1
Turbulent dissipation rate	σ_ε	1.3
Constant parameters of k-w turbulence model		
–	α	0.12
–	β_0	0.072
–	β_0^*	0.09
Turbulent kinetic energy	σ_k^*	0.5
Specific turbulent dissipation rate	σ_ω	0.5

The component of deformation rate P_k is:

$$\mathrm{P}_k=\mu_T\left[\nabla\mathbf{u}:\left(\nabla\mathbf{u}+(\nabla\mathbf{u})^T\right)-\frac{2}{3}(\nabla\cdot\mathbf{u})^2\right]-\frac{2}{3}\rho k\nabla\cdot\mathbf{u} \tag{26}$$

where «:» means tensor convolution.

4.2. The $k-\omega$ Model of Turbulent Flow

The turbulence viscosity ratio μ_T for the $k-\omega$ model is:

$$\mu_T=\rho\frac{k}{\omega} \tag{27}$$

where ω is relative dissipation rate, Hz.
The resultant equations of the $k-\omega$ model are the following [75]:

The convection–diffusion equation for turbulent kinetic energy k:

$$\rho\frac{\partial k}{\partial t}+\rho(\mathbf{u}\cdot\nabla)k=\nabla\cdot[(\mu+\sigma_k^*\mu_T)\nabla \mathrm{k}]+\mathrm{P}_k-\rho\beta^* k\omega \tag{28}$$

The equation of relative dissipation rate ω:

$$\rho\frac{\partial \omega}{\partial t}+\rho(\mathbf{u}\cdot\nabla)\omega=\nabla\cdot[(\mu+\sigma_\omega\mu_T)\nabla\omega]+\alpha\frac{\omega}{k}\mathrm{P}_k-\rho\beta\omega^2 \tag{29}$$

where α; $\beta=\beta_0 f_\beta$; $\beta^*=\beta_0^* f_\beta$; σ_ω; σ_k^*; β_0; $f_\beta=\frac{1+70\chi_\omega}{1+80\chi_\omega}$; and $\chi_\omega=\left|\frac{\Omega_{ij}\Omega_{jk}S_{ki}}{(\beta_0^*\omega)^3}\right|$—are model parameters, and their values are listed in Table 4.

$\Omega_{ij}=\frac{1}{2}\left(\frac{\partial\bar{u}_i}{\partial x_j}-\frac{\partial\bar{u}_j}{\partial x_i}\right)$ is the tensor of average rotation rate;

$S_{ij}=\frac{1}{2}\left(\frac{\partial\bar{u}_i}{\partial x_j}-\frac{\partial\bar{u}_j}{\partial x_i}\right)$ is the tensor of average deformation rate:

Furthermore, the wall boundary conditions were formed for the $k-\omega$ model [75].

4.3. The Computational Model of the Object under Study

The model of interruption chamber of a dead-tank SF6 gas CB. The dimensions of the model were approximately the same as those of the real CB. Elements of the chamber, which were not utilized in gas dynamics analysis, were not taken into account (Figure 8).

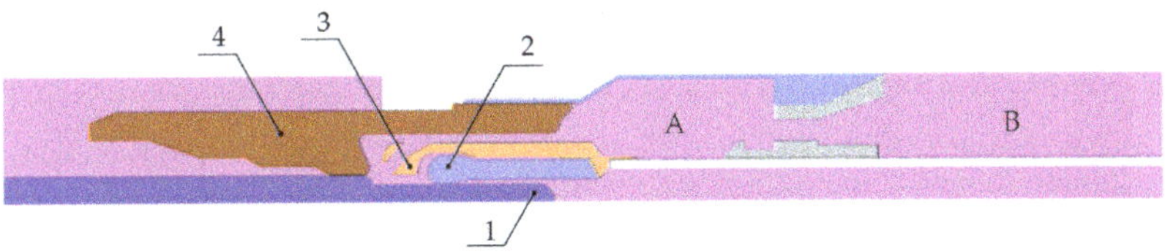

Figure 8. Computational model for numerical calculation (axial symmetry): 1—fixed arcing contact; 2—moving arcing contact; 3—auxiliary PTFE nozzle; 4—main PTFE nozzle; A—above-piston volume; B—under-piston volume.

Parameters of the turbulence models k-e and k-w for numerical analysis can be found in Table 4.

The grid of the numerical model was triangular (the triangulation method). The rectangular domain was used at the boundary layer. Grid parameters are shown in Table 5. The calculation was carried out using the finite element method in combination with the arbitrary Lagrangian Eulerian method (ALE). The latter was used to solve gas dynamics equations in the third coordinate system, thus called mesh frame.

Table 5. Mesh and time dependent solver parameters.

Number of Elements	Vertex Elements	Edge Elements	Average Element Quality	Automatic Remeshing	Relative Tolerance	Tolerance Factor	Termination Technique	Max Iterations
4737	92	1090	0.4474	0.08	0.1	1	Tolerance	20

This study was distinguished by taking into account the movement of the mobile parts of the object under study in a gas dynamics simulation. The module moving mesh was used to simulate the movement of a piston and the movable contact.

The k-e model included viscous effects that were considered for the sliding wall; consequently, there was a boundary layer. No slip was the default boundary condition to model solid walls. A no-slip wall was a wall where the fluid velocity relative to the wall velocity was zero. For a stationary wall, it meant that $\mathbf{u} = 0$. The constraint could be mathematically formulated for the problem with a moving wall:

$$\mathbf{u}_{rel} \cdot n = 0 \tag{30}$$

$$\mathbf{u_{rel}} = \mathbf{u} - \mathbf{u_{tr}} \tag{31}$$

$$\mathbf{K} - (\mathbf{K} \cdot \mathbf{n})\mathbf{n} = 0 \tag{32}$$

$$\mathbf{K} = \mu\left(\nabla \mathbf{u}_{rel} + (\nabla \mathbf{u}_{rel})^T\right)\mathbf{n} \tag{33}$$

where $\mathbf{n}$—boundary normal, with direction outside the region; $\mathbf{u_{rel}}$ is relative velocity; and $\mathbf{u_{tr}}$ is translation velocity.

Turbulence parameters k and ε were subject to homogenous Neumann boundary conditions:

$$\nabla k \cdot \mathbf{n} = 0 \tag{34}$$

$$\nabla \varepsilon \cdot \mathbf{n} = 0 \tag{35}$$

During the calculation, due to the movement of the boundaries of the object, the grid of the computational domain was strongly deformed, which reduced its quality and led to the appearances of discontinuities, inconvergences of equations, etc. In order to eliminate possible errors, it was customary to use the Automatic Remeshing function, which rebuilt the computational mesh when it reached a critically low quality.

4.4. The Proposed Model of Interaction between SF6 Gas Flow and Arc

The model is based on the additional element, which is either the line (axisymmetric) or cylinder (three-dimensional) in the contact gap. The arc column measured temperature is assigned to the line. The temperature is obtained from the experiment of 10 kA symmetrical short-circuit breaking [76].

The temperature change is shown in Figure 9a. The duration of the application of temperature to the moving line was given with the assumption that the arc burned for two power frequency half cycles (i.e., 20 ms). In this case, it was decided to neglect the change in temperature in the vicinity of the zero current. This assumption was very rough because the problem of skipping current zero was extremely important [77–79]. Studies devoted to the analysis of the occurrence of thermal and/or electrical breakdowns due to the passage of zero current gave a complete picture of arcing. However, in this study, the emphasis was on modeling the processes of arcing until the moment of arc extinction, and the processes of restoring electrical strength were not studied.

At the line boundaries (Figure 9b), the temperature of the arc stem was set according to [76] with a current cut of 10 kA. A feature of the model was the fact that the line moved along with the moving contact until the contacts opened (moving mesh). After opening the contacts, the upper point of the line was fixed, and it lengthened after the moving contact. In a three-dimensional picture, this line was represented by a cylindrical arc, on the surface of which the temperature was experimentally set. Thus, gas heating occurs in the contact gap as the contacts are being separated.

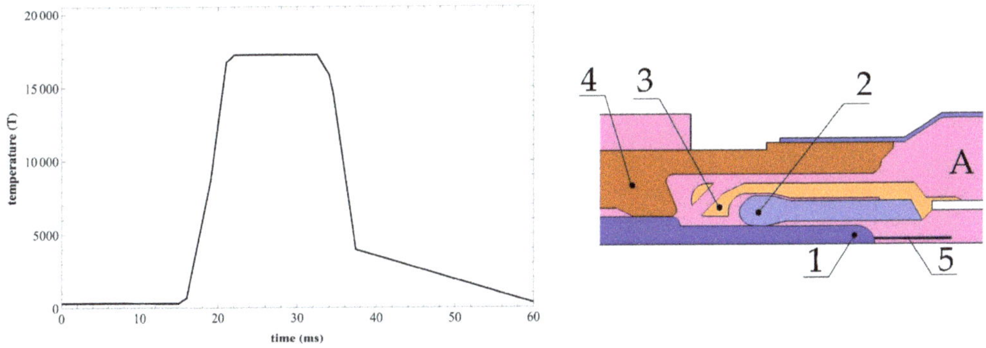

(a) (b)

Figure 9. (**a**) Temperature change according to [76]; (**b**) Arc model with additional line 5 between arcing contacts for adaptive heat release (along the moving contact system): 1—fixed arcing contact; 2—moving arcing contact; 3—auxiliary PTFE nozzle; 4—main PTFE nozzle; A—above-piston volume; B—under-piston volume.

When adding heat sources to the model, the heat balance equation was added to the solution of the above equations for gas dynamics:

$$\rho C_p \frac{\partial T}{\partial t} + \rho C_p \mathbf{u} \cdot \nabla T + \nabla \cdot \mathbf{q} = Q + Q_p + Q_{vd}, \tag{36}$$

where C_p is the specific heat capacity at constant pressure, $J/(kg{\cdot}K)$; $\mathbf{q}$—heat flow due to thermal conductivity, W/m^2; Q—heat source other than viscous dissipation, W/m^3; Q_p is the work performed by changing the pressure and is the result of heating during adiabatic compression, as well as some thermoacoustic effects, W/m^3; Q_{vd} is viscous dissipation in liquid, W/m^3.

$$Q_p = \alpha_p T \left(\frac{\partial p_A}{\partial t} + \boldsymbol{u} \cdot \nabla p_A \right), \tag{37}$$

where α_p is the coefficient of thermal expansion, $1/K$:

$$\alpha_p = -\frac{1}{\rho} \left(\frac{\partial \rho}{\partial T} \right)_p \tag{38}$$

For heat transport turbulence consideration, the Kays–Crawford model was used (the default in COMSOL Multiphysics). The viscous dissipation, in this case:

$$Q_{\mathrm{vd}} = \tau : \nabla \mathbf{u} + Q_{turb} \tag{39}$$

where τ is the viscous stress tensor, Pa.

Heat flux by conduction (RANS turbulence model):

$$\boldsymbol{q} = -(k + k_T) \nabla T \tag{40}$$

with the turbulent thermal conductivity defined as:

$$k_T = \frac{\mu_T C_p}{Pr_T} \tag{41}$$

where Pr_T—Prandtl number, according to [80].

In this case, the boundary conditions for the first kind of Dirichlet were written in the form of Equations (41) and (42) on the boundary where there was no heat flux across the boundary (in general form):

$$-\mathbf{n}\cdot\mathbf{q} = 0 \tag{42}$$

However, taking the wall treatment into account, the boundary condition is rewritten as:

$$-\mathbf{n}\cdot\mathbf{q} = \mathbf{q}_{\mathrm{wf}} = \rho C_p \mathbf{u}_\tau \frac{T_w - T}{T^+} \tag{43}$$

where $\mathbf{q}_{\mathrm{wf}}$—the heat flux between the fluid with temperature T and a wall with temperature T_w; C_p—is the fluid heat capacity; $\mathbf{u}_\tau$—is the friction velocity given by wall treatment (Equations (29)–(32)); T_w—wall temperature (on the boundaries which are "temperature source"); and T^+– is the dimensionless temperature and is given by [81].

The temperature on the boundaries that represent the experimental arc temperature profile (according to [76]):

$$T = T_0 \tag{44}$$

4.5. Calculation Results

Table 6 presents the results of calculation of two turbulent models ($k-\varepsilon$ and $k-\omega$) of the main gas-dynamic parameters: pressure in the under-piston volume p; gas velocity in the nozzle u; mass flow rate G. For convenience, all results are given to contact separation, which are considered in the analytical calculation.

Table 6. Gas dynamic parameters obtained numerically.

l, mm	t, ms	$p_{k-\varepsilon}$, MPa	$p_{k-\omega}$, MPa	$u_{k-\varepsilon}$, m/s	$u_{k-\omega}$, m/s	$G_{k-\varepsilon}$, kg/s	$G_{k-\omega}$, kg/s
0	0	0.420	0.420	0	0	0	0
6	5.70	0.441	0.441	0	0	0	0
12	8.50	0.462	0.462	0	0	0	0
18	10.80	0.484	0.484	6.0	6.0	0	0
24	12.95	0.507	0.507	31.0	30.0	0.367	0.360
30	15.00	0.523	0.523	59.0	55.5	0.699	0.656
36	16.50	0.533	0.533	83.0	74.5	0.855	0.787
42	17.95	0.552	0.554	105.0	89.5	0.973	0.902
48	19.13	0.572	0.575	121.0	103.0	1.005	0.936
54	20.20	0.590	0.596	121.0	105.0	1.143	1.092
60	21.30	0.607	0.613	123.0	106.0	1.380	1.301
66	22.35	0.626	0.631	105.0	103.0	1.550	1.442
72	23.42	0.644	0.665	119.0	124.0	1.595	1.520
78	24.53	0.659	0.666	126.0	111.0	1.693	1.650
84	25.70	0.673	0.681	119.0	118.0	1.888	1.838
90	27.00	0.682	0.690	145.0	129.0	2.083	2.007
96	28.55	0.676	0.687	98.0	105.0	2.196	2.105
102	30.93	0.645	0.658	105.0	105.0	2.303	2.327
108	33.90	0.590	0.600	83.0	95.0	2.391	2.486
114	37.46	0.511	0.514	65.0	75.0	2.036	2.151
120	45.00	0.383	0.378	20.0	21.0	0.727	0.710

Furthermore, the results of the calculations are presented in the form of fields of profiles of speeds, pressures, and temperatures, as the contact system of the arcing device moved when the short circuit currents were turned off. In the profile field, the gas and the corresponding changes in dependent variables (gas velocity, pressure, temperature) are shown in color, and the solid elements of the arc extinguisher involved in the movement (nozzles, contacts, pre-piston areas) are shown in white.

Figure 10 shows the fields of velocity profiles for different positions of the moving part without taking into account the arc (turbulence model $k-\varepsilon$). It can be seen that the gas velocity inside the moving contact cavity exceeded the gas velocity in the nozzle at some moments. However, even in this case, the main volume of gas flowed out through a large nozzle because the cross-sectional area of the nozzle was greater than the cross-sectional area inside the moving contact.

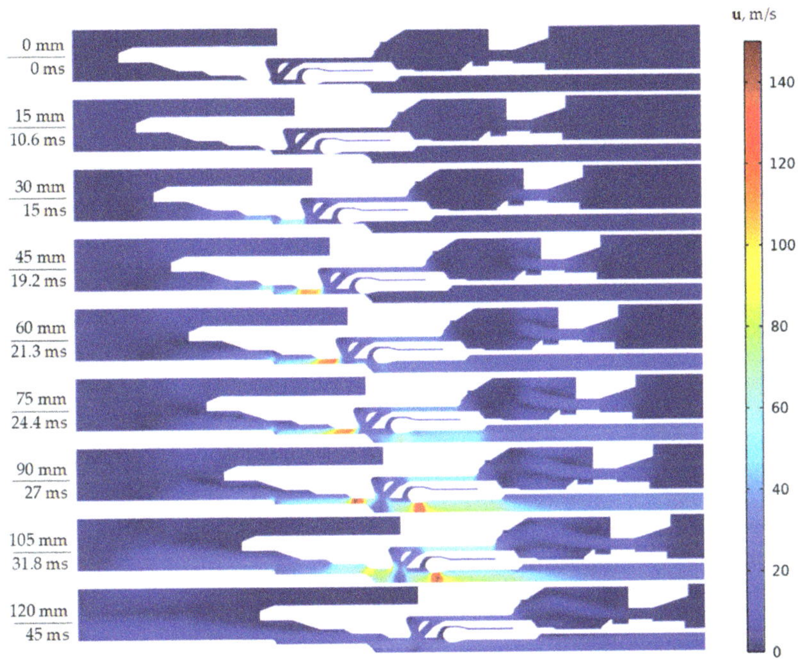

Figure 10. Velocity field without taking arc into account for turbulent model $k-\varepsilon$ (on the left—the contact separation and time).

Figure 11 shows the change in gas pressure at different positions of the moving part without taking into account the arc (turbulence model $k-\varepsilon$). An increase in the pressure in the under-piston and above-piston volumes could be noticed (assumption: it was assumed that there is no valve between them). With a real shutdown of the short-circuit current, the valve between the areas would begin to close, and the arc extinguishing was mainly carried out by blowing from the above-piston volume. Furthermore, the figure basically shows that the pressure in the contact gap was pumped after the opening of the large nozzle (contact separation is 90 mm).

Figure 11. Pressure field without taking arc into account for turbulent model $k-\varepsilon$ (on the left—the contact separation and time).

Figure 12 shows the change in pressure in the under-piston volume for both analytical and the numerical calculations, taking into account the turbulent models $k-\varepsilon$ and $k-\omega$, respectively. According to Figure 12, it can be seen that the pressure in the under-piston volume changed smoothly—almost linearly almost until the very end of the contact separation. This happened due to the neglect of both turbulent SF6 gas flows and the impact of the shutdown arc on this flow. According to the results obtained numerically, it is seen that the pressure peak occurred at the moment of the deceleration of the contact separation, which was $t = 27$ ms. The decrease in pressure could be explained by the fact that the gas compression rate slowed down, whereas the mass flow continued to increase.

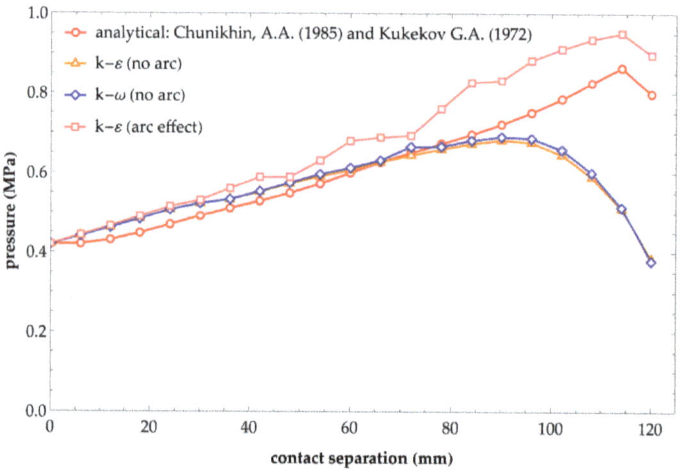

Figure 12. Pressure change in the under-piston volume obtained analytically (according to [15,18]) and numerically for turbulent models $k-\varepsilon$ and $k-\omega$.

Figure 13 shows the gas mass flow change through the section of the under-piston volume. It can be seen that the flow curve was tied to the cross section of the regions that the SF6 gas passed on the way to the contact gap, as in the analytical calculation. The differences were caused by less linear transitions with an increase in the cross-sectional area of the SF6 gas movement in the direction of the contact system due to turbulence. This meant that the analytical method gave only a superficial understanding of gas dynamics during the CB breaking or making. Considering the turbulent flows (without arc affect) and turbulent heat transport (with arc affect), even with low relative tolerance, gave a different nature of the interrupter's switching process. The proposed numerical model might have been used for the influence estimation of the thermal effect of the switching arc on the nozzles, as well as for the estimation of both the thermal and electric breakdowns.

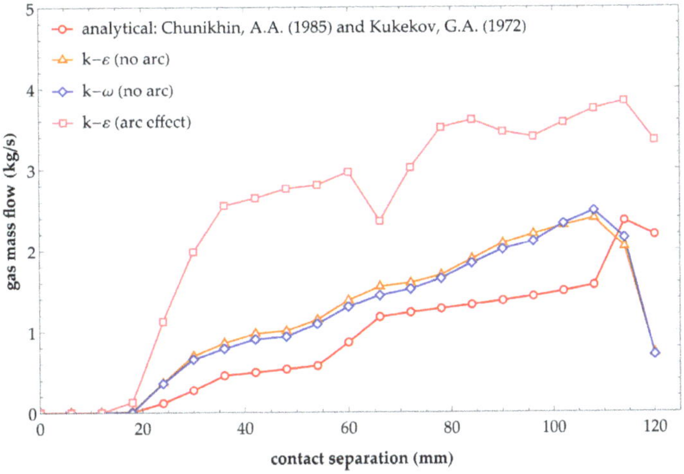

Figure 13. SF6 gas mass flow through cross section of the under-piston volume for analytical (according to [15,18]) and numerical calculations.

Figure 14 shows the gas velocity change in the narrowest part of the nozzle (confuser) versus time and contact separation, respectively.

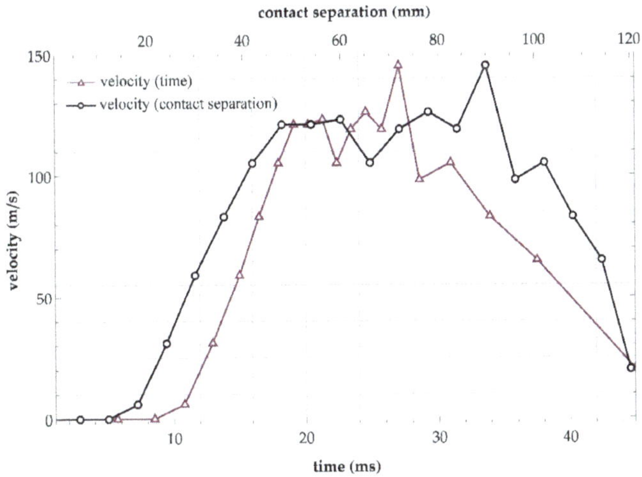

Figure 14. Velocity changes in the smallest cross-section of the nozzle depending on the time and contact separation.

Up to a certain point in time ($t_{crit} = 19.13$ ms), the gas velocity increased, after which some constancy could be observed. Therefore, we could assume that the speed had reached its critical value. This meant that the time t_{crit} could be considered a critical point, and the time interval up to this point was a subcritical gas outflow regime. The interval after the critical point and before the moment of braking was a supercritical regime of gas outflow. Velocity fluctuations occurred due to additional outflow of gas into the area of the opened section inside the moving contact.

Figure 15 shows the gas velocity field, taking into account the setting of the arc in the form of a thermal heating source. It can be seen that the gas velocity during opening reached supersonic values (the speed of sound in SF6 gas under normal conditions was $130-135$ m/s and increased with increasing pressure of the medium). For the process of arc extinguishing, this was a favorable factor because there was an intensification in the cooling of the plasma channel of the cut-off arc. However, one could also notice that a local area was formed along the stationary arcing contact, where the velocities prevailed. This fact was due to the assumption made in the calculation model—the main part of the arc-extinguishing movable contact was cut out. Thus, it was considered that the main blast was directed through a large nozzle (disconnection of high short-circuit currents).

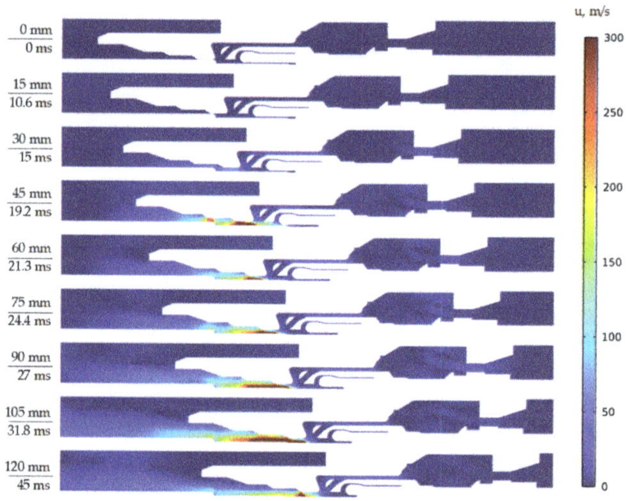

Figure 15. Velocity field taking arc into account for turbulent model $k - \varepsilon$ (on the left—the contact separation and time).

Figure 16 shows the gas pressure field, taking into account the setting of the arc in the form of a thermal heating source. When such a short-circuit current breaking took place, the pressure increased by about 2.0–2.5 times from the nominal value (0.5 MPa) and reached approximately 1.0–1.5 MPa for the self-blast-type interrupters [17].

Figure 17 shows a picture of the temperature profile in the arc quencher during shutdown. In contrast to the arc-free mode, there was a clear change in the SF6 flow in the contact gap due to the occurrence of flow stagnation areas, where the SF6 gas flow, passing through a small nozzle, met the SF6 gas flow going through a large nozzle. However, their directions were opposite to each other. The local heating of the area along the fixed arcing contact was due to the assumption made above.

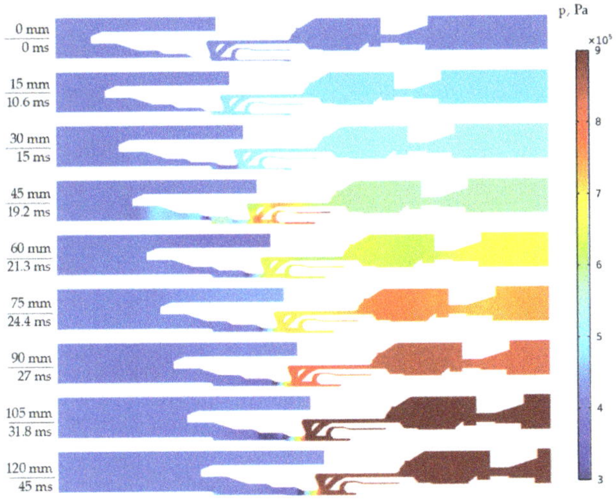

Figure 16. Pressure field taking arc into account for turbulent model $k-\varepsilon$ (on the left—the contact separation and time).

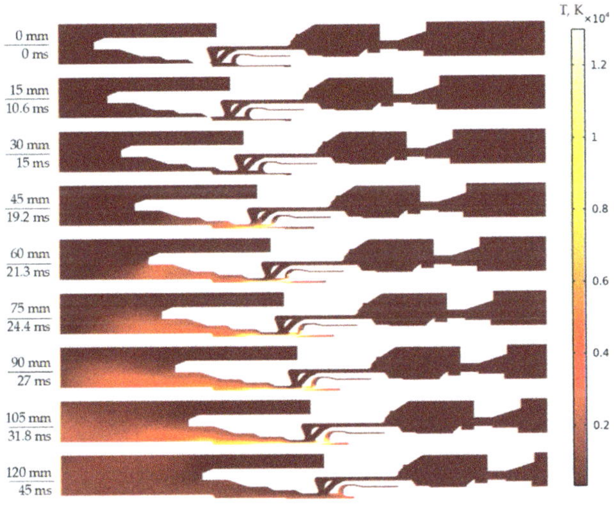

Figure 17. Temperature field (on the left—the contact separation and time).

Due to the consideration of the arc action, the pressure in the contact gap changed and has a less decreasing character, in contrast to the arc-free mode.

5. Conclusions

In the developed model described above, the temperature characteristics of the arc shaft were obtained from the experiment of switching off the symmetrical short-circuit current of 10 kA [76]. The proposed model of the switching arc was implemented in numerical software using the finite element method based on a moving-mesh technique (ALE method). The ALE method allowed us to take the contact separation curve of the CB into consideration and to make the developed model adaptive. In order to make a comparison with experimental data on breaking higher short-circuit currents, the temperature change was adapted according to [73] for an arcing time of 20 ms (two half cycles of power frequency). The accuracy of the developed model was determined in comparison with experiments on breaking the symmetrical current of 25 kA by a self-blast interrupter from the study [82]. The comparison was carried out by the means of calculating the coefficient of determination R^2 by changing the pressure in the under-piston volume (Figure 18). Due to the consideration of the arc effect, the pressure in the contact gap changed and had a less decreasing character, in contrast to the arc-free case.

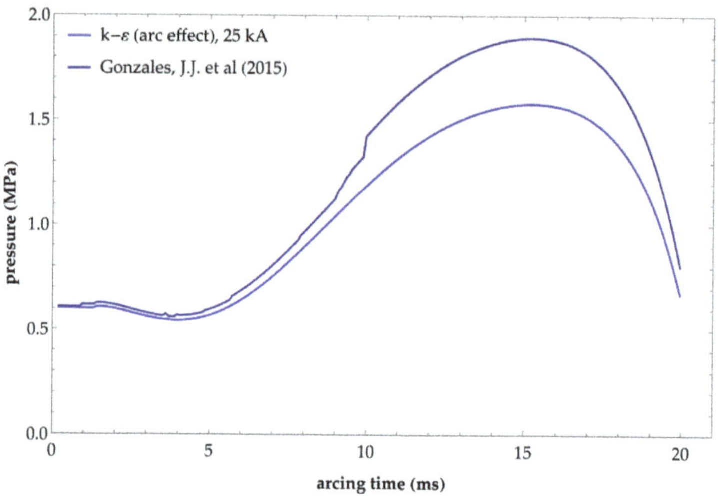

Figure 18. Pressure change in under-piston volume (comparison with [82]).

From Figure 18, it can be seen that the developed model in the form of a temperature heating source had a good quality in comparison with the experiment [82], with a determination coefficient $R^2 = 0.997$.

According to the results of the calculations, it is advisable to note the assumptions and critical parameters of the model:

1. The absence of a valve between the under-piston and above-piston volumes, which clearly affects the velocity profile of the SF6 flow moving into the contact gap.

2. Cut out the main part of the arcing moving contact, leading to a one-way blow (through a large nozzle). Therefore, the model was suitable for breaking large short-circuit currents.

3. The construction of the computational grid was performed by the user (user controlled) and not by the built-in tools of the numerical software complex (physics controlled). As a result, the computational mesh on the boundary layers had the same quality as in the main computational domain. In other words, the calculation of gas dynamics was carried out quite roughly.

4. The buoyancy forces were not taken into account in the calculations of gas dynamics.

5. The assumption of Mach numbers <0.3 was used. However, in the peak phases of the arcing, the SF6 flow velocity exceeded the supersonic flow. Thus, it was more correct to use models that were used for the sonic and supersonic flows' descriptions.

6. To increase the convergence, the accuracy parameter of the moving grid (relative tolerance) was taken as equal to 0.1, which made it possible to consider the physical processes only in the first approximation.

7. The contact separation curve was taken from the arc-free case ("no-load") and was considered unchanged even when the arc was taken into account. However, when the arc was extinguished, the contact separation curve "collapses" [83], which clearly affected the entire process of the gas dynamics, including the gas outflow regime.

8. Despite the fact that the radiation effect was significant within such high temperatures of the switching arc column (>13,000 K), the assumption of the neglecting this effect was used.

As prospects for further research directions, in addition to considering the above assumptions, the following can be noted:

1. Taking all the physical processes of interaction the SF6 gas flow and switching arc (creation of magnetohydrodynamic, hydrokinetic, and kinetic models that make it possible to trace the movement of the conductive medium—the arc plasma in SF6 gas), with the aim of their possible syntheses and the creation of universal approaches to the simulation of the switching arc.
2. Adapt the developed model for "near zero" arc extinguishing processes, taking into account the processes of restoring the electrical strength of the contact gap, the residual conductivity of the arc, and assessing the likelihood of thermal and electrical breakdowns, etc.
3. An improvement of the methods for calculating the equations of gas dynamics in order to minimize computational costs.
4. Analyze the occurrence of computational instability in the calculations of gas dynamics and develop prerequisites for combating this problem without losing the accuracy of the simulation.

Author Contributions: Conceptualization, V.V.P., A.I.K. and Y.V.P.; methodology, A.I.K. and V.V.P.; software, V.V.P. and Y.V.P.; validation, A.I.K. and V.V.P.; writing—original draft preparation, A.I.K., V.V.P. and Y.V.P.; writing—review and editing, A.I.K. and V.V.P.; visualization, V.V.P.; supervision, A.I.K. All authors have read and agreed to the published version of the manuscript.

Funding: The research was carried out within the state assignment with the financial support of the Ministry of Science and Higher Education of the Russian Federation (subject No. FEUZ-2022-0030 Development of an intelligent multi-agent system for modeling deeply integrated technological systems in the power industry).

Conflicts of Interest: The authors declare no conflict of interest.

References

1. Küchler, A. *High Voltage Engineering. Fundamentals—Technology—Applications*; Springer-Verlag GmbH Germany: Schweinfurt, Germany, 2018; 650p. [CrossRef]
2. *GOST 52565-2006*; Alternating-Current Circuit Breakers for Voltages from 3 to 750 kV. General Specifications. National standard of Russia. Rosstandart: Moscow, Russia, 2007; 91p. (In Russian)
3. *IEC 62271-100*; High-Voltage Switchgear and Controlgear—Part 100: Alternating-Current Circuit-Breakers. IEC: Geneva, Switzerland, 2008; 695p.
4. *IEC 62271-110*; High Voltage Switchgear and Controlgear—Part 110: Inductive Load Switching, Ed. 2.0. IEC: Geneva, Switzerland, 2017; 60p.
5. *IEEE Std C37.015–2017 (Revision of IEEE Std C37.015-2009)*; Guide for the Application of Shunt Reactor Switching. IEEE: New York, NY, USA, 2018; 62p. [CrossRef]
6. Zalesskiy, A. *Electric Arc*; Gosenergoizdat: Leningrad, Russia, 1963; 267p. (In Russian)
7. Engelsht, V.; Gurovich, V.; Desyatkov, G. *Electric Arc Column Theory*; AN USSR, Siberian department, Thermal physics institute: Novosibirsk, Russia, 1990; 373p. (In Russian)
8. Agafonov, G.; Babkin, I. *High Voltage Electrical Apparatus with SF6 Insulation*; Energoatomizdat: St. Petersburg, Russia, 2002; 728p. (In Russian)
9. Tonkonogov, E. *The Design of Electrical Apparatus. High Voltage SF6 Circuit Breakers*; Izdatelstvo of Peter the Great St. Petersburg Polytechnic University: St. Petersburg, Russia, 2008; 160p. (In Russian)
10. Averyanova, S.A. *Theory of Arc Extinguishing in Electrical Apparatuses. Interaction of the Electric Arc with the Gas Flow in High Voltage Circuit Breakers*; Izdatelstvo of Peter the Great St. Petersburg Polytechnic University: St. Petersburg, Russia, 2015; 68p. (In Russian)
11. Poltev, A. *Design and Calculation of SF6 High Voltage Apparatus*; Energiya, Leningrad department: Leningrad, Russia, 1979; 240p. (In Russian)
12. Eroshenko, S. *Calculation of Short Circuit Currents in Power Systems*; Izdatelstvo of Ural Federal University: Ekaterinburg, Russia, 2019; 104p. (In Russian)
13. Khalyasmaa, A.I.; Eroshenko, S.A.; Zinovyev, K.A.; Bolgov, V. Improvement of Short-Circuit Calculation Results Reliability for Large Electric Power Systems. In *2019 Electric Power Quality and Supply Reliability Conference and 2019 Symposium on Electrical Engineering and Mechatronics, PQ and SEEM 2019*; IEEE: Kärdla, Estonia, 2019; pp. 1–6. [CrossRef]

14. Il'in, A.S. *Mathematical Modeling of Thermodynamic Processes of Arc Extinguishing in SF6 flow in Electrical Apparatus*; Candidate of Technical Science Dissertation, Ural Federal University: Ekaterinburg, Russia, 2012. (In Russian)
15. Chunikhin, A.; Zhavoronkov, M. *High Voltage Apparatus*; Energoatomizdat: Moscow, Russia, 1985; 432p. (In Russian)
16. Smeets, R.; Van Der Sluis, L.; Kapetanović, M.; Peelo, D.; Janssen, A. *Switching in Electrical Transmission and Distribution Systems*, 1st ed.; John Wiley & Sons, Ltd.: Chichester, UK, 2015; 425p.
17. Kapetanović, M. *High Voltage Circuit Breakers*; Faculty Electrotech. Eng., Univ. Sarajevo: Sarajevo, Bosnia and Herzegovina, 2011; 648p.
18. Kukekov, G. *High Voltage AC Circuit Breakers*, 2nd ed.; revised; Energiya, Leningrad Department: Leningrad, Russia, 1972; 336p. (In Russian)
19. Cassie, A.M. A new theory of rupture and circuit severity. *CIGRE Rep.* **1939**, *102*, 588–608.
20. Mayr, O. Beitrage zur Theorie des Statischen und des Dynamischen Lichtbogens. *Arch. Für Elektrotechnik* **1943**, *37*, 588–608. (In German) [CrossRef]
21. Browne, T.E. A study of A-C. arc behavior near currents zero by means of mathematical models. *AIEE Trans.* **1948**, *67*, 147–153. [CrossRef]
22. Browne, T.E. An Approach to Mathematical Analysis of A-C Arc Extinction in Circuit Breakers. *AIEE Trans.* **1959**, *77*, 1508–1517. [CrossRef]
23. Ragaller, K. *Current Interruption in High-Voltage Networks*, 1st ed.; Springer: New York, NY, USA, 1978; 360p. [CrossRef]
24. Hermann, W.; Kogelschatz, U.; Niemeyer, L.; Ragaller, K.; Schade, E. Experimental and theoretical study of a stationary high-current arc in a supersonic nozzle flow. *J. Phys. D (Appl. Phys.)* **1974**, *7*, 1703–1723. [CrossRef]
25. Hermann, W.; Ragaller, K. Theoretical description of the current interruption in HV gas blast breakers. *IEEE Trans. Power Appar. Syst.* **1977**, *96*, 1546–1555. [CrossRef]
26. Tuma, D.T.; Lowke, J.J. Prediction of properties of arcs stabilized by forced convection. *J. Appl. Phys.* **1975**, *46*, 3361–3367. [CrossRef]
27. Lowke, J.J.; Ludwig, H.C. A simple model for high-current arcs stabilized by forced convection. *J. Appl. Phys.* **1975**, *46*, 3352–3360. [CrossRef]
28. El-Akkari, F.R.; Tuma, D.T. Simulation of transient and zero current behavior of arcs stabilized by forced convection. *IEEE Trans. Power Appar. Syst.* **1977**, *96*, 1784–1788. [CrossRef]
29. Swanson, B.W. Nozzle arc interruption in supersonic flow. *IEEE Trans. Power Appar. Syst.* **1977**, *96*, 1697–1706. [CrossRef]
30. Il'in, A.S. Numerical simulation of arc quenching processes in a high-voltage SF6 circuit breaker and comparison of results with real tests. *Sci. Tech. Bull. Povolzhye* **2011**, *5*, 140–146. (In Russian)
31. Il'in, A.S. Model of arc quenching processes in high voltage circuit breaker. *Electrotekhnika* **2011**, *12*, 36–42. (In Russian)
32. Park, S.H.; Kim, H.K.; Bae, C.Y.; Jung, H.K. Evaluation on Short Line Fault Breaking Performance of SF6 Gas Circuit Breaker Considering Effects of Ablated Nozzle Vapor. *IEEE Trans. Magn.* **2009**, *45*, 1836–1839. [CrossRef]
33. Park, J.H.; Kim, K.H.; Yeo, C.H.; Kim, H.K. CFD Analysis of Arc-Flow Interaction in a High-Voltage Gas Circuit Breaker Using an Overset Method. *IEEE Trans. Plasma Sci.* **2014**, *42*, 175–184. [CrossRef]
34. Schavemaker, P.H.; Van Der Sluis, L. An Improved Mayr-Type Arc Model Based on Current-Zero Measurements. *IEEE Trans. Power Deliv.* **2000**, *15*, 580–584. [CrossRef]
35. Smeets, R.P.P.; Kertész, V. Evaluation of High-Voltage Circuit Breaker Performance with a Validated Arc Model. *IEEE Proc. Gener. Transm. Distrib.* **2000**, *147*, 121–125. [CrossRef]
36. Ahmethodžić, A.; Kapetanović, M.; Sokolija, K.; Smeets, R.P.P.; Kertész, V. Linking a Physical Arc Model with a Black Box Arc Model and Verification. *IEEE Trans. Dielectr. Electr. Insul.* **2011**, *18*, 1029–1037. [CrossRef]
37. Ohtaka, T.; Kertesz, V.; Smeets, R.P.P. Novel Black-Box Arc Model Validated by High-Voltage Circuit Breaker Testing. *IEEE Trans. Power Deliv.* **2018**, *33*, 1835–1844. [CrossRef]
38. Sinkevich, O.A.; Stahanov, I.P. *Plasma Physics. Stationary Processes in a Partially Ionized Gas*; Graduate school: Moscow, Russia, 1991; 191p. (In Russian)
39. Cherednichenko, V.S.; Anshakov, A.S.; Kuzmin, M.G. *Plasma Electrotechnological Installations*; Izdatelstvo of Novosibirsk State University: Novosibirsk, Russia, 2009; 508p. (In Russian)
40. Boulos, M.I.; Fauchais, P.L.; Pfender, E. *Handbook of Thermal Plasmas*; Springer: Cham, Switzerland, 2020; p. 1500. [CrossRef]
41. Zhong, L.; Cressault, Y.; Teulet, P. Evaluation of Arc Quenching Ability for a Gas by Combining 1-D Hydrokinetic Modeling and Boltzmann Equation Analysis. *IEEE Trans. Plasma Sci.* **2019**, *47*, 1835–1840. [CrossRef]
42. Zhong, L.; Gu, Q.; Zheng, S. An Improved Method for Fast Evaluating Arc Quenching Performance of a Gas Based on 1D Arc Decaying Model. *Phys. Plasmas.* **2019**, *26*, 103507. [CrossRef]
43. Zhong, L.; Wang, J.; Wang, X.; Rong, M. Comparison of Dielectric Breakdown Properties for Different Carbon-Fluoride Insulating Gases as SF6 Alternatives. *AIP Adv.* **2018**, *8*, 085122. [CrossRef]
44. Ivanov, M.F.; Galburt, V.A. *Numerical Simulation of Gas and Plasma Dynamics by Particle Methods*; Izdatelstvo of Moscow Institute of Physics and Technology: Moscow, Russia, 2000; 168p. (In Russian)
45. Klimontonovich, Y.L. *Kinetic Theory of Non-Ideal Gas and Non-Ideal Plasma*; Science: Moscow, Russia, 1975; 352p. (In Russian)
46. Fridman, A.; Kennedy, L.A. *Plasma Physics and Engineering*, 3rd ed.; CRC Press: Boca Raton, FL, USA, 2021; 724p. [CrossRef]

47. Muratović, M.; Kapetanović, M.; Ahmethodzić, A.; Delic, S.; Suh, W.B. Nozzle Ablation Model: Calculation of Nozzle Ablation Intensity and Its Influence on State of SF6 Gas in Thermal Chamber. In Proceedings of the 2013 IEEE International Conference on Solid Dielectrics (ICSD), Bologna, Italy, 30 June–4 July 2013; pp. 692–697. [CrossRef]
48. Park, J.H.; Ha, M.J. Experimental and Numerical Studies of Nozzle Ablation and Geometric Change in Real Gas Circuit Breakers. *IEEE Trans. Power Deliv.* **2022**, *37*, 4506–4514. [CrossRef]
49. Kuroda, M.; Urai, H.; Terada, M.; Ishii, T.; Kojima, Y.; Yokomizu, Y. Evaluation of Dielectric Interruption Performance in Gas Circuit Breaker with Ablated PTFE/BN Vapor. In Proceedings of the 2019 5th International Conference on Electric Power Equipment—Switching Technology: Frontiers of Switching Technology for a Future Sustainable Power System, ICEPE-ST 2019, Kitakyushu, Japan, 13–16 October 2019; pp. 551–554. [CrossRef]
50. Jianying, Z.; Zhijun, W.; Bo, Z.; Yongqi, Y.; Yapei, L. Research on Parameters Optimization of High Voltage Circuit Breaker Nozzle Based on Image Recognition and Deep Learning. *IEEJ Trans. Electr. Electron. Eng.* **2021**, *16*, 496–504. [CrossRef]
51. Kwak, C.S.; Kim, H.K.; Lee, S.H. Bezier Curve-Based Shape Optimization of SF6 Gas Circuit Breaker to Improve the Dielectric Withstanding Performance for Both Medium and Maximum Arcing Time. In Proceedings of the ICEPE-ST 2017—4th International Conference on Electric Power Equipment-Switching Technology, Xi'an, China, 22–25 October 2017; Volume 2017-Decem, pp. 61–65. [CrossRef]
52. Bang, B.H.; Lee, Y.S.; Choi, J.U.; Ahn, H.S.; Park, S.W. Prediction and Improvement of Dielectric Breakdown between Arc Contacts in Gas Circuit Breaker. In Proceedings of the 2013 2nd International Conference on Electric Power Equipment—Switching Technology, ICEPE-ST 2013, Matsue, Japan, 20–23 October 2013; pp. 1–4. [CrossRef]
53. Homaee, O.; Gholami, A. Prestrike Modeling in SF6 Circuit Breakers. *Int. J. Electr. Power Energy Syst.* **2020**, *114*, 105385. [CrossRef]
54. Zhang, H.; Yao, Y.; Wang, Z.; Zhang, B.; Hao, X.; Liu, Y.; Du, Y. Application of Arc Breaking Simulation in Development of Extra High Voltage SF6 Circuit Breaker. In Proceedings of the 16th IET International Conference on AC and DC Power Transmission (ACDC 2020), Online, 2–3 July 2020; pp. 842–845. [CrossRef]
55. Dhotre, M.T.; Ye, X.; Seeger, M.; Schwinne, M.; Kotilainen, S. CFD Simulation and Prediction of Breakdown Voltage in High Voltage Circuit Breakers. In Proceedings of the 2017 IEEE Electrical Insulation Conference, EIC 2017, Baltimore, MD, USA, 11–14 June 2017; pp. 201–204. [CrossRef]
56. Ha, M.J.; Kim, J.; Yeo, C.H.; Kweon, K.Y. Influence of PTFE Ablation on the Performance of High Voltage Self-Blast Circuit Breaker. In *Transmission and Distribution Conference and Exposition: Asia and Pacific, T and D Asia 2009*; IEEE: Piscataway, NJ, USA, 2009. [CrossRef]
57. Iordanidis, A.A.; Franck, C.M. Simulation of Ablation ARCS in Realistic Nozzles. In *GD 2008—17th International Conference on Gas Discharges and Their Applications*; IEEE: Piscataway, NJ, USA, 2008; pp. 209–212.
58. Zhang, J.; Lan, J.; Tian, L. Influence of DC Component of Short-Circuit Current on Arc Characteristics during the Arcing Period. *IEEE Trans. Power Deliv.* **2014**, *29*, 81–87. [CrossRef]
59. Ahmethodžić, A.; Kapetanović, M.; Gajić, Z. Computer Simulation of High-Voltage SF6 Circuit Breakers: Approach to Modeling and Application Results. *IEEE Trans. Dielectr. Electr. Insul.* **2011**, *18*, 1314–1322. [CrossRef]
60. Choi, Y.K.; Shin, J.K. Arc Gas-Flow Simulation Algorithm Considering the Effects of Nozzle Ablation in a Self-Blast GCB. *IEEE Trans. Power Deliv.* **2015**, *30*, 1663–1668. [CrossRef]
61. Iordanidis, A.A.; Franck, C.M. Self-Consistent Radiation-Based Simulation of Electric Arcs: II. Application to Gas Circuit Breakers. *J. Phys. D Appl. Phys.* **2008**, *41*, 135206. [CrossRef]
62. Zhang, J.L.; Yan, J.D.; Fang, M.T.C. Investigation of the Effects of Pressure Ratios on Arc Behavior in a Supersonic Nozzle. *IEEE Trans. Plasma Sci.* **2000**, *28*, 1720–1724. [CrossRef]
63. Park, Y.; Song, T. Plasma Arc Simulation of High Voltage Circuit Breaker with a Hybrid 2D/3D Model. In Proceedings of the 2022 6th International Conference on Electric Power Equipment-Switching Technology (ICEPE-ST), Seoul, Republic of Korea, 15–18 March 2022; Volume 4, pp. 190–193. [CrossRef]
64. Golovin, S.V. Partially Invariant Solutions of the Magnetohydrodynamics Equations. Doctor of Physical-Mathematical Science Dissertation, Novosibirsk State University, Novosibirsk, Russia, 2009. (In Russian).
65. Lie, S. *Vorlesungen über Continuierliche Gruppen mit Geometrischen und Anderen Anwendungen*; B.G. Teubner: Leipzig, Germany, 1893; 805p. (In German)
66. Ovsyannikov, L.V. *Group Properties of Differential Equations*; AN USSR, Siberian department: Novosibirsk, Russia, 1962; 239p. (In Russian)
67. Versteeg, H.K.; Malalasekera, W. *An Introduction to Computational Fluid Dynamics*, 2nd ed.; Pearson Education Ltd.: Harlow, UK, 2007; 520p.
68. Najm, H.N. Uncertainty Quantification and Polynomial Chaos Techniques in Computational Fluid Dynamics. *Annu. Rev. Fluid Mech.* **2009**, *41*, 35–52. [CrossRef]
69. Loycanskiy, L.G. *Fluid and Gas Mechanics*, 7th ed.; revised; Drofa: Moscow, Russia, 2003; 840p. (In Russian)
70. Batchelor, G.K. *An Introduction to Fluid Dynamics*; Cambridge University Press: Cambridge, UK, 2012; 658p. [CrossRef]
71. Averyanova, S.A. *Numerical Simulation of Gas Flow in the Arcing Device of a High-Voltage Circuit Breaker*; Candidate of Physical-Mathematical Science Dissertation, Peter the Great St. Petersburg Polytechnic University: St. Petersburg, Russia, 2005. (In Russian)
72. Swanson, B.W.; Roidt, R.M. Thermal Analysis of an SF6 Circuit Breaker ARC. *IEEE Trans. Power Appar. Syst.* **1971**, *PAS-91*, 381–389. [CrossRef]

73. Pei, Y. Computer Simulation of Fundamental Processes in High Voltage Circuit Breakers Based on an Automated Modelling Platform. Ph.D. Thesis, The University of Liverpool, Liverpool, UK, November 2014.
74. Liu, J. Modelling and Simulation of Air and SF6 Switching Arcs in High Voltage Circuit Breakers. Ph.D. Thesis, The University of Liverpool, Liverpool, UK, May 2016.
75. Wilcox, D.C. *Turbulence Modeling for CFD*, 3rd ed.; DCW Industries: La Cañada, CA, USA, 2006; 522p.
76. Bai, S.; Luo, H.; Guan, Y.; Liu, W. Arc Shape and Arc Temperature Measurements in SF6 High-Voltage Circuit Breakers Using a Transparent Nozzle. *IEEE Trans. Plasma Sci.* **2018**, *46*, 2120–2125. [CrossRef]
77. Chernoskutov, D.; Popovtsev, V.; Sarapulov, S. Analysis of SF6 Circuit Breakers Failures Related to Missing Current Zero. Part I. In *2020 Ural Smart Energy Conference (USEC)*; IEEE: Ekaterinburg, Russia, 2020; pp. 51–54. [CrossRef]
78. Chernoskutov, D.; Popovtsev, V.; Sarapulov, S. Analysis of SF6 Circuit Breakers Failures Related to Missing Current Zero. Part II. In *2020 Ural Smart Energy Conference (USEC)*; IEEE: Ekaterinburg, Russia, 2020; pp. 55–58. [CrossRef]
79. Thomas, R. Three Phase Controlled Fault Interruption Using High Voltage SF6 Circuit Breakers. Ph.D. Thesis, The University of Liverpool, Göteborg, Sweden, 2007.
80. Kays, W.M. Turbulent Pratidtl Number—Where Are We? *J. Heat Transfer.* **1994**, *116*, 284–295. [CrossRef]
81. Lacasse, D.; Turgeon, É.; Pelletier, D. On the Judicious Use of the k-ε Model, Wall Functions and Adaptivity. *Int. J. Therm. Sci.* **2004**, *43*, 925–938. [CrossRef]
82. Gonzalez, J.J.; Freton, P.; Reichert, F.; Petchanka, A. PTFE Vapor Contribution to Pressure Changes in High-Voltage Circuit Breakers. *IEEE Trans. Plasma Sci.* **2015**, *43*, 2703–2714. [CrossRef]
83. Chernoskutov, D.V. *Increasing the Switching Capacity of High-Voltage Electrical Equipment*; Candidate of Technical Science Dissertation, Ural Federal University: Ekaterinburg, Russia, 2017. (In Russian)

axioms

Article

Steam Gasification in a Fluidized Bed with Various Methods of In-Core Coal Treatment

Nikolay Abaimov [1,*], Alexander Ryzhkov [1], Vladimir Tuponogov [1], Leonid Simbiriatin [2], Alexey Dubinin [1], Lu Ding [3] and Sergey Alekseenko [1,4]

1 Ural Power Engineering Institute, Ural Federal University Named after the First President of Russia B.N. Yeltsin, Str. Mira, 19, 620002 Yekaterinburg, Russia; a.f.ryzhkov@urfu.ru (A.R.); v.g.tuponogov@urfu.ru (V.T.); a.m.dubinin@urfu.ru (A.D.); aleks@itp.nsc.ru (S.A.)
2 "Promyshlennyj Perlit", 4b/35 V. Vysotsky Street, 620072 Yekaterinburg, Russia; leonid.simbiryatin@gmail.com
3 Key Laboratory of Coal Gasification and Energy Chemical Engineering of the Ministry of Education, East China University of Science and Technology, Shanghai 200237, China; dinglu101@163.com
4 Kutateladze Institute of Thermophysics SB RAS, 1 Academician Lavrentiev Avenue, 630090 Novosibirsk, Russia
* Correspondence: nick.sum41@mail.ru; Tel.: +7-906-815-08-28

Abstract: The aim of this work is to study coal steam gasification with various methods of coal in-core treatment in FB using a newly developed thermodynamic calculation method. A calculational study of subbituminous coal steam non-catalytic gasification was carried out using four different methods of coal in-core treatment in single-vessel multisectional fluidized-bed gasifiers. A semi-empirical model based on the entropy maximization thermodynamic method and "restricted equilibria" based on previously obtained experimental data has been developed. Based on thermodynamic calculations, the effect of the leading thermochemical processes and operating parameters of the fluidized bed (temperature, fluidization number, steam/coal ratio feed rate) was revealed. New information was obtained regarding the composition of char and syngas at the gasifier outlet, the syngas heating value, and the cold gas efficiency of the steam gasification of Borodinskiy subbituminous coal char. The results indicate the possibility of significantly accelerating and improving non-catalytic steam gasification in fluidized bed gasifiers through the appropriate organization of in-core coal treatment. Based on the results obtained, the following recommendation is made—when designing multi-section and multi-vessel steam-blown gasifiers, the ratio of residence times should be set in favor of increasing the coal residence time in the steam-blown carbonization zone. Structurally, this can be achieved by increasing the volume and/or area of the steam-blown carbonization section (vessel).

Keywords: engineering calculations; mathematical modeling; improvement of technical devices; coal; syngas; steam gasification; bubbling fluidized bed; thermodynamic

MSC: 74A15; 81T80; 62P30

Citation: Abaimov, N.; Ryzhkov, A.; Tuponogov, V.; Simbiriatin, L.; Dubinin, A.; Ding, L.; Alekseenko, S. Steam Gasification in a Fluidized Bed with Various Methods of In-Core Coal Treatment. *Axioms* **2023**, *12*, 587. https://doi.org/10.3390/axioms12060587

Academic Editor: Leonid Plotnikov

Received: 4 May 2023
Revised: 10 June 2023
Accepted: 11 June 2023
Published: 13 June 2023

1. Introduction

Steam gasification in a fluidized bed (FB) is, theoretically, an ideal way to produce high-quality, medium heating value syngas for energy and chemical technologies in a non-catalytic low-temperature process that ensures the unit's slag-free operation at a satisfactory conversion rate [1].

Steam gasification in FB was initially developed in the 1970s with the aim of obtaining medium heating value syngas from coal (about 10 MJ/m^3) for a high-capacity integrated gasification combined cycle (IGCC) and as a feedstock for large chemical enterprises. The process, in accordance with technological requirements, was carried out, as a rule, under a pressure of up to 4 MPa in the low temperature range of 650–1100 °C [2]. Recently, steam gasification in a bubbling FB (BFB) has gained renewed popularity with the development of distributed generation on a local raw fuel base (low-grade coals and biomass), which needs highly efficient energy and chemical sources of low-pressure syngas. Interest in coal is present both in emerging [3–5] and developed economies [1,6–8].

One direction in the modern transition to alternative energy sources is the use of hydrogen, which can be obtained in its pure form from solid fuel syngas [5,9]. Syngas with a high hydrogen content is also used in power plants with solid oxide fuel cells [10].

A characteristic feature of steam gasification gases in FB is, as a rule, an increased content of CO_2, reaching 28.3% for coal and 25–32% for biomass [11–14] and high hydrogen content (up to 53–66%). Such gases cannot be used for chemical production, but are suitable for hydrogen production, because they almost do not require a water gas shift reaction (WGSR), although they need deep purification due to the content of pyrolysis ballast products. There are data on the production of syngas with relatively close concentrations of H_2 and CO (47 and 38%) at a low content of CO_2 (3.3%) [2].

In this work the equilibrium composition of the reaction products was calculated by the entropy maximization method (EMM) using "restricted equilibria". However, in its pure form, the EMM method, as well as its analogue, the Gibbs energy minimization method, is not suitable for predicting the composition of syngas in a low-temperature FB reactor [15]. This is due to the overly brief gas residence time, which is 2–3 orders of magnitude less than necessary [16]. As a result of such calculations, the content of CO_2, CH_4, and C_2H_4 is usually far from their experimental equivalents [17]. Even in entrained-flow gasifiers at relatively high temperatures (up to 1200 °C), syngas composition can differ markedly from the equilibrium composition, as determined by the WGSR equilibrium constant [18].

Models with "restricted equilibria" allow for the implementation of user-defined constraints. A similar approach is commonly used in modeling in Aspen Plus [19]. Bed gasifiers are also modeled using a multi-zone model based on non-stoichiometric equilibrium models and a redistribution coefficient with a bypass of pyrolysis products through the oxidation zone [20]. According to this approach, only some of the gasification reactions are in equilibrium. At the same time, exact energy balances are achieved only if the experimental gas composition is used as an input parameter for modeling [19].

The research [21] was completed via a study of the influence on the equilibrium of additional parameters such as the quantity of steam, the pressure or the kind of biomass using the Gibbs energy minimization method. In [22], a one-dimensional unsteady state model is developed for simulation of biomass gasification in a bubbling fluidized bed. The effect of biomass feeding position is investigated, and the performance of a reactor under non-isothermal conditions is compared with its performance under isothermal operation. A comprehensive process model is proposed to simulate the steam gasification of biomass in a bubbling fluidized bed reactor using the Aspen Plus simulator [23]. In this work, it was found that the steam-to-biomass ratio is directly proportional to an increase in the content of hydrogen and carbon monoxide, while gas yield and carbon conversion efficiency enhance significantly with increasing temperature. In [24] researchers investigate the fluidized bed gasification of several pure and blended feedstock prepared in the form of pellets: oak bark, two bark/wheat straw blends (85/15 and 50/50% wt) and lignin residue remaining from bioethanol production.

The aim of this work is to study coal steam gasification with various methods of coal in-core treatment in FB using a newly developed thermodynamic calculation method.

The objectives of the research include:

1. Development of a semi-empirical model for calculating steam gasification in FB.
2. Refinement of the steam gasification model based on the experimental results obtained earlier.
3. Carrying out thermodynamic calculational studies of subbituminous coal char steam gasification in the field of technologically justified FB parameters.

2. Methods and Materials

The most common steam gasification technology in FB, combining autothermal and allothermal modes, is a multi-stage multi-section technology with external circulation (Figure 1). Such a multi-vessel technology makes it possible to spatially separate the heating, pyrolysis, and gasification zones by fuel and gas, while maintaining intensive heat and mass transfer both within the zones and between them [25].

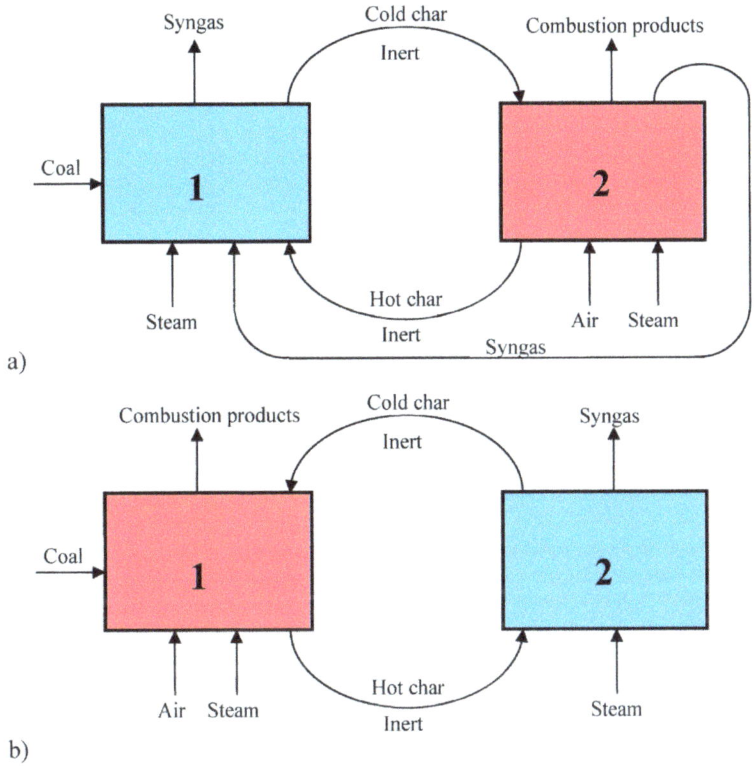

Figure 1. Schemes of media movement in double-vessel gasifiers with external circulation: (**a**) DFB-I; (**b**) DFB-II.

In most cases, a double-vessel FB (DFB) technology is used, usually combining a coal pyrolyzer with BFB or fast FB (FFB) and a gasifier (either an FB char furnace or a circulating FB (CFB)).

Single-vessel FB (SFB) reactors without internal sectioning for the steam gasification of coal can be used under special conditions (an allothermic regime, prepared fuel, catalytic packing, etc.) [26]. Exploratory studies of autothermal SFBs with steam–oxygen- and steam–air-blown processes are ongoing [27,28]. However, the use of single-vessel gasifiers is more developed in entrained-flow high-temperature steam-oxygen-blown technologies [29], or in special gasifiers with low-temperature plasma [30–32].

In some cases, gasification is carried out in one vessel via a two-section FB (SDFB) and a three-section FB (STFB) with internal sectioning and internal circulation (Figure 2). Single-vessel gasifiers with internal sectioning SDFB and STFB allow, as do multi-vessel gasifiers, to separate the heating, pyrolysis, and gasification zones and control the operation of the gasifier in autothermal mode [1,33,34].

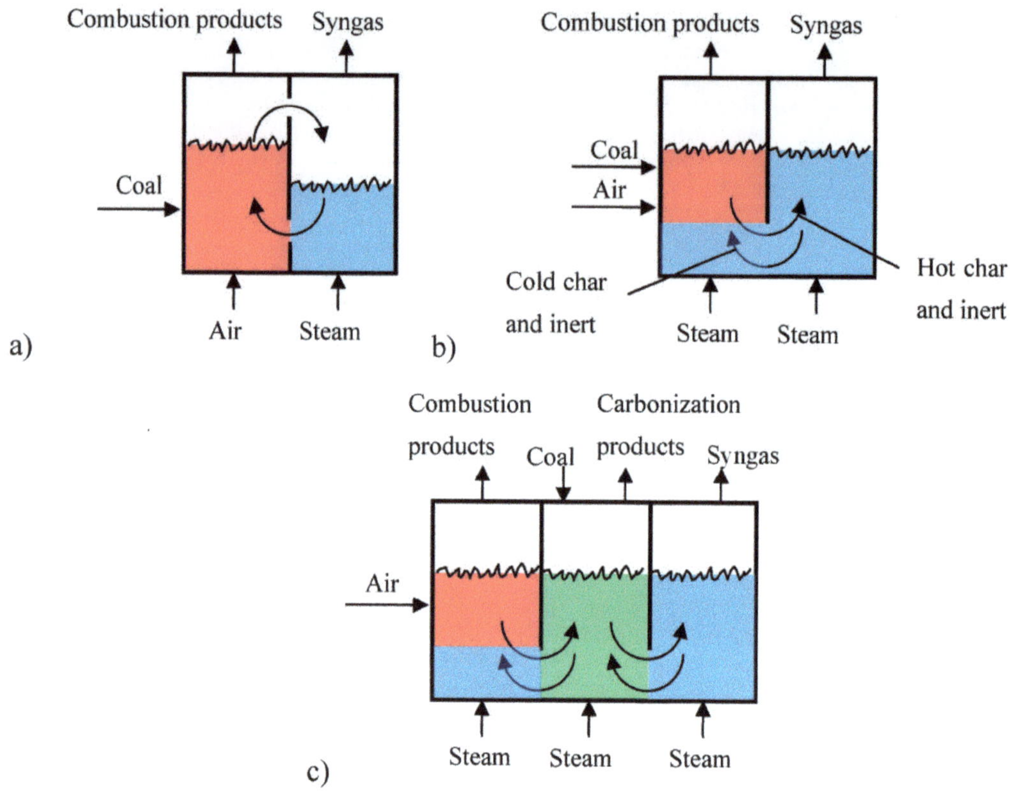

Figure 2. Schemes of media movement in single-vessel gasifiers with internal sectioning: (**a**) SDFB-I; (**b**) SDFB-II; (**c**) STFB.

Multi-section gasifiers operate according to one of two schemes. In Group I gasifiers, syngas is withdrawn from the same gasifier into which all initial reagents (fuel, oxidizer, etc.) are supplied, as is the case in single-vessel gasifiers without internal sectioning. A typical example is the DFB-I and TFB-I gasifiers [2,27,28,31,32,35], where raw coal is gasified. In these, the selection of syngas output from the pyrolysis section, which is supplied with coal, steam, and heated ash (pos. 4 of Table 1) or heated char and heated gas (pos. 5 of Table 1), is carried out. In such gasifiers, the syngas consists of the products of fresh coal and char steam pyrolysis.

Table 1. Steam gasification performance in a fluidized bed and in other systems.

No.	Gasifier	Fuel	Reagent		P, MPa	T, °C	Syngas Composition				Q^d, MJ/m³	Ref
			Section 1	Section 2			H_2	CO	CO_2	CH_4		
1	DFB–II, allothermal, 2 vessels, lab scale	Brown coal	Pyrolysis gas coal	Steam	4	742/790	57.5	19.8	20.8	1.9	12.9 (10.6 (gas 1)	[2]
2	DFB-II, allothermal, lab scale	Brown coal	Steam coal	Pyrolysis gas /steam	4	660/666	67.1	3	28.3	1.6	9.58	[2]
3	"Cogas", TFB-II, 5 vessels, autothermal, pilot scale	Bituminous coal subbitumious coal	Generator gas coal	Steam (Section 2) air (Section 3)	0.1	300–800 800 800–900	49.5	32.5	15.6	0	9.44	[2]

Table 1. *Cont.*

No.	Gasifier	Fuel	Reagent		P, MPa	T, °C	Syngas Composition				Q^d, MJ/m³	Ref
			Section 1	Section 2			H_2	CO	CO_2	CH_4		
4	Agglomerating, DFB-I, autothermal, 2 vessels, pilot scale	Bituminous coal	Steam coal	Air	0.8	850–930 1100	47.9–66.2	0–38.6	3.3–28.2	1.6–1.9	up to 10	[2]
5	"Westinghouse", TFB-I, 2 vessels, 3 sections, autothermal, pilot scale	Coal	Generator gas + coal (Section 1) generator gas + char (Section 2)	Steam/air (3rd section, vessel 2)	1.1–1.7	700–900 (vessel1) 1000–1100 (vessel 2)	14.4	19.2	9.4	2.8	5.2 (vessel 1)	[2]
6	DFB-II, 2 vessels, allothermal, catalytic (Na(OH)$_2$), lab scale	Coal	N_2, Ar coal	Steam	0.1	700 800 900	57 56 55	8 10 18	32 30 25	3 3 3	-	[36]
7	DFB-II, 2 vessels, allothermal, lab scale	Coal D SS	Steam coal	Steam	0.1	750 850 950 750 850 950	61.2 58.7 57.8 61 62.5 60	14.7 16.2 19 16 15 19.1	15.6 17.1 19.4 12 10.2 12.9	4.7 8.1 2.9 2.1 2.4 3.7	-	[5]
8	SFB-I, 1 vessel, autothermal, pilot scale	Anthracite	Steam/O_2 coal		2.5	995	38.49	26.35	23.6	4.89	-	[27]
9	SFB-I, 1 vessel, autothermal, pilot scale	Coal	Steam/air coal		0.1	950	15–20	15–20	10–12	1–2	-	[28]
10	Fixed bed, 2 vessels, allothermal, lab scale	Bituminous coal	Pyrolysis gas coal	Steam	0.1	950 1000 1100	59.6 52.5 52.9	22.4 37.6 39.6	16.00 8.18 5.3	1.33 1.19 0.77	-	[3]
11	Plasma gasifier, lab scale	Brown coal	Steam coal		-	2427–3177	46.8–51	39.3–46	0	0	-	[35]
12	Plasma gasifier, lab scale	Bituminous coal	Steam/air air coal		-	3077–3577 2417–2577	17.0–61.2 16.8–17.9	21.5–45.8 32.4–38.1	0	0	-	[30]
13	Microwave, plasma gasifier, pilot scale	Brown coal	Steam coal		-	1640	39.8	32	18.2	0	-	[31]
14	Microwave plasma gasifier lab scale	Brown coal	Steam/air coal		-	5727	36–49	19–24	24–46	0	-	[32]
15	TFB-II, 1 vessel, 3 sections, autothermal lab scale	Bituminous coal	Steam coal (Section 1)	steam/air кокс (Section 2) steam char (Section 3)	0.1	925–950 *	58.9–75.3	13.2–29.6	6.9–14.7	0–4.5	9.7–11.5	[33]
		Semicoke	Steam semicoke (Section 1)	steam/air semicoke (Section 2) steam semicoke (Section 3)	0.1	925–950 **	47.5–48.6	47.1–48.8	1.8–4.7	0.7–0.8	11.7	

Table 1. *Cont.*

No.	Gasifier	Fuel	Reagent		P, MPa	T, °C	Syngas Composition				Q^d, MJ/m³	Ref
			Section 1	Section 2			H_2	CO	CO_2	CH_4		
16	DFB-II, 1 vessel, 2 sections, autothermal lab scale	Bituminous coal	Steam/air coal (Section 1)	Steam char (Section 2)	0.1	700–1000 ***	22.2–42.8	15.9–36.1	11.9–22.1	0.4–4.1	9.0–10.4	[34]
17	Fixed bed, 1 vessel, Allothermal, lab scale	Bituminous coal	Steam coal	-	~0.1	900–1200	28–58	19–29	42–17	10–1	220–230 kJ/mol	[37]

*—dry gas, in wet gas H_2O = 9.1–13.2. **—dry gas, in wet gas H_2O = 15.8–18.3. ***—wet gas H_2O = 8.5–35.5.

In group II gasifiers (DFB-II, TFB-II), syngas is withdrawn from the section into which steam, and heated char are fed, while coal is fed into the pyrolyzer for carbonization and heating. In allothermic processes (pos. 1 and 2 from Table 1), heating in the pyrolyzer is carried out with steam, syngas, and helium (in a built-in heat exchanger) [2]. In autothermal schemes, heating is performed with burning volatiles, part of the raw coal and char.

Group I gasifiers produce syngas saturated with hydrocarbons with a higher heating value (about 12 MJ/m^3). Group II gasifiers produce somewhat leaner (9–10 MJ/m^3) and purer syngas (with a lower content of hydrocarbons).

2.1. Experiment

To carry out the planned studies, we used data obtained earlier in two previously developed [33,34] single-vessel FB gasifiers with two and three sections (SDFB-II and STFB-II). The sections are interconnected by internal flows through which the inert material circulates, while heat and fuel particles are transferred throughout the volume of the reaction space. For modeling, steam gasification modes were selected for four different cases of thermochemical in-core treatment of coal, carried out at different temperatures and steam flow rates.

The work of the sections is based on the general properties of FB, of which the main property is their representation as ideal mixing reactors in the solid phase and ideal displacement in the gas phase [38].

The fuel used was subbituminous Borodinskiy coal and semicoke from Borodinskiy coal (Table 2): the particle size distribution was typical for bubble fluidization, with a particle size of 1–1.5 mm. As an inert layer material, a well-flowing corundum with a density of 3900 kg/m^3 and average particle sizes of 0.32 and 0.5 mm was used. Such materials, belonging to group B according to the Geldart classification, form a stationary BFB when blown at relatively low speeds (<2–3 m/s). The high density of inert particles made it possible to use sufficiently small fractions in operating conditions without entrainment from the apparatus.

2.1.1. SDFB-II

The SDFB-II plant consists of a vessel and an inner pipe dividing it into combustion and gasification sections (Figure 3). Raw coal is fed into the oxygen zone of the combustion section, where it heats up and releases the bulk of the rapidly flammable volatiles matter. The resulting gases are distilled by an upward flow, and char particles descend into the oxygen-free zone of the gasifier, fluidized by steam and the products of its interaction with char: they then flow into the central pipe of the steam gasification section. The problem of steam overheating to the required temperature of 700–1000 °C, which is unattainable in thermal power plants with indirect/indirect steam heating, is solved in FB by the most efficient method of direct heating of low-potential steam (~200 °C), circulating hot inert material. The removal of ash, together with some of the inert material, is carried out from the upper gasification section. The flow diagram of media in SDFB-II is shown in Figure 3.

Table 2. Ultimate and proximate analysis of Borodinskiy coal, semicoke and ash composition.

Parameter	Coal	Semicoke	Ash	
LHV, MJ/kg	22.676	26.857	Slagging start temperature, °C	950
W^r, %	7.11	2.51	SiO_2, %	46.8
A^r, %	13.39	14.41	Al_2O_3, %	12.9
C^{daf}, %	71.26	90.48	TiO_2, %	0.6
H^{daf}, %	4.88	1.81	Fe_2O_3, %	7.9
N^{daf}, %	1.21	0.97	CaO, %	25.8
O^{daf}, %	22.26	6.74	MgO, %	5
S^{daf}, %	0.39	-	K_2O, %	0.5
V^{daf}, %	48.89	10.46	Na_2O, %	0.5

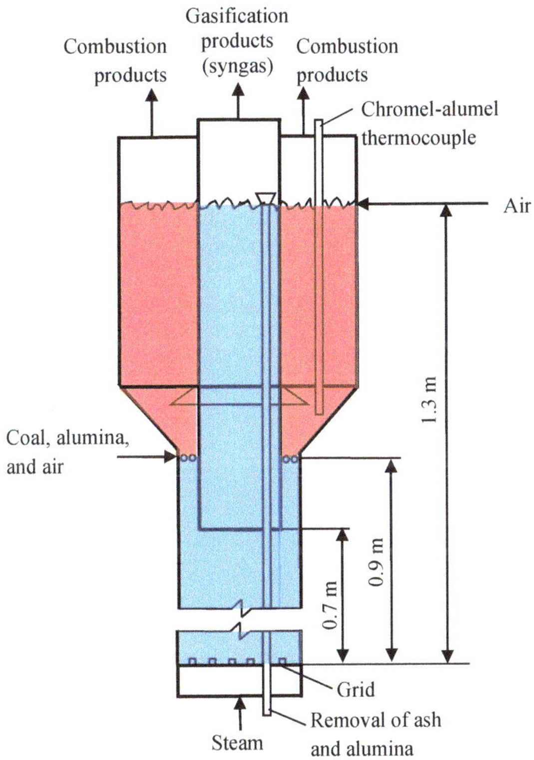

Figure 3. SDFB-II plant diagram.

As shown in Figure 3, at the point of air and coal supply into the gasifier, the newly formed gases are carried away by the upwards gas flow, while the particles of char and ash move downward towards the gas distribution grate. The temperature at the exit from the combustion zone approaches the temperature of the FB (700–1000 °C). This contributes to the development of adverse reactions [39]. Part of the resulting products from side reactions (NO_x, SO_2, CO_2) interact with particles of char and ash according to Equations (1) and (2), [40,41] with the formation of N_2, CO, $CaSO_4$:

$$NO + C = 0.5N_2 + CO \quad (1)$$

$$SO_2 + CaO + 0.5O_2 = CaSO_4 \quad (2)$$

The ash particles, together with the trapped sulfur, are removed through the gasification section. The total gas flow is distributed between the gasification section and the combustion zone in the proportion: s = $(d/D)^2$ in the gasification section and (1-s) in the combustion section.

2.1.2. STFB-II

The STFB-II consists of a vessel and two inner tubes arranged concentrically, one inside the other (Figure 4). The gasifier is divided by these pipes into three sections: central—steam pyrolysis–carbonization; outdoor—steam–air combustion; and an annular gap between them—steam gasification. High superheated steam pyrolysis is a well-known technology widely used to produce activated carbon [42].

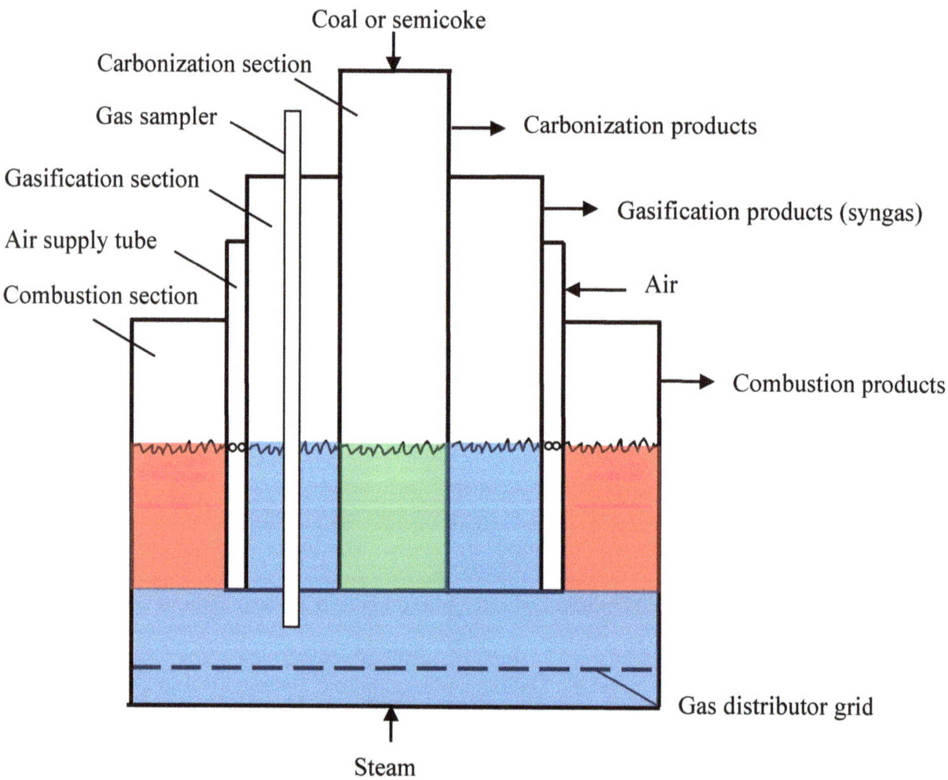

Figure 4. STFB-II gasifier.

The products from each section are removed separately through pipes, while the heat from the combustion zone is carried throughout the gasifier by circulating particles. Raw fuel is supplied to the oxygen-free zone through nozzle, where char with a stable graphite-like carbon structure is formed in the reducing environment. An increase in the reactivity of such a product is mainly due to an increase in the growth of quantitative indicators (the reactive surface of particles during the crushing or opening of internal pores). The char formed in the pyrolysis–carbonization section enters the gasification and combustion sections.

The operating conditions of the gasifiers and the experimental syngas compositions are given in Tables 3–5.

Table 3. Operating conditions for the gasifier.

Parameter	SDFB-II	STFB-II
Gasifier diameter	0.28 m	0.18 m
Bed height	1.3 m	1.0 m
Bed temperature	7000–1000 °C	925–950 °C
Gasification section diameter	0.134–0.213 m	0.08/0.03 m
Coal mean particle diameter	1×10^{-3} m	1.2×10^{-3} m
Density of coal	1250 kg/m^3	1250 kg/m^3
Coke bed concentration	0.6 kg on 1 kg of mixture	0.12 kg on 1 kg of mixture
Alumina average particle diameter	0.5×10^{-3} m	0.32×10^{-3} m
Alumina density	3900 kg/m^3	3900 kg/m^3
Alumina minimum fluidization velocity	0.294 m/s	0.127 m/s
Air flow rate	30–100 m^3/h	20–30 m^3/h
Steam flow rate	17 kg/h	4.4 kg/h
Steam velocity	0.4 m/s	0.28 m/s

Table 4. Investigation program.

Parameter	Case 1	Case 2	Case 3	Case 4
Gasifier	SDFB-II		STFB-II	
Fuel	Coal			Semicoke
Temperature. °C	700	1000	950	

Table 5. Experimental syngas composition.

Component	Case 1	Case 2	Case 3	Case 4
H_2	22.0	42.3	45	39.9
CO	15.9	35.9	17.3	39.7
CO_2	21.9	11.5	10.8	1.4
CH_4	3.9	0.5	0.4	0.7
N_2	0.3	0.1	0	0
H_2O	35.5	8.6	26.5	18.3
O_2	-	-	-	-

In the operating mode at bed temperatures of 700–1000 °C, the superheated steam velocities in the bed near the gas distributor grid were 0.4 m/s in the SDFB-II and 0.28 m/s in the STFB-II. Along the height of the gasifier, the rate of the mixture of steam and gasification products increases due to an increase in the steam conversion products. With complete steam conversion, the gas velocity at the exit from the bed increases by a factor of 2, while the dimensionless velocity $W = u_g/u_{mf}$ reaches 2.72 and 4.4 for SDFB-II and STFB-II, respectively. At the same time, for large particles of coal and semicoke, the gas velocity slightly exceeds the minimum fluidization velocity, due to which there is no accumulation of large particles on the gas distributor grid. Under such hydrodynamic conditions, fuel inert particles are intensively mixed by gas bubbles, creating a high temperature uniformity over the volume of the gasifier sections and a diffusion flow of non-volatile carbon particles from the coal loading are to the ash unloading area in the gasification section.

2.2. Modelling

As shown in the previous section, multiple reaction zones are formed in multi-section gasifiers. Depending on the layout, two to three main zones can be distinguished—the pyrolysis zone, the gasification zone, and the combustion zone; however, these zones do not fully correspond to the corresponding sections. Therefore, a multizone thermodynamic model has been developed to calculate the operation parameters of a multi-sectional gasifier. Similar models apply for single section gasifiers [19]. However, the multi-section gasifier has its own specifics in the form of two or three separate gas outlets (char gas, syngas, and combustion products).

The developed model is based on the following assumptions. The gasifier operates at a steady state at a pressure of 101.13 kPa. The supply air is dry at 25 °C and 101.13 kPa. The ash reaction was not considered. Syngas is an ideal gas consisting of CO, CO_2, H_2, CH_4, H_2O, and N_2. The combustion products consist of CO_2, H_2O, N_2, and O_2. Coal and char contain C, H, N, O, S, H_2O, and ash (inert).

The char composition was determined from the following conditions and assumptions:

(1) in the gasification section, only char is gasified, without raw fuel impurities;

(2) there is no moisture and volatile matter in the char;

(3) the main components of char are fixed carbon and ash.

The equilibrium composition of the reaction products was calculated by the entropy maximization method (EMM) using "restricted equilibria". A detailed description of the practical implementation of the EMM method and its validation are presented in [35,43].

It follows from the laws of thermodynamics that, in a state of equilibrium, the entropy of an isolated system is maximum. Therefore, the problem of calculating the equilibrium composition can be reduced to finding the coordinates of the conditional maximum entropy. The principle of maximum entropy is valid for any equilibrium system, regardless of the path by which the system has reached equilibrium (according to the second law of thermodynamics):

$$S = \sum_{i=1}^{k} S_i^{(p_i)} \cdot n_i + \sum_{l=1}^{L} S_l \cdot n_l = \sum_{i=1}^{k} \left(S_i^0 - R_0 \ln \frac{R_0 T n_i}{v} \right) \cdot n_i + \sum_{l=1}^{L} S_l^0 \cdot n_l, \quad (3)$$

where S—entropy (J/(kg K)); $S_i^{(p_i)}$—entropy of the i-th component of the gas phase (J/(mol K)) at the partial pressure of its equilibrium state $p_i = R_0 T n_i / v$ (Pa); n_i—content of the i-th gaseous component in the system (mol/kg); S_l—entropy of the condensed phase l, which depends only on temperature; v—the specific volume of the system; $S_i{}^0$—standard entropy of the i-th component of the gas phase at temperature T and pressure 0.1 MPa; and R_0—universal gas constant (J/(mol K)).

Determining the parameters of the equilibrium state consists in finding the values of all dependent variables, including the numbers of moles of components and phases at which entropy reaches its maximum. When finding an extremum, additional connections are imposed on the values of the unknown unknowns, reflecting the conditions for the system's existence: the constancy of the total internal energy (because the system is isolated by condition), the constancy of the mass of chemical elements for a closed system and the condition of general electrical neutrality:

$$-U + \sum_{i=1}^{k+L} U_i \cdot n_i = 0; b_j = \sum_{i=1}^{k+L} a_{ji} n_i, j = 1,2,\ldots m; \sum_{i=1}^{k} a_{ji} n_i = 0, \quad (4)$$

where U—internal energy (J/(kg K)); U_i—internal energy of the i-th component (J/mol K); a_{ji}—stoichiometric coefficients; m—the number of chemical elements in the system; and b_j—the content of the j-th element in the system.

As a result, to find the composition and properties of an arbitrary composition corresponding to the state of the maximum entropy of a conditionally isolated system, it is necessary to solve a nonlinear system of equations:

$$G_i - R_0 \ln \frac{R_0 T}{v} - R_0 \ln n_i + \sum_{j=1}^{m} a_{ji}\lambda_j + a_{ei}\lambda_e = 0, (i = 1,2,\ldots k); \tag{5}$$

$$\left(G_l + \sum_{j=1}^{m} a_{jl}\lambda_j\right) \cdot n_l = 0, (l = 1,2,\ldots L); \tag{6}$$

$$\sum_{i=1}^{k+L} a_{ji}n_i - b_j = 0, (j = 1,2,\ldots m); \tag{7}$$

$$\sum_{i=1}^{k} a_{ei}n_i = 0; \tag{8}$$

$$R_0 T \sum_{i=1}^{k} n_i - pv = 0; \tag{9}$$

$$\sum_{i=1}^{k+L} U_i n_i - U = 0, \tag{10}$$

where G_i—Gibbs energy of the i-th component (J/(mol K)) and λ_j—the Lagrange multiplier of the j-th component.

The system of Equations (3)–(10) is solved iteratively. The thermochemical and thermodynamic characteristics of individual substances are taken from [44].

However, the reaction of coal with water vapor within the temperature range of 700–1000 °C at atmospheric pressure and in the absence of a catalyst proceeds slowly [36]. The acceleration of the process is possible by organizing preliminary chemothermal preparation (pyrolysis, carbonization), aimed at reducing the content of hydrogen and heteroatoms (nitrogen, oxygen, sulfur) and the formation of new reaction centers, as well as at increasing the specific reaction surface and at increasing the carbon concentration in the feed gasification material. An important factor influencing the degree of devolumization/degassing and the volume and structure of pores is the gas medium. Carrying out pyrolysis in an oxidizing environment is accompanied by a known activation effect, depending on the type of oxidizing agent.

In an atmosphere of atmospheric oxygen with an equivalent ratio greater than one (in the combustion zone), the devolumization process will be accompanied by the external burning of particles without the oxidative study of the opening pores. A system of thin deep pores will be worked out in the steam, and these can have a positive effect on the conversion rate in the gasification section and on the degree of approximation of the syngas parameters to thermodynamic equilibrium.

To consider the kinetic limitations in determining the composition of the syngas, it is necessary to select the weight coefficient of the reactions taking place from the experimental data.

With a relatively high concentration of fixed carbon in the reactor zone, characteristic of FB, the process of its gasification is determined by the competitive course of the following two heterogeneous reactions of water gas:

$$C + H_2O = CO + H_2; +117 \text{ kJ/mol}; \tag{11}$$

$$C + 2H_2O = CO_2 + 2H_2; +75 \text{ kJ/mol}. \tag{12}$$

Simultaneously with them, the third heterogeneous Boudoard reaction also proceeds:

$$C + CO_2 = 2CO; +159 \text{ kJ/mol}. \tag{13}$$

However, this reaction does not participate in the competition due to its secondary role: CO_2 (the reagent of this reaction) appears in the system only after reaction (12) has taken place. These reactions are traditionally used to describe steam gasification [16,17,45].

To assess the competition between the rates of reactions (11) and (12), the coefficient K is used, which shows by how much reaction (12)'s contribution is greater than reaction (11)'s contribution in the format (12)/(11).

The combustion section (in the DFB-II reactor) or carbonization section (in the TFB-II reactor) is supplied with fuel with a known composition, however, the gasification section does not receive raw fuel but the product of its conversion (carbonization)—char, with an unknown composition that varies depending on carbonization conditions (temperature, consumption of carbonizing media (steam and/or air), etc.).

The algorithm for carrying out thermodynamic calculations for developing models and determining the char composition in the gasification section is shown in Figure 5.

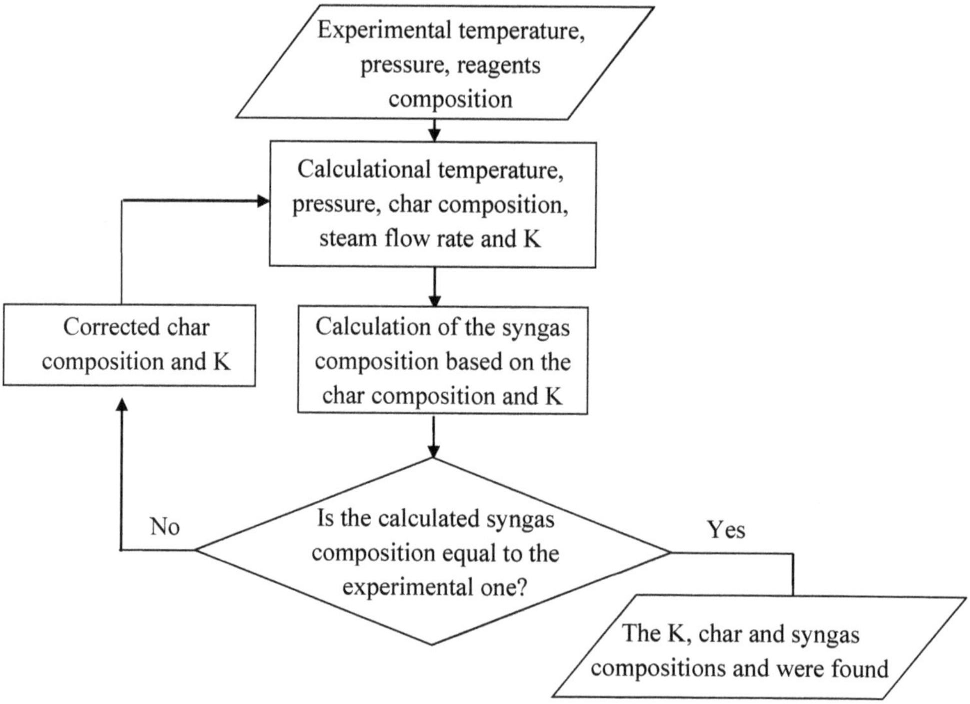

Figure 5. Algorithm for calculating the process of steam gasification in FB.

3. Results and Discussion

3.1. Statistical Evaluation of Results

Based on experimental tests to validate the calculated results obtained, a statistical processing of the data was carried out to confirm the adequacy of the developed model. Figure 6 shows the results of the statistical processing of the concentrations of the main syngas components using the SDFB-II gasifier as an example.

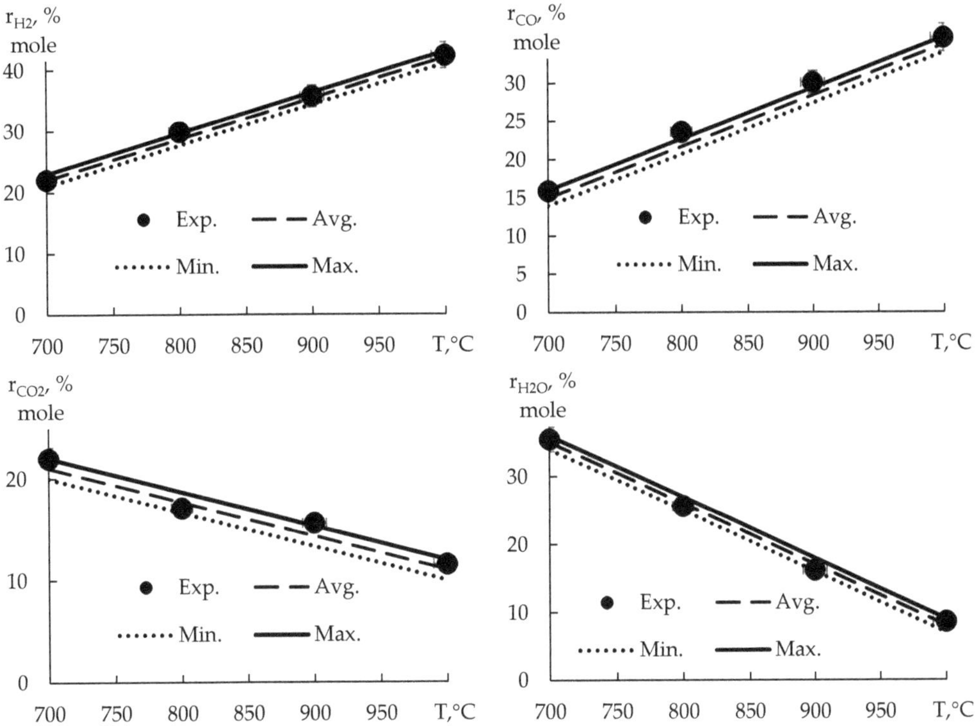

Figure 6. Statistical evaluation of calculation results using experimental data.

As can be seen from Figure 6, the developed model gives results close to the experimental ones. The minimum and maximum values differ from the average by about 1%. The statistical evaluation of the calculation results made it possible to validate the developed model and confirm its satisfactory applicability for solving such problems.

3.2. Dominant Reactions and Their Ratio

The results of multivariate calculations according to the developed algorithm are given in Table 6. Table 6 shows the calculated syngas compositions in three versions:

Table 6. Estimated compositions of syngas.

r, %	Case 1				Case 2				Case 3				Case 4			
	A	B	C	Exp.	A	B	C	Exp.	A	B	C	Exp.	A	B	C	Exp.
H_2	47	47	23	22.0	51	49	43	42.3	42	43	41.4	45	47	46	39.8	39.9
CO	29	24	17	15.9	38	46	36	35.9	20	24	24.3	17.3	16	32	39.8	39.7
CO_2	9	13	23	21.9	4	2	12	11.5	9	9	8.5	10.8	7	7	0	1.4
CH_4	3	1	1	3.9	0	0	0	0.49	0	0	0	0.4	0	0	0	0.7
N_2	0	0	0	0.3	0	0	0	0.14	0	0	0	0	0	0	0	0
H_2O	9	15	7	35.5	7	3	9	8.59	29	24	25.7	26.5	16	15	20.3	18.3

A—original version, without taking into account the char composition and coefficient K;

B—variant taking into account the char composition, but without taking into account coefficient K;

C—variant taking into account the char composition and coefficient K (Table 7);

Table 7. Estimated compositions of syngas.

	Case 1	Case 2	Case	Case 4
K	1/2.45	1/1.18	1/1	1/0

Exp.—experimental version.

The convergence of coefficients K from case 1 to cases 2 and 3 indicates a tendency to equalize the degree of completion of the reactions (11) and (12), which occurs due to a change in their rates that occurs with the achievement of parity in case 3 and the subsequent disparity in case 4.

Figure 7 shows that syngas composition (experiment and modelled, as calculated according to version C) generally does not coincide with the equilibrium A and B. An analysis of the ratio of the syngas components found using versions B and C is shown in Figure 6. The experimental (Exp.) and calculated (C) compositions generally do not coincide with the compositions of A and B. The coincidence of B unrestricted and C restricted equilibria is achieved only in case 3, where syngas composition is controlled by WGSR (14):

$$CO + H_2O = CO_2 + H_2; -42{,}9\ kJ/mol. \quad (14)$$

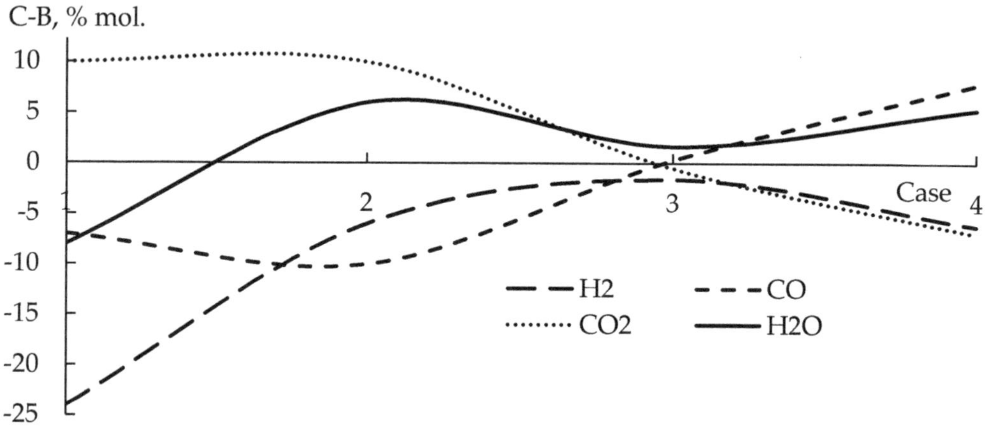

Figure 7. Comparison of calculated results B and C.

In other cases, kinetic restrictions do not allow for the obtaining of syngas with an equilibrium composition. In cases 1 and 2, CO_2 and H_2O are the most abundant in the nonequilibrium mixture, while in case 4, there is an excess of CO and a deficit of CO_2.

In cases 1 (to a greater extent) and 2 (to a lesser extent), the char reactivity is reduced due to oxidative (air) pyrolysis, which occurs in the combustion section. In case 1, the reaction speed is also reduced due to the relatively low temperature of 700 °C. The low reaction rate leads to the dominance of reaction (12), an increase in CO_2, and a decrease in CO. In these cases, the syngas composition obtained by the dominant reaction (12) is first calculated, and then the unreacted carbon reacts with CO_2 according to reaction (13). In case 3, the syngas composition obtained in the experiment and the calculation practically coincide. This indicates the occurrence of reactions (11) and (12) at the same rate, and the achievement of thermodynamic equilibrium at a temperature of 950 °C due to the high char reactivity. In case 2, syngas was obtained at almost the same temperature; however, due to the use of air carbonization, instead of steam, it has an unequal contribution from reactions (11) and (12), which confirms the advantage of steam as a carbonization medium. From the experimental data of case 4, we can see that the syngas H_2 and CO are almost the same, while the CO_2 is negligible, which indicates that reaction (11) proceeds at a much higher rate than reaction (12). This is explained by the greater reactivity of the source fuel (semicoke). The calculation for this case was carried out without considering reaction (12), therefore $CO_2 = 0$. In case 4, the situation is the reverse to that of cases 1 and 2, since steam carbonization char initially has increased reactivity, and additional steam carbonization in TFB-II further increases it. In this case, reaction (11) dominates so strongly ($CO_2 \approx 0$, $H_2 = CO$) that reaction (12) can be neglected. Syngas at the outlet again turns out to be non-equilibrium.

As can be seen, the syngas composition calculated from the global equilibrium according to variant A is close to the experimental composition only in case 3. In the other cases, the action of kinetic restrictions does not allow one to obtain the composition of the syngas from the equilibrium of all reactions.

Depending on the competitive course of reactions (11) and (12), the resulting composition approaches one of three gasses:

Gas 1. In the lower temperature range (600–700 °C) with low reaction fuel, reaction (12) is faster than (11). Syngas composition is mainly determined by equation (12), with the characteristic ratio $H_2/CO_2 = 2$.

Gas 2. In the upper temperature range (900–1000 °C) with low-reactivity fuel, reaction (11) is connected to reaction (12), and they begin to proceed at approximately the same rate. Additionally, due to high temperatures, the water gas shift reaction according to reaction (14) is accelerated.

Gas 3. In the upper temperature range with a highly reactive fuel, reaction (12) ceases to determine syngas composition, and the main contribution begins to be made by reaction (11), with the known ratio $H_2/CO = 1$.

The lowest heating value—gas 1—is the most suitable for producing pure hydrogen and reaches the highest concentration of hydrogen. The lowest concentration of hydrogen at the equality $H_2/CO = 1$ will reach the highest heating value—gas 3—which is more suitable for organic syntheses.

Quantitative confirmation of the above explanations can be found in the classical literature on gasification, starting with [45]; however, this describes the results of experiments under special conditions, far removed from those found in industry. The proposed method makes it possible to obtain the calculated syngas composition close enough to the experimental one. The deviation of the calculated data from the experimental data lies in the allowable range of 5–15%. The presented data show the effectiveness of the method of "restricted equilibria" for the predictive study of steam gasification in both low-temperature and high-temperature regimes.

3.3. *Char Composition*

Char composition was also determined from the above algorithm; however, it does not affect the ratio of the syngas components as much as coefficient K. The char composition (along with the temperature and steam flow rate) is needed to calculate syngas composition. If the temperature and steam flow are determined by the gasifier operating mode, then the reacted char composition is not directly controlled. In our cases, this can be determined only by the algorithm described above, the calculation results for which are shown in Figure 8.

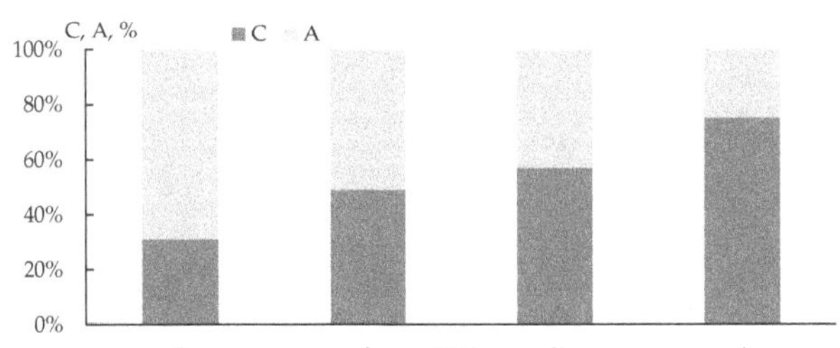

Figure 8. Reacted char composition.

Figure 7 shows that the content of reacted carbon in char increases from case 1 to case 4, and that the ash content decreases without an unambiguous dependence on temperature.

3.4. *Temperature Influence*

Figure 9 shows the calculation results of the temperature dependences of syngas composition using models describing cases 1–4. The results were obtained using the algorithm (Figure 5) and the obtained char composition (Figure 8). The temperature ranges for cases 1 and 2 were taken from the experiment. For cases 3 and 4, a narrow temperature range of ±50 °C was chosen, since the change in K must be considered over a wide temperature range.

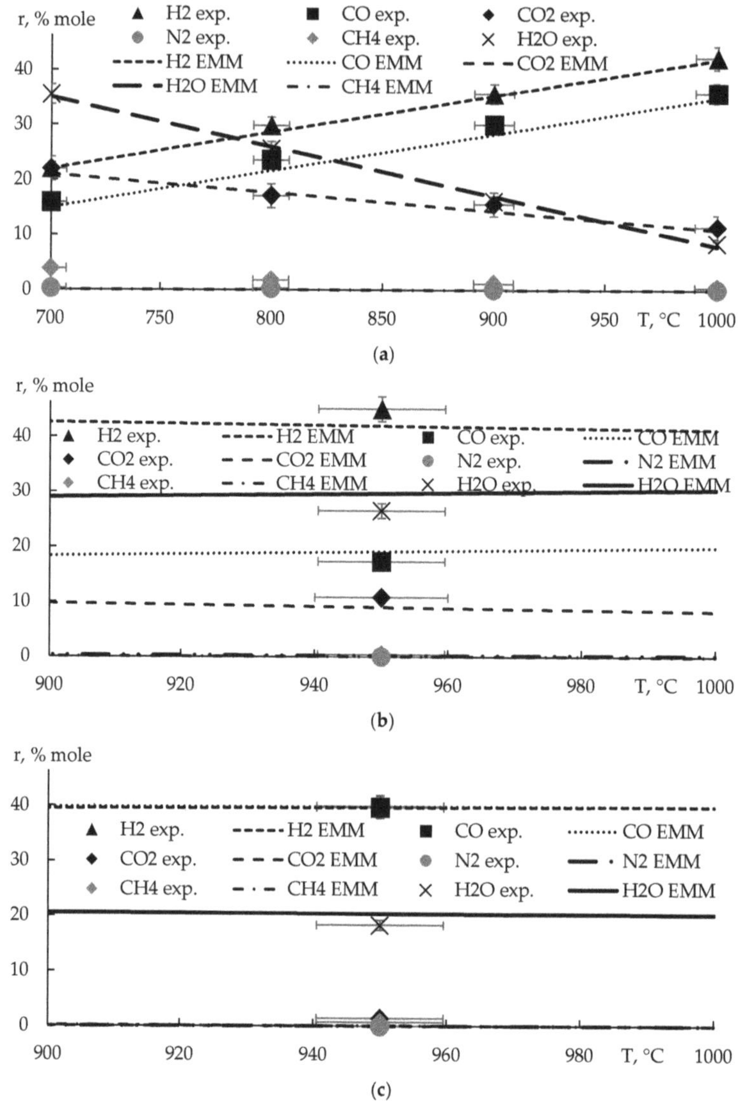

Figure 9. Dependence of syngas composition on temperature in the case of: (**a**) 1 and 2; (**b**) 3; (**c**) 4.

Figure 8 shows the results obtained for SDFB-II. A change in temperature from 700 °C (case 1) to 1000 °C (case 2), which creates differences in the composition and char reactivity, causes a change in the contribution of reactions (12) and (11) from 2.45/1 for case 1 to 1.18/1 for case 2. In the interval between 700 and 1000 °C, the weighting factors change in accordance with Table 8. To obtain these, additional experimental data from [34] were used. Thermodynamic analysis showed that, because of changes to the method of fuel preparation and to the increase of its reactivity at 950 °C, it was possible to achieve an equilibrium syngas composition of the gas 3 type in case 3. In case 2, however, equilibrium was not achieved, even at 1000 °C.

Table 8. Coefficient K for SDFB-II at different temperatures.

	Temperature, °C			
	700	**800**	**900**	**1000**
K	2.45/1	2.02/1	1.61/1	1.18/1

Such a monotonic convergence of coefficients K indicates a tendency to equalize the degree of completion of reactions (11) and (12), which occurs due to the equalization of their rates with increasing temperature. However, as noted above, the equalization of the rates of reactions (11) and (12) in case 2 is not achieved.

Such a redistribution with an increase in temperature by 300 °C causes a doubling of the concentrations of the syngas components: H_2 and CO—towards an increase, and CO_2—towards a decrease. A similar nature of the change in the composition of syngas with temperature is often observed in experimental studies of the steam reforming of coal and biomass in a fluidized bed [1,34]

With a deeper preparation of fuel for gasification, performed in cases 3 and 4 (STFB-II), the contribution of reaction (12) continues to decrease to 1 (case 3) and then to zero (case 4). In this case, the sensitivity of syngas composition to temperature drops sharply (almost to zero in case 4).

To determine the temperature limits of the low sensitivity of the steam reforming gas composition to temperatures in the range of 700–1000 °C, additional experimental studies are required.

Most often, the low temperature sensitivity of the steam reforming syngas composition to temperature is encountered at a higher temperature range—such as in entrained-flow steam-oxygen reactors, plasma processes, etc.—which seem to occur in case 4 [29,35]. The presented data show the flexibility and effectiveness of the method for the predictive study of low-temperature steam reforming in the studied regimes.

3.5. Steam Flow Rate Influence

Figure 10 shows the calculation of the wet (r) and dry (r^d) syngas composition, as well as its heat of combustion depending on the dimensionless steam consumption (g_s) kg of steam/kg of fuel (in our case, char). The lower limit of steam flow rate is selected based on the minimum steam flow required to gasify all of the char carbon, so that no free carbon remains in the system. The calculation for determining this steam flow for all cases was carried out according to equilibrium without restrictions (version C, K = 1/1). The maximum steam flow rate on the graphs was obtained by doubling the minimum value. At the same time, it should be considered that char composition and K can vary depending on steam flow rate. For example, the rate of reaction (12) can increase with an increase in steam consumption and, vice versa, decrease with its decrease. The syngas heating value was determined by the formula:

$$Q = \frac{12.63 \cdot r_{CO} + 10.78 \cdot r_{H_2} + 35.83 \cdot r_{CH_4}}{100} \tag{15}$$

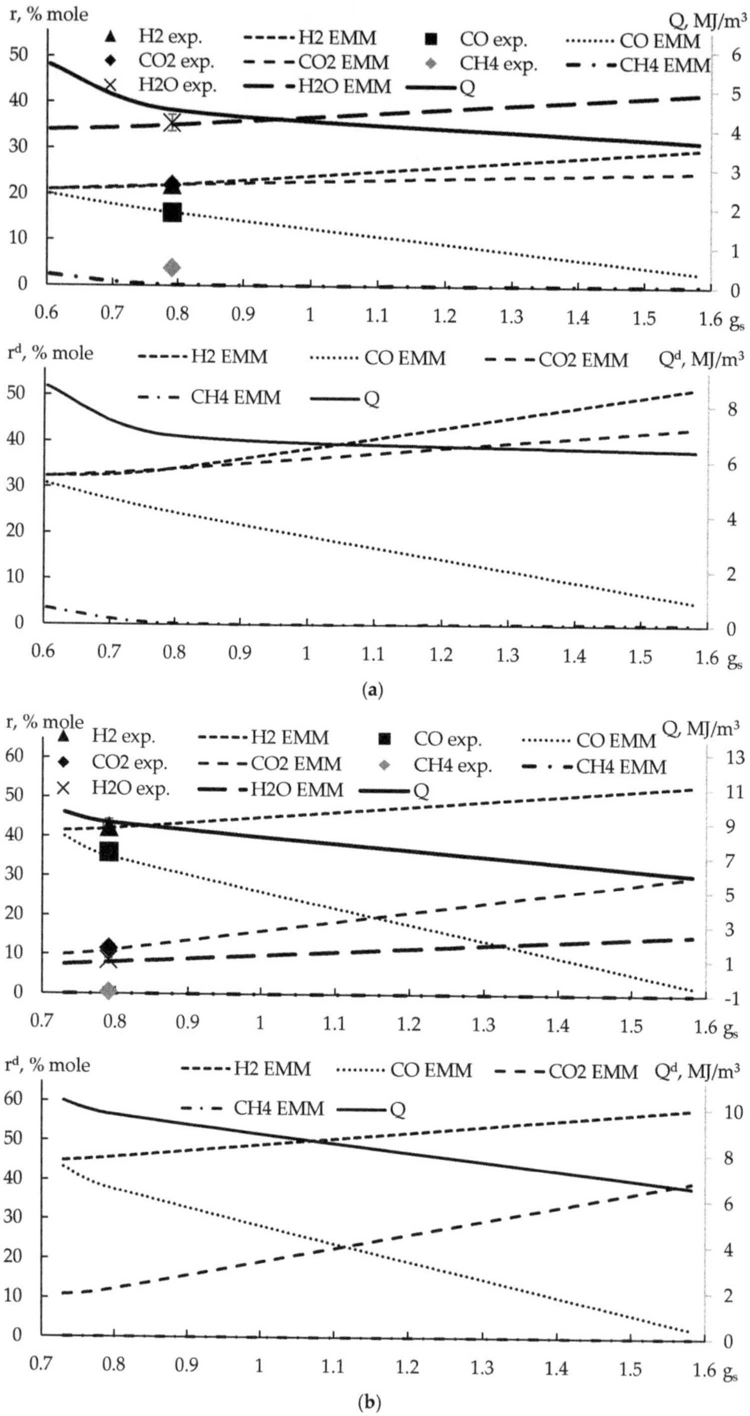

Figure 10. *Cont.*

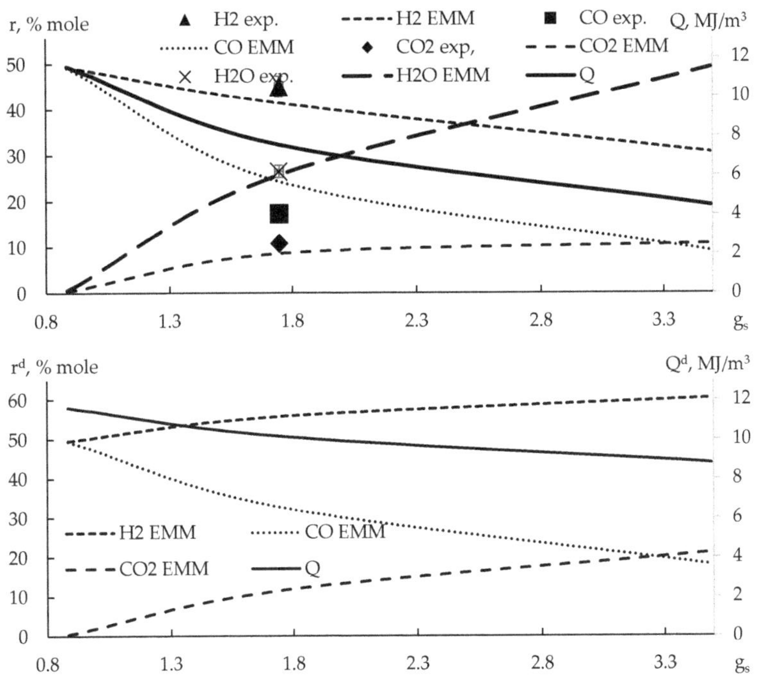

(**c**)

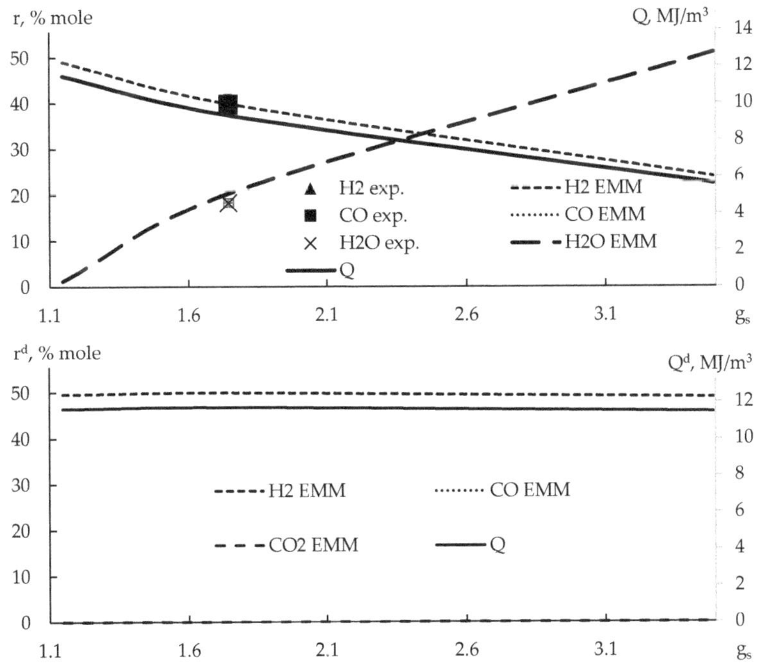

(**d**)

Figure 10. Dependence of wet (r) and dry (r^d) syngas composition on the relative steam consumption in the case of: (**a**) 1; (**b**) 2; (**c**) 3; (**d**) 4.

Figure 9 shows that, with increasing steam consumption, the concentrations of CO_2 and H_2 increase due to reaction (14), while the content of H_2O increases due to the dilution of syngas with steam. This is true in all cases except for case 4, because reaction (14) is impossible due to the absence of CO_2 and reaction (12). The heating value in all cases has its maximum at a minimum steam flow. With an increase in steam consumption, the heating value decreases. This is due to the increased concentration of H_2 and the decreased proportion of CO, given that the heating value of CO is greater than that of H_2. Dependencies of the same kind are also observed in experimental studies, such as in [46]. At the minimum steam flow, syngas composition is close to gas 3, and at the maximum steam flow, to gas 1.

Table 9 compares the obtained results with data from other steam-blown gasifiers. Rows 4–7 show the calculated data of the experimental and optimized modes obtained on the SDFB-II and STFB-II. Optimization was carried out according to the heating value of the syngas. The only parameter that significantly increases the calorific value of syngas in the cases under study is the steam flow rate, namely its reduction (Section 3.4). Therefore, the optimized modes are those in which the steam flow rate is reduced to the minimum required to avoid free carbon in the system.

Table 9. Steam-blown gasifiers.

No.	Gasifier	Fuel	g_s, kg/kg of Fuel	P, MPa	T, °C	Dry Syngas Composition				Q^d, MJ/m^3
						H_2	CO	CO_2	CH_4	
1	Entrained-flow [29]	Oil palm residues	1–8.4	0.1	1000	42–55	25–12	14–25	6–5	9.9–9.2
2	Fixed bed [8]	Palm kernel shells	0.85	0.1	850	48.2	28.2	15.9	7.6	11.5
3	Fluidized bed [46]	Wood	0.83–1.2	0.1	840	45.8–48.2	21.1–19.9	19.2–19.4	6.9–5.9	10.1–9.8
					750	38.3–44.3	22.2–17.9	20.7–22	8–6.6	9.8–9.4
			1.2	0.1–0.25		44.3–45	17.9–11.8	22–26.6	6.6–9	9.4–9.6
4	Case 1 (before optimization/after optimization)	Subbituminous coal	0.79/0.61	0.1	700	34/32	25/31	34/32	0.3/3.5	6.9/8.6
5	Case 2 (before optimization/after optimization)	Subbituminous coal	0.79/0.73	0.1	1000	46/45	38/43	12/11	0/0	9.7/10.3
6	Case 3 (before optimization/after optimization)	Subbituminous coal	1.7/0.9	0.1	950	56/50	33/49	1/0	0/0	10.1/11.6
7	Case 4 (before optimization/after optimization)	Semicoke	1.7/1.1	0.1	950	50/50	50/50	0/0	0/0	11.6/11.6

It can be seen from Table 9 that, due to optimization, the heating value of dry syngas increased by 25, 6, and 15% in cases 1, 2, and 3, respectively. In case 4 the heating value remained unchanged, because at K = 0/1 the composition of the syngas does not depend on the steam flow rate. The largest increase in the heating value occurred in case 1 due to a significant increase in the concentration of CH_4, which is undesirable in systems with cold gas cleaning due to its ability to condense. The heating value in the optimized case 3 and in case 4 exceeds the values obtained in the literature, and at the same time does not contain CH_4, as in [46].

Thus, the proposed model allows one to optimize the gasifiers. Sensitivity analysis in such systems can be performed using the matrix decomposition technique [47]. The development of more complex models will require the use of the Statistical and Dynamic System Criteria [48].

3.6. Pressure Influence

The influence of pressure on syngas composition is shown in Figure 11. The pressure range is selected as 0.1–4 MPa, based on the range of operating pressures encountered in the installation with the FB. Pressure can affect the composition of the reacting char and K, but additional experimental studies are needed to evaluate this effect.

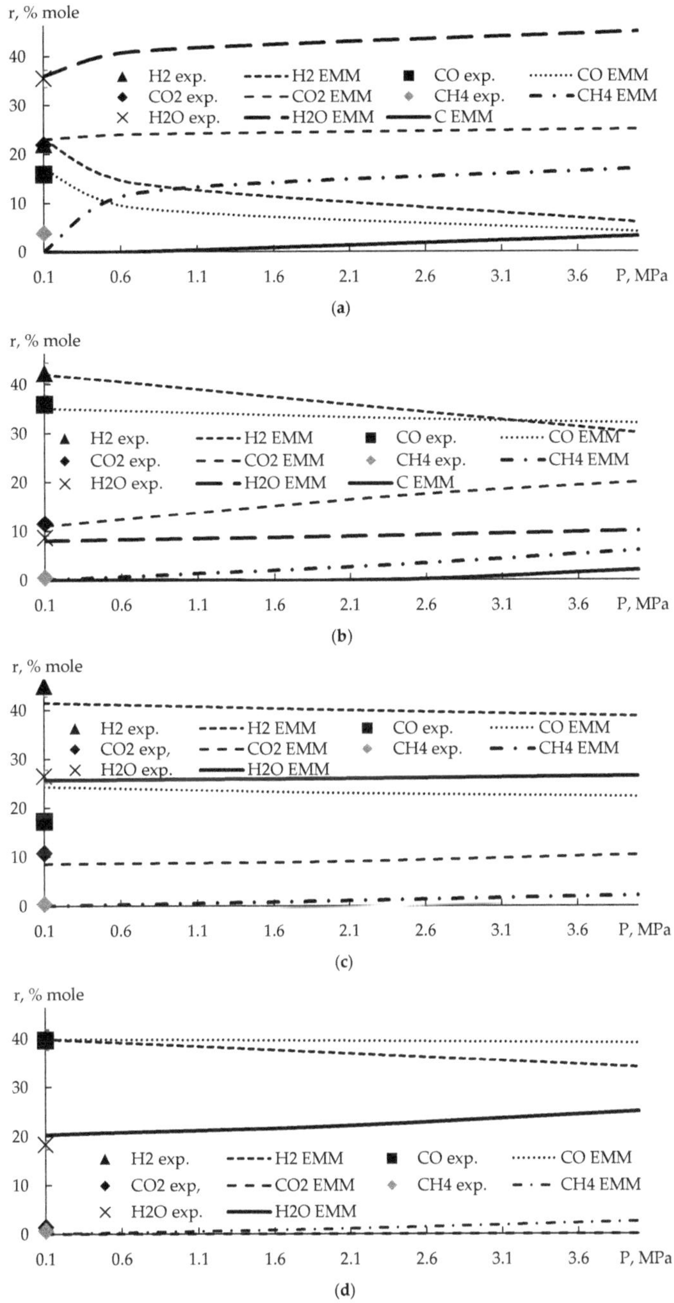

Figure 11. Dependence of syngas composition and heating value on the relative steam flow rate in the case of: (**a**) 1; (**b**) 2; (**c**) 3; (**d**) 4.

Figure 11 shows that in cases 1 and 2, at pressures greater than 0.7 and 2.35 MPa, free carbon begins to be present in the system, which is explained by a shift in the equilibrium of heterogeneous reactions (11)–(13) to the side with a smaller volume (number of moles) according to the Le Chatelier–Brown principle. In case 2, the pressure of the appearance of free carbon in the system is higher, since the temperature is higher; according to the Le Chatelier–Brown principle, the balance of endothermic reactions shifts towards the formation of reagents. In cases 3 and 4, free carbon does not appear in the system in the studied pressure range, as the reactivity of the fuel in these cases is higher. In all cases, the same trend is observed—the proportion of heavy molecules (CH_4, CO_2, H_2O) increases, while the proportion of light molecules (CO and H_2) decreases, something which is also explained by the Le Chatelier–Brown principle. In general, this character of the dependence of syngas composition on pressure in gasifiers with FB is quite characteristic and has been experimentally confirmed [46].

3.7. Comparison with Literature Data

Table 10 shows a comparison of syngas component ratios from the literature data and the performed calculation. In some gasifiers CO or $CO_2 \approx 0$, in this case the table shows "-".

Table 10. Comparison of syngas component ratios.

No.	Gasifier	H_2/CO	H_2/CO_2	CO/CO_2	Q^d, MJ/m^3	Gas Type (from Section 3.1)
1	DFB-II, allothermal, 2 vessels, lab scale [2]	2.9	2.8	1	10.6	2
2	DFB-II, allothermal, lab scale [2]	22.4	2.4	0.1	9.58	1
3	"Cogas", TFB–II, 5 vessls, autothermal, pilot scale [2]	1.5	3.2	2.1	9.44	3
4	Agglomerating, DFB-I, autothermal, 2 vessels, pilot scale [2]	-	14.5	0	10	2
5	"Westinghouse", TFB-I, 2 vessels, 3 sections autothermal, pilot scale [2]	0.75	1.5	2.1	5.2	1
6	DFB-II, 2 vessels, allothermal, catalytic ($Na(OH)_2$), lab scale [36]	7.1	1.7	0.25	7.1	2
7	DFB-II, 2 vessels, allothermal, lab scale [5]	4.2	3.9	0.9	11.2	2

Table 10. *Cont.*

No.	Gasifier	H_2/CO	H_2/CO_2	CO/CO_2	Q^d, MJ/m^3	Gas Type (from Section 3.1)
8	Case 1 (before optimization/after optimization)	1.4/1	1/1	0.7/1	6.8/7.3	2/1
9	Case 2 (before optimization/after optimization)	1.2/1	3.8/4	3.2/4	9.8/10.3	2/3
10	Case 3 (before optimization/after optimization)	1.7/1	56/-	33/-	10.2/11.6	2/3
11	Case 4 (before optimization/after optimization)	1/1	-/-	-/-	11.7/11.7	3/3

As can be seen from Table 10, most of the gasifiers from the literature review produce gas 2 and two of them produce gas 1. This is explained by the low char reactivity, which is prepared from coal in these gasifiers. As a rule, the coal treatment in these gasifiers takes place in an oxidizing atmosphere of air oxygen or steam, which leads to significant differences in the carbonized product (char) reactivity and the syngas composition of char steam-blown gasification. As a result, under the same temperature regime and residence time in the gasification section in cases 2, 3, and 4, syngas was obtained with different degrees of approximation to thermodynamic equilibrium. Oxygen reacts relatively quickly with carbon, so the conversion occurs on the particle surface, reducing its area and reactivity. Therefore, the char conversion in cases 1 and 2 does not reach thermodynamic equilibrium ($K > 1/1$), the syngas tends to gas 2, and in the case of optimization it approaches (but does not reach) gases 1 and 3. The steam reacts slower than oxygen, therefore, it has time to penetrate the pores, react in them, and increase their size, particle area and reactive ability. Therefore, the char conversion in case 3 reaches thermodynamic equilibrium ($K = 1/1$) according to the gas 2 already in the non-optimized mode. Additionally, the conversion of semicoke, which has undergone additional carbonization even before loading into the gasifier, is ultimately observed, again, to be non-equilibrium, but with the dominance of reaction (11), $K = 1/0$, which determines the composition of gas 3.

Based on the results obtained, the following recommendations were made—when designing multi-section and multi-vessel steam-blown gasifiers, the ratio of residence times should be set in favor of increasing the coal residence time in the steam-blown carbonization zone. Structurally, this can be achieved by increasing the volume and/or area of the steam-blown carbonization section (vessel).

4. Conclusions

(1) A new efficient, reliable, and simple method for calculating syngas composition at the gasifier outlet using the EMM method with the "restricted equilibria" approach has been developed. This makes it possible to visually interpret the actual data, clearly determine the limits of applicability of the reactions and confidently predict the linear development of the process in the confidence range of the variation of the system parameters.

(2) The effectiveness of in-core thermochemical fuel treatment in improving the syngas quality by increasing the reactivity of the steam-blown processed fuel was determined. This makes it possible to significantly guide the fast conversion processes in the FB closer to thermodynamic equilibrium and achieve, in a controlled process, the syngas compositions required by the chemical industry, such as gas 3 ($H_2/CO = 1$, $CO_2 = 0$) and gas 1 ($H_2/CO_2 = 2$, $CO = 0$) without resorting to extreme technologies (plasma) or expensive fillers (catalysts). Such a high quality of syngas is achieved even without the use of steam gasification in multi-vessel FBs. Oxygen reacts relatively quickly with carbon, so the conversion occurs on the surface of the particle, reducing its area and reactivity. Steam reacts more slowly than oxygen, so it has time to penetrate the pores, react in them, and increase their size, particle area and reactive ability. The char obtained from semicoke (case 4) had the longest carbonization, which began even before loading into the gasifier, so it has the maximum reactivity and thus makes it possible to obtain gas 3.

(3) Thermodynamic studies of the process of the steam-blown gasification of subbituminous coal char were carried out in the field with technologically justified parameters of a combustion chamber in terms of temperature (700–1000 °C), pressure (0.1–4 MPa) and steam consumption (0.6–1.7 kg/kg of fuel). Increasing the heating value (optimization) is achieved by reducing the steam flow rate to that which was stoichiometrically necessary for the absence of free carbon from the system. The maximum increase in the heating value is achieved in case 1 (10%), and the minimum (no increase) in case 4. Gas 3 is only reached in case 4 before optimization and in all cases after optimization due to oxygen-free conditions for CO_2 formation.

Author Contributions: Conceptualization, L.S. and L.D.; methodology, N.A.; software, N.A.; validation, N.A.; formal analysis, N.A. and A.R.; investigation, N.A. and A.R.; resources, A.D.; data curation, A.R.; writing—original draft preparation, V.T., N.A. and A.R.; writing—review and editing, A.R.; visualization, N.A.; supervision, S.A.; project administration, S.A.; funding acquisition, S.A. All authors have read and agreed to the published version of the manuscript.

Funding: Research funding from the Ministry of Science and Higher Education of the Russian Federation (Ural Federal University Program of Development within the Priority-2030 Program) and the National Natural Science Foundation of China (22278142) is gratefully acknowledged.

Institutional Review Board Statement: Not applicable.

Informed Consent Statement: Not applicable.

Data Availability Statement: Not applicable.

Conflicts of Interest: The authors declare no conflict of interest.

Nomenclature

D	gasifier diameter, m
d	gasification section diameter, m
g_s	dimensionless steam consumption, kg of steam/kg of fuel
P	pressure, Pa
Q	heating value, MJ/m^3
R_0	universal gas constant, J/(mol K)
r	volume concentration, %
S	entropy J/(kg K)
T	temperature, °C
u_g	superficial gas velocity, m/s
u_{mf}	minimum fluidization gas velocity, m/s
BFB	bubbling fluidized bed
CFB	circulating fluidized bed
DFB	double-vessel fluidized bed
EMM	entropy maximization method
FB	fluidized bed
FFB	fast fluidized bed
IGCC	integrated gasification combined cycle
SFB	single-vessel fluidized bed
SDFB	two-section fluidized bed
STFB	three-section fluidized bed
WGSR	water gas shift reaction

References

1. Karl, J.; Pröll, T. Steam gasification of biomass in dual fluidized bed gasifier: A review. *Renew. Sustain. Energy Rev.* **2018**, *98*, 64–78. [CrossRef]
2. Schilling, H.-D.; Bonn, B.; Kraus, U. *Coal Gasification*; Graham and Trotman: London, UK, 1981; p. 175.
3. Wang, D.; Li, S.; He, S.; Gao, L. Coal to substitute natural gas based on combined coal-steam gasification and one-step methanation. *Appl. Energy* **2019**, *240*, 851–859. [CrossRef]

4. Chavan, P.; Datta, S.; Saha, S.; Sahu, G.; Sharma, T. Influence of high ash indian coals in fluidized bed gasification under different operating conditions. *Solid Fuel Chem.* **2012**, *46*, 108–113. [CrossRef]
5. Shevyrev, S.A.; Mazheiko, N.E.; Yakutin, S.K.; Strizhak, P.A. Investigation of characteristics of gas and coke residue for the regime of quasi- and non-stationary steam gasification of coal in a fluidized bed: Part 1. *Energy* **2022**, *251*, 123938. [CrossRef]
6. Aigner, I.; Pfeifer, C.; Hofbauer, H. Co-gasification of coal and wood in a dual fluidized bed gasifier. *Fuel* **2011**, *90*, 2404–2412. [CrossRef]
7. Kern, S.; Pfeifer, C.; Hofbauer, H. Gasification of lignite in a dual fluidized bed gasifier—Influence of bed material particles size and the amount of steam. *Fuel Process. Technol.* **2013**, *111*, 1–13. [CrossRef]
8. Barco-Burgos, J.; Carles-Bruno, J.; Eicker, U.; Saldana-Robles, A.L.; Alcantar-Camarena, V. Hydrogen-rich syngas production from palm kernel shells (PKS) biomass on a downdraft allothermal gasifier using steam as a gasifying agent. *Energy Convers. Manag.* **2021**, *245*, 114592. [CrossRef]
9. Feng, J.; Yan, S.; Zhang, R.; Gu, S.; Qu, X.; Bi, J. Characteristics of Co–Ca catalyzed coal hydrogasification in a mixture of H2 and CO2 atmosphere. *Fuel* **2022**, *324*, 124486. [CrossRef]
10. Li, B.; Lyu, Z.; Zhu, J.; Han, M.; Sun, Z. Study on the operating parameters of the 10 kW SOFC-CHP system with syngas. *Int. J. Coal Sci. Technol.* **2021**, *8*, 500–509. [CrossRef]
11. Antonelli, J.; Lindino, C.A.; Bariccatti, R.A.; Souza, S.N.M.; Lenz, A.M. Lead adsorption and subsequent gasification with Pinus elliottii waste. *Manag. Environ. Qual. Int. J.* **2017**, *28*, 839–850. [CrossRef]
12. Dubinin, A.M.; Shcheklein, S.E. Mini coal-fired CHP plant on the basis of synthesis gas generator ($CO+H_2$) and electrochemical current generator. *Int. J. Hydrog. Energy* **2017**, *42*, 26048–26058. [CrossRef]
13. Dubinin, A.M.; Tuponogov, V.G.; Kagramanov, Y.A. Air-based coal gasification in a two-chamber gas reactor with circulating fluidized bed. *Therm. Eng.* **2017**, *64*, 46–52. [CrossRef]
14. Valin, S.; Bedel, L.; Guillaudeau, J.; Thiery, S.; Ravel, S. CO_2 as a substitute of steam or inert transport gas in a fluidised bed for biomass gasification. *Fuel* **2016**, *177*, 288–295. [CrossRef]
15. Pletka, R.; Brown, R.C.; Smeenk, J. Indirectly heated biomass gasifcation using a latent heat ballast—1: Experimental evaluations. *Biomass Bioenergy* **2001**, *20*, 297–305. [CrossRef]
16. Murakami, T.; Xu, G.; Suda, T.; Matsuzawa, Y.; Tani, H.; Fujimori, T. Some process fundamentals of biomass gasification in dual fluidized bed. *Fuel* **2007**, *86*, 244–255. [CrossRef]
17. Puig-Arnavat, M.; Bruno, J.C.; Coronas, A. Review and analysis of biomass gasification models. *Renew. Sustain. Energy Rev.* **2010**, *14*, 2841–2851. [CrossRef]
18. Kobayashi, N.; Tanaka, M.; Piao, G.; Kobayashi, J.; Hatano, S.; Itaya, Y.; Mori, S. High temperature air-blown woody biomass gasification model for the estimation of an entrained down-flow gasifier. *Waste Manag.* **2009**, *29*, 245–251. [CrossRef]
19. Biagini, E.; Barontini, F.; Tognotti, L. Development of a bi-equilibrium model for biomass gasification in a downdraft bed reactor. *Bioresour. Technol.* **2016**, *201*, 156–165. [CrossRef]
20. Liao, C.; Summers, M.; Seiser, R.; Cattolica, R.; Herz, R. Simulation of a pilot-scale dual-fluidized-bed gasifier for biomass. *Environ. Prog. Sustain. Energy* **2014**, *33*, 732–736. [CrossRef]
21. Detournay, M.; Hemati, M.; Andreux, R. Biomass steam gasification in fluidized bed of inert or catalytic particles: Comparison between experimental results and thermodynamic equilibrium predictions. *Powder Technol.* **2011**, *208*, 558–567. [CrossRef]
22. Agu, C.E.; Pfeifer, C.; Eikeland, M.; Tokheim, L.-A.; Moldestad, B.M.E. Detailed One-Dimensional Model for Steam-Biomass Gasification in a Bubbling Fluidized Bed. *Energy Fuels* **2019**, *33*, 7385–7397. [CrossRef]
23. Nguyen, N.M.; Alobaid, F.; Epple, B. Process Simulation of Steam Gasification of Torrefied Woodchips in a Bubbling Fluidized Bed Reactor Using Aspen Plus. *Appl. Sci.* **2021**, *11*, 2877. [CrossRef]
24. Valin, S.; Ravel, S.; Pons de Vincent, P.; Thiery, S.; Miller, H.; Defoort, F.; Grateau, M. Fluidised Bed Gasification of Diverse Biomass Feedstocks and S. Blends—An Overall Performance Study. *Energies* **2020**, *13*, 3706. [CrossRef]
25. Leckner, B. Developments in fluidized bed conversion of solid fuels. *Therm. Sci.* **2016**, *20*, S1–S18. [CrossRef]
26. Midilli, A.; Kucuk, H.; Topal, M.E.; Akbulut, U.; Dincer, I. A comprehensive review on hydrogen production from coal gasification: Challenges and Opportunities. *Int. J. Hydrog. Energy* **2021**, *46*, 25385–25412. [CrossRef]
27. Li, G.; Liu, Z.; Liu, F.; Weng, Y.; Ma, S.; Zhang, Y. Thermodynamic analysis and techno-economic assessment of synthetic natural gas production via ash agglomerating fluidized bed gasification using coal as fuel. *Int. J. Hydrog. Energy* **2020**, *45*, 27359–27368. [CrossRef]
28. Datta, S.; Sarkar, P.; Chavan, P.D.; Saha, S.; Sahu, G.; Sinha, A.K.; Saxena, V.K. Agglomeration behavior of high ash Indian coals in fluidized bed gasification pilot plant. *Appl. Therm. Eng.* **2015**, *86*, 222–228. [CrossRef]
29. Ogi, T.; Nakanishi, M.; Fukuda, Y.; Matsumoto, K. Gasification of oil palm residues (empty fruit bunch) in an entrained-flow gasifier. *Fuel* **2013**, *104*, 28–35. [CrossRef]
30. Messerle, V.E.; Ustimenko, A.B.; Lavrichshev, O.A. Comparative study of coal plasma gasification: Simulation and experiment. *Fuel* **2016**, *164*, 172–179. [CrossRef]
31. Uhm, H.S.; Na, Y.H.; Hong, Y.C.; Shin, D.H.; Cho, C.H. Production of hydrogen-rich synthetic gas from low-grade coals by microwave steam-plasmas. *Int. J. Hydrog. Energy* **2014**, *39*, 4351–4355. [CrossRef]
32. Hong, Y.C.; Lee, S.J.; Shin, D.H.; Kim, Y.J.; Lee, B.J.; Cho, S.Y.; Chang, H.S. Syngas production from gasification of brown coal in a microwave torch plasma. *Energy* **2012**, *47*, 36–40. [CrossRef]

33. Dubinin, A.M.; Munts, V.A.; Baskakov, A.P.; Choinzonov, B.L. Gasification of Irsha-Borodino coal in a fluidized-bed reactor. *Solid Fuel Chem.* **1983**, *17*, 117–120.
34. Dubinin, A.M.; Tuponogov, V.G.; Filippov, D.V. Determining maximum capacity of an autothermal fluidized-bed gas generator. *Therm. Eng.* **2009**, *56*, 421–425. [CrossRef]
35. Messerle, V.E.; Ustimenko, A.B.; Lavrichshev, O.A. Plasma coal conversion including mineral mass utilization. *Fuel* **2017**, *203*, 877–883. [CrossRef]
36. Qiu, P.; Du, C.; Liu, L.; Chen, L. Hydrogen and syngas production from catalytic steam gasification of char derived from ion-exchangeable Na- and Ca-loaded coal. *Int. J. Hydrog. Energy* **2018**, *43*, 12034–12048. [CrossRef]
37. Kriengsak, S.N.; Buczynski, R.; Gmurczyk, J.; Gupta, A.K. Hydrogen production by high-temperature steam gasification of biomass and coal. *Environ. Eng. Sci.* **2009**, *26*, 739–744. [CrossRef]
38. Kunii, D.; Levenspiel, O. *Fluidization Engineering*, 2nd ed.; Butterworth-Heinemann: Waltham, MA, USA, 1991; p. 490.
39. Wang, X.; Li, Y.; Zhang, W.; Zhao, J.; Wang, Z. Simultaneous SO_2 and NO removal by pellets made of carbide slag and coal char in a bubbling fluidized-bed reactor. *Process Saf. Environ. Prot.* **2020**, *134*, 83–94. [CrossRef]
40. Pérez-Astray, A.; Adánez-Rubio, I.; Mendiara, T.; Izquierdo, M.; Abad, A.; Gayán, P.; de Diego, L.; García-Labiano, F.; Adánez, J. Comparative study of fuel-N and tar evolution in chemical looping combustion of biomass under both iG-CLC and CLOU modes. *Fuel* **2019**, *236*, 598–607. [CrossRef]
41. Ding, N.; Zhang, C.; Luo, C.; Zheng, Y.; Liu, Z. Effect of hematite addition to $CaSO_4$ oxygen carrier in chemical looping combustion of coal char. *RSC Adv.* **2015**, *5*, 56362–56376. [CrossRef]
42. Mondal, S.; Sinha, K.; Aikat, K.; Halder, G. Adsorption thermodynamics and kinetics of ranitidine hydrochloride onto superheated steam activated carbon derived from mung bean husk. *J. Environ. Chem. Eng.* **2015**, *3*, 187–195. [CrossRef]
43. Gorokhovski, M.; Karpenko, E.I.; Lockwood, F.C.; Messerle, V.E.; Trusov, B.G.; Ustimenko, A.B. Plasma technologies for solid fuels: Experiment and theory. *J. Energy Inst.* **2005**, *78*, 157–171. [CrossRef]
44. Chase, M.W.; Davies, C.A.; Downey, J.R.; Frurip, D.J.; McDonald, R.A.; Syverud, A.N. JANAF thermochemical tables. *J. Phys. Chem. Ref. Data* **1985**, *14* (Suppl. 1), 535.
45. Rambush, N.E. *Modern Gas Producers*; Benn Bros. Ltd.: London, UK, 1923; p. 545.
46. Mayerhofer, M.; Mitsakis, P.P.; Meng, X.; de Jong, W.; Hartmut Spliethoff, H.; Gaderer, M. Influence of pressure; temperature and steam on tar and gas in allothermal fluidized bed gasification. *Fuel* **2012**, *99*, 204–209. [CrossRef]
47. Praks, P.; Lampart, M.; Praksová, R.; Brkić, D.; Kozubek, T.; Najser, J. Selection of Appropriate Symbolic Regression Models Using Statistical and Dynamic System Criteria: Example of Waste Gasification. *Axioms* **2022**, *11*, 463. [CrossRef]
48. Yang, Q.; Peng, X. A Fast Calculation Method for Sensitivity Analysis Using Matrix Decomposition Technique. *Axioms* **2023**, *12*, 179. [CrossRef]

axioms

Article

Thermo-Optical Measurements and Simulation in a Fibre-Optic Circuit Using an Extrinsic Fabry–Pérot Interferometer under Pulsed Laser Heating

Artem N. Kotov *, Aleksandr A. Starostin, Vladimir I. Gorbatov and Pavel V. Skripov

Institute of Thermal Physics, Ural Branch, Russian Academy of Sciences, Yekaterinburg 620016, Russia; astar2006@mail.ru (A.A.S.); gorbatov@ursmu.ru (V.I.G.); pavel-skripov@bk.ru (P.V.S.)
* Correspondence: artem625@mail.ru

Abstract: Advantages of using an external Fabry–Pérot interferometer (EFPI) as a high-speed local temperature deformation sensor are demonstrated for the fibre-optic circuit combining a powerful laser beam for surface heating with a low-power probing radiation. The difference in the formation of the heating and probing radiation provides a simple basis for varying the gap between the fibre end and the surface in order to change the ratio between the heating and EFPI measuring areas. Using an example of modelling the laser heating by radiation from a standard single-mode fibre, we demonstrate the possibility of employing the EFPI to measure the temperature deformation of the surface on a quasi-isothermal area with the temperature close to the maximum at gap values of more than 100 μm. With the condition of preliminary calibration, the proposed scheme can be used to evaluate the heat treatment of the surface with the speed of the applied photodetector. The practical possibilities of the method are demonstrated on examples of heating some metal and semiconductor samples by laser pulses of microsecond duration.

Keywords: thermo-optical effects; Fabry–Pérot interferometer; Gaussian beam model; thermal diffusivity

MSC: 80-05; 78A55

check for updates

Citation: Kotov, A.N.; Starostin, A.A.; Gorbatov, V.I.; Skripov, P.V. Thermo-Optical Measurements and Simulation in a Fibre-Optic Circuit Using an Extrinsic Fabry–Pérot Interferometer under Pulsed Laser Heating. *Axioms* **2023**, *12*, 568. https://doi.org/10.3390/axioms12060568

Academic Editor: Giovanni Mascali

Received: 6 May 2023
Revised: 1 June 2023
Accepted: 6 June 2023
Published: 8 June 2023

1. Introduction

As a general rule, destructive processes leading to the failure of materials begin from the surface. Thus, the solution to the problem of improving the quality of machines and devices largely depends on the technological quality assurance of the surface layer of parts, including their geometric characteristics and physicochemical properties [1]. When using concentrated energy flows for the surface treatment of materials, it is necessary to convert the energy of the source into thermal energy with a certain efficiency. The application of this energy produces changes in the structural phase state of the surface layer to achieve the desired quality. The quality of surface heat treatment mainly depends on the rigour with which the specified heating and cooling modes are applied. Popular pulsed laser processing methods provide unique opportunities for thermally hardening the surface layers of parts [2–4]. One of the main advantages of laser processing is the high spatial resolution of the impact on the material.

This method provides the formation of structures or impact on functional elements with geometric dimensions characteristic of microtechnology (10^{-4}–10^{-6} m). Thus, it becomes necessary to develop a means of measuring fast processes during pulsed heat treatment of such microelements. The diverse physical processes taking place during laser processing depend on the power density of the laser radiation on the surface. Pulsed laser heating begins with rapid thermal expansion of the surface layer of the base material to induce mechanical stresses in the heated layer [4–6]. Thermal expansion occurs in a thin surface layer having a thickness of the order of $(a\tau_d)^{1/2}$, where a is the thermal diffusivity of the surface layer material, while τ_d is the duration of the radiation pulse (for a microsecond pulse $(a\tau_d)^{1/2}$ is on the order of units of micrometres). The force arising from the accelerated displacement of the centre of mass of this layer compresses it during the action of the leading edge of the pulse and stretches it during the action of the trailing edge. The radial thermal expansion of the elastic layer leads to the occurrence of thermal compression stresses and surface deformation. In order to perform the specified heat treatment modes, it is necessary to monitor the maximum values of temperature, stress and strain achieved in the centre of the irradiated area. Similar problems arise when using laser methods in the field of thermal control of the properties of materials and products. For example, the reaction of a material upon exposure to a short laser pulse is investigated in the well-known pump–probe method [7–10]. In particular, the thermophysical properties of the material or defects in its structure are determined by changing the temperature of its surface [10]. For non-contact measurements of rapidly changing temperatures, optical sensors are widely used based on a number of thermo-optical effects [11], see below:

- Thermal radiation of a heated body;
- Temperature dependence of the refractive index;
- Temperature change of the absorption coefficient;
- Temperature change of the reflection coefficient as a result of changes in refraction and absorption, as well as thermal deformation of the surface.

However, for these effects, the thermo-optical coefficients are small, and measurements are possible either at large temperature differences or in special cases, for example, at the edge of the material's own optical absorption. For increasing the sensitivity of optical circuits for thermo-optical measurements, interferometers can be used. However, the bulky circuits generally used for this purpose involve large-sized optics and inconvenient settings.

In this connection, the use of fibre optics in laser processing opens up new technological possibilities, especially when using robots. Accordingly, there is a demand for the development of methods for the remote control of thermal processes by fibre-optic means [7–9]. By configuring low-coherence interferometry with fibre optics, sensors can be miniaturised and placed at some distance from the electronics [7,12]. Chemically inert, compact and durable optical fibre integrated in the process of production is affordable due to the relatively inexpensive optical components. The purpose of the present work is to demonstrate the possibilities of combining low-coherence interferometry with fibre optics to create compact circuits for the high-speed control of thermal processes during pulsed laser processing.

2. Materials and Methods

2.1. Thermal Mirror and Interferometry Method

Photothermal methods are based on the various secondary effects that occur following the absorption of a non-stationary light flux by a sample [13,14]. In the case of photo reflection from a "thermal mirror", a spatial resolution of up to 1 μm is provided [15–20]. The method consists of detecting changes in the reflection coefficient of the sample as a result of heating. The most significant changes occur as a result of thermo-optical and thermoelastic effects. In order to determine the local thermoelastic properties, an interferometer is included in the experimental setup. This is used to detect the displacement of the sample surface due to thermal expansion caused by an increase in temperature. The main feature of the used optical scheme consists of the difference in the sizes of the heating and probing rays. In this case, the heating beam has a diameter several times larger than the probing beam [18]. The narrow probing beam reads information from the central part of the heating spot at temperature and deformation parameters close to the maximum values. The scheme of using optical fibre for radiation transmission is shown in Figure 1.

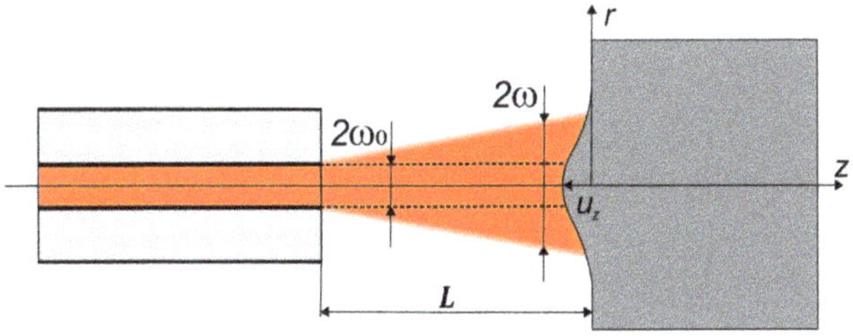

Figure 1. Geometric diagram of the procedures for heating and testing the sample surface.

The change in the optical path due to the thermoelastic deformation of the surface is determined from the interferometric signal using semi-analytical and numerical tools. As a result of the influence of both thermo-optical and thermoelastic effects, a signal can typically contain both contributions. However, by using an interferometer as an amplifier and detector of the phase of the reflected signal, it becomes possible to isolate the effects of surface displacement and radiation phase changes during reflection.

Temporal and spatial distributions of temperature and deformations under local heating are considered elsewhere [6,15–20]. In our case, we apply the model [19,20] with a Gaussian power distribution in a heating beam having radius ω:

$$I(r,t) = \frac{E}{\pi\omega^2}\exp\left(-\frac{r^2}{\omega^2}\right)f(t) \quad (1)$$

where r is the radial distance from the heating centre; E is the pulse energy; $f(t)$ is the normalised dependence of the pulse amplitude on time; and ω is the radius of the Gaussian light beam.

With a surface absorption of radiation, the size of the heated site is much larger than the absorption depth; thus, it is possible to proceed from the model of a local surface heat source using:

$$W(r,t) = \frac{(1-R)}{c_p\rho_0}I(r,t)\delta(z) \quad (2)$$

where R, c_p, and ρ_0 are the average reflection coefficient, specific heat capacity and density of the material, respectively, and $\delta(z)$ is the delta function.

Temperature change $T(r, t)$ for a thermally insulated surface can be calculated with:

$$T(r,t) = \frac{2E_0}{c_p\rho_0}\int\limits_0^t \frac{\exp\left(-\frac{z^2}{4a\Delta t}\right)\exp\left(-\frac{r^2}{\omega^2+4a\Delta t}\right)}{(4\pi a\Delta t)^{1/2}\pi(\omega^2+4a\Delta t)}f(t')dt' \quad (3)$$

where $E_0 = E(1-R)$; $\Delta t = t - t'$; $z = 0$—depth from the heating surface.

Associated thermal deformations along the normal to the heated surface:

$$\begin{aligned} u_z = {} & -u_0\exp(-r^2/\omega^2)F(t) - u_0\tfrac{2}{\pi}\int\limits_0^t\int\limits_0^{\pi/2}\frac{\exp\left(-\frac{r^2\cos^2\beta/\omega^2}{1+\Delta t/\tau}\right)}{(1+\Delta t/T)^{3/2}(\Delta t/\tau)^{1/2}} \\ & \times\left(\frac{r^2\cos^2\beta/\omega^2}{(1+\Delta t/\tau)}-1\right)\frac{F(t')}{\tau}\cos^2\beta d\beta dt', \end{aligned} \quad (4)$$

where $u_0 = 2(1+\nu)\alpha_T\frac{E}{c_p\rho_0\pi\omega^2}$; $F(t) = \int\limits_0^t f(t')dt'$; u_0—quasi-one-dimensional mode amplitude; $\tau = \omega^2/4a$—characteristic thermal time; α_T—coefficient of linear thermal expansion; and ν—Poisson's ratio.

As shown in [19], surface displacements due to thermal expansion are retarded with respect to temperature changes. Figure 2 shows an example of relative temperature $T(0, t)$ and deformation u_z (0, t) changes in the centre of the heating spot. The retardation of deformation with respect to temperature depends on the size of the heating spot and the thermal diffusivity of the sample.

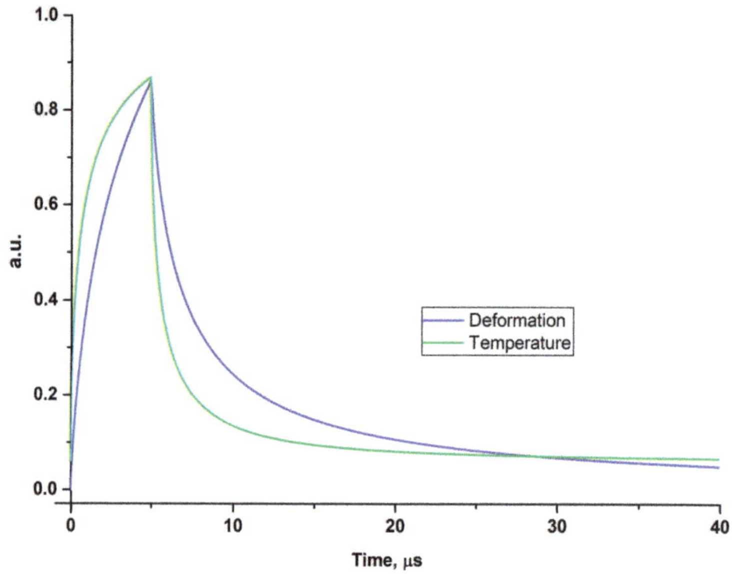

Figure 2. Retardation of thermal deformation in the centre of the heating spot with respect to temperature change.

2.2. Interferometer

When irradiated by means of a light guide, a natural gap is formed between the end of the light guide and the surface of the sample. If the planes of the end face and surface are approximately parallel, then a Fabry–Pérot interferometer is generated in the gap. The Fabry–Pérot interferometric fibre-optic sensor is one of the most commonly used types of fibre-optic sensors due to its versatility, simplicity, speed, accuracy and immunity to environmental noise [21]. The extrinsic Fabry–Pérot interferometer (EFPI) is a fibre-optic modification of the Fabry–Pérot interferometer. The fibre-optic EFPI is formed by the fibre end face (having a typical reflection coefficient of about 3.5% for a wavelength range of 1.55 μm) and the outer surface [22]. As a rule, the intensity of the reflected light registered in the EFPI demonstrates an oscillating relationship to the interferometer baseline due to interferometric effects [23]. The short base of such an interferometer makes it very convenient for measuring various physical quantities. Sensors based on a fibre-optic EFPI have become widely used in science and industry due to their low cost, small dimensions, electromagnetic protection and high performance over a wide range of applications [24]. There are two main classes of measurement methods using EFPI: those that track only the deviations of the interferometer signal and those used to analyse the spectral function of the interferometer (interferometry in the wavelength region). Although deviation tracking approaches offer relatively high speed and resolution, their main disadvantage is the uncertainty of the initial value. The slower method of recording the transmission spectrum of the interferometer has a high absolute accuracy and a wide dynamic range of measurement [25]. In our case, we apply the first method to track the deviations of the interferometer signal from the operating point when the reflecting thermally deformable surface is displaced.

The operation of EFPI as a displacement sensor of the reflecting surface involves two mechanisms: interference and intensity.

The interference-based mechanism is described by the Fabry–Pérot interference of reflected beams between the two surfaces of the gap [26]. A low-coherence version of the interferometer is obtained if the average reflection coefficient $R = (R_1 \cdot R_2)^{1/2} << 1$ for reflections from both sides of the gap L. Then, multiple reflections can be disregarded and the interference equation for the probing beam is expressed in a known form associated with the harmonic approximation of the Airy function:

$$I_r(t) = I_0 \cdot (1 - R)^2 \cdot \left[1 + 2R \cdot \cos\left(\frac{4\pi n L(t)}{\lambda} + \varphi(t)\right)\right] \tag{5}$$

where $I_r(t)$—dependence of the radiation intensity on the time at the EFPI output; I_0—the initial radiation intensity at the EFPI input; n—refractive index of the medium in the interferometer gap ($n \approx 1$); $L(t)$—dependence of the gap size on time; λ—wavelength of the probing radiation; $\varphi(t)$—dependence of the phase shift of the radiation reflected from the sample on time; and η—attenuation coefficient of the radiation intensity due to beam divergence in the interferometer gap.

The mechanism of attenuation of the radiation intensity in the EFPI gap is associated with the expansion of the radiation beam at the output of the fibre in accordance with its numerical aperture N_A. The propagation of radiation behind the fibre occurs within the framework of the Gaussian beam model with a constriction at the end of the fibre. The width of the Gaussian beam is characterised by a radius ω, which varies (diverges) along the beam axis and corresponds to the radius of the constriction ω_0 at its minimum at the end of the fibre. In the case under consideration, ω_0 is given by half the diameter of the mode spot of the light guide. The approach based on the Gaussian beam model assumes that the beam passes the length of the interferometer gap twice: in the forward and reverse directions, it remains Gaussian, while at the boundary with the fibre, it has a radius exceeding ω_0. The Rayleigh beam length L_0 is determined by the distance at which the beam expands twice. The relative fraction η of the power of the probing radiation reflected from the sample, which will be "captured" by the fibre in the form of radiation of the main mode, will be determined by the integral of the overlap of the mode field and the radiation incident at the end [27]:

$$\eta = 1/(1+(L/L_0)^2) \tag{6}$$

For a standard single-mode SMF-28 light guide, the diameter of the mode field is about 10 μm. At a numerical aperture of 0.12, the divergence of the heating radiation will increase the diameter of the heating spot by approximately 10 μm for every 42 μm of the gap ($L_0 \approx 42$ μm). For a gap value of $L \approx 100$ μm, we have: $2\omega \approx 34$ μm and $\eta \approx 0.15$. At the same time, the information platform of the interferometer on the treated surface corresponds to the field $2\omega_0 \approx 10$ μm of the main mode of the light guide. In this case, it follows Formula (1), whereby the heating power on the information platform changes by less than 8%. Therefore, it is possible to select such a gap value L at which the heating intensity will be sufficient for processing, and the interferometer readings will correspond with an acceptable attenuation η to a quasi-isothermal site in the centre of the heating spot.

2.3. *Laser Surface Heating Control Unit with Fibre-Optic EFPI*

The optical scheme of the installation contains only fibre-optic elements (see Figure 3). An add-drop multiplexer is used to combine and separate the heating and probing rays. The separation of the reflections of the probing radiation from the sample and the end of the light guide in the EFPI interferometer is carried out by a fibre-optic circulator. From the output of the circulator, interfering rays are fed to a pulsed photodetector. The function of a lock-in amplifier is performed by a digital oscilloscope with external synchronisation from the pulse of the heating laser pump source. Semiconductor laser diodes with a fibre-optic output are used as radiation sources. Pulsed heating of the sample is provided by a 0.1–0.2 W pulse-pumped laser diode operating at a wavelength of 1470 nm. The source of probing radiation, which is based on a distributed feedback (DFB) laser diode, operates at a wavelength of 1530 nm with a continuous power of 0.001 W. The use of the DFB laser diode fulfils the requirement of a constant wavelength of probing radiation to stabilise the operating point of the interferometer for a measurement cycle.

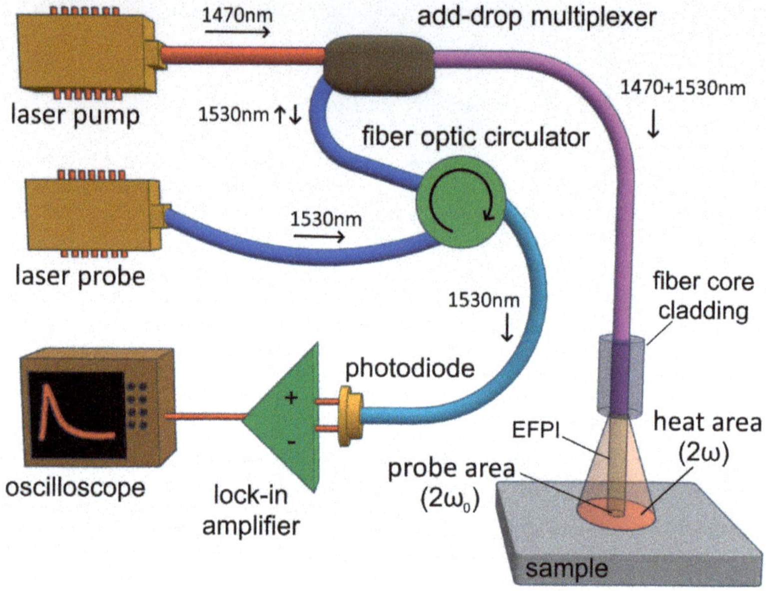

Figure 3. Experimental setup of a laser surface heating with an EFPI fibre-optic interferometer.

Preliminary adjustment of the interferometer operating point is carried out by adjusting the gap using a piezoelectric actuator. When installing the operating point in the middle of the linear section of the interferometer, sensitivity to surface displacements of about 10 mV/nm was achieved with a noise intensity of no more than 1 mV. The average value of the recorded signals was 10–30 mV at the length of the characteristic linear section, at about 3 V [7].

3. Results

Experimental Section

The experiments were carried out on small-sized samples of metals and semiconductors, having a surface area measured in units to tens of square millimetres. The size of the heated spot on the sample surface was up to 100 μm, while the measuring area of the EFPI interferometer in the centre of the heating spot corresponded to the size of a single light-guide mode of about 10 μm. According to estimates available elsewhere [19,20], the average values of temperature and deformation differ from the actual values by less than 10% when heated by a Gaussian beam at a site three times larger than that of the measuring site. The piezoelectric-driven measuring cell was designed in such a way so as to permit experiments to be conducted across a wide temperature range [7]. After setting the average temperature of the sample, the interferometer operating point was adjusted. Then, a series of heating pulses was applied with simultaneous control of the interferometer signal. The signal observed on the oscilloscope contained heating and cooling sections. The amplitude and speed of transients depended on the duration of the heating pulse, the properties of the sample and the size of the heating spot. Figure 4 shows the characteristic dependences of the recorded signals for the selected samples of metals and semiconductors. Higher values of the thermal diffusivity of metals contribute to faster processes of heat relaxation and, accordingly, higher values of the rate of change of the recorded signals.

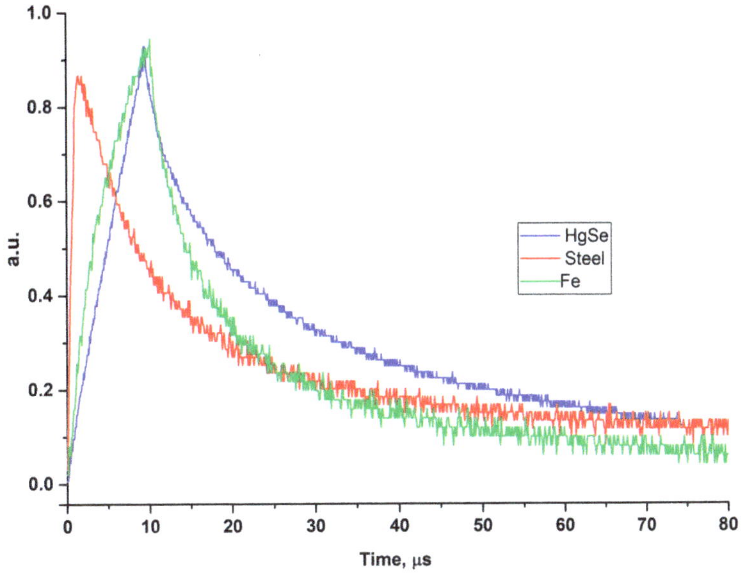

Figure 4. Experimental data of laser heating with EFPI interferometer for different samples. Typical response signals to laser heating with EFPI interferometer for samples under investigation.

4. Discussion

The formation of the interferometer signal is possible both when the heated surface is deformed and when its optical characteristics change with temperature. In both cases, there is a change in the phase characteristic of reflections in the interferometer. The amplitude at the level of 1% of the magnitude of the linear section of the interferometer characteristic of the observed signals suggests the additive influence of both factors. The experimental data presented in Figure 5 show the relationship between the shape of the recorded signals and the normalised calculated dependences of temperature change and deformation in the centre of the heating spot.

Our first hypothesis with regard to the direct dependence of the recorded signal on temperature changes was not confirmed. On the contrary, the second hypothesis on the existence of a relationship between the signal change and the heated surface thermal deformations has been confirmed by the mathematical model of the surface displacement.

Figure 5 shows the better correspondence between experimental data and the model of the thermal deformation process, with respect to the temperature change.

To obtain a better agreement between the calculation result and the experiment, the procedure of the iterative fitting of the deformation curve u_z to the experimental data were carried out. The target fitting criterion was to minimise the standard deviation of the difference between values calculated and revealed in the experiment.

In the used model, the main influence on the signal shape is assumed to be due to changes in the thermal diffusivity of the sample and the size of the heated area. Deformations of the sample are delayed relative to temperature changes. Comparing experimental data with model data, it is possible to estimate the influence of each factor on the formation of the resulting signal. Figure 6 depicts the corresponding dependences, which demonstrate the predominance of the mechanism of thermal deformations in the case of metal samples and the total effect of both factors in the case of semiconductors.

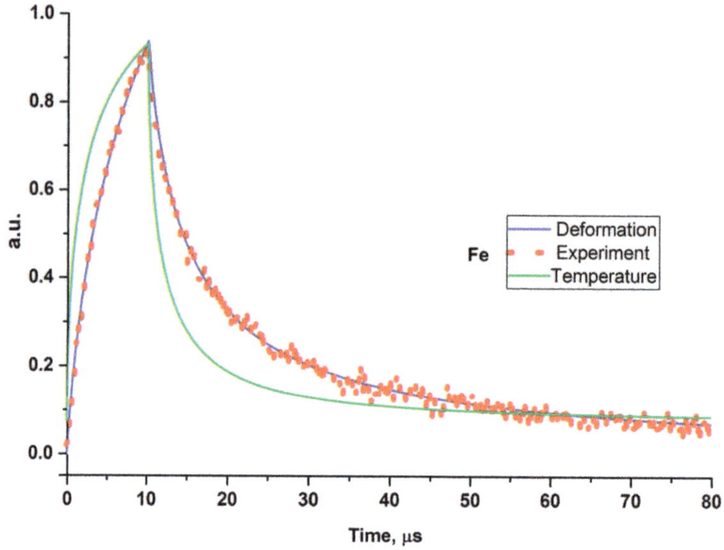

Figure 5. Comparison of normalised calculated dependences of temperature change and deformation in the centre of the heating spot with experimental data.

The reflected optical signal is recorded as voltage fluctuations of the photodetector. However, the results (Figures 4–6) are given in relative units. The signal scale significantly depends on the heating conditions of the sample, which are also affected by the settings of the measuring system and the local optical properties of the material. The normalised signals obtained in a series of successive experiments can be used to study the mechanical effects of thermal deformation and the dynamics of thermal processes. The results presented in the graphs were obtained by accumulating and averaging 256 pulsed experiments with appropriate synchronisation. Deviations in a series of measurements did not exceed a few percent. An additional advantage of signal averaging and accumulation is the possibility of improving the quality of the obtained experimental data. Thermo-optical signals have a small amplitude (appeared to be in the range of 10–50 mV) with a noise level of about 5 mV. Accumulation with averaging allows an increase in the signal-to-noise ratio by over 10 times.

Statistical results for the studied materials are presented in Table 1. The data show correspondence between the model prediction and the experimental data. To calculate the standard deviation, a preliminary calculation of the difference between the values of the strain u_z and the normalised reflection signal for each time interval of the experiment was performed. Then, for the obtained data, the standard deviation, expressed as a percentage, was calculated. The Pearson criterion was calculated for two discrete datasets for the strain curves u_z and the experimental data.

Table 1. Correlation statistics and fitting parameter data.

Parameter	Nb	ZnSeNi	Steel	HgSe	Fe
Standard deviation, %	2.192	3.341	1.593	8.935	2.068
Pearson correlation coefficient	0.9945	0.9856	0.9973	0.9062	0.9960
Thermal diffusivity a, m^2/s	$2.48 \cdot 10^{-5}$	$1.05 \cdot 10^{-5}$	$4 \cdot 10^{-6}$	$1.18 \cdot 10^{-6}$	$2.2 \cdot 10^{-5}$

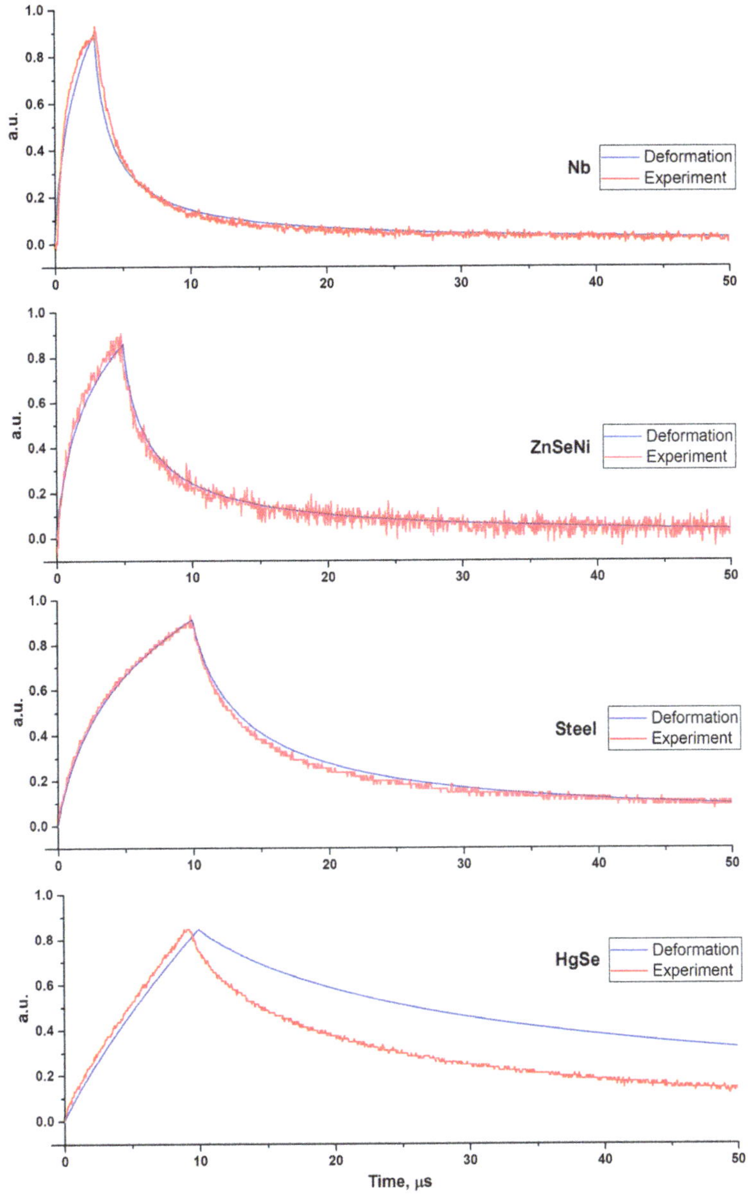

Figure 6. Experimental results and normalised calculation of deformation in the centre of the heating spot for different samples.

Experiments carried out with various materials revealed a strong dependence of the amplitude of the recorded signal on the size of the gap between the light guide and the sample in accordance with Formulas (5) and (6). Nevertheless, relative changes in the gap over the linear section of the characteristic are experimentally controlled with respect to the position of the interferometer baseline. The absolute value of the gap can be determined by analysing the spectral characteristics of the interferometer at different probing radiation wavelengths. In this work, relative measurements of the thermo-optical signal were carried out at one wavelength. In addition, the amplitude and shape of the signal depended on the relative position of the studied area on the sample surface. Thus, the thermo-optical signal carries information on the changes in the optical and thermophysical properties of the sample surface [13,28]. Prompt receipt and analysis of this information allow the intensity of the impact (in particular, the thermal one) on the sample to be regulated for various purposes, including the creation and preservation of selected structures on the surface, as well as their modification and destruction.

5. Conclusions

In the course of the study, the possibility of assessing the intensity of pulse heat treatment processes on the microsites of the surface of samples by means of low-coherence interferometry was revealed. This was implemented in a compact fibre-optic circuit with a simultaneous supply of heating and probing radiation through a single-mode light guide. The structure of fibre optics in the pump–probing circuit with an extrinsic Fabry–Pérot interferometer for a high-speed thermo-optical control of thermal processes during pulsed laser heating was investigated. By adjusting the EFPI gap, it is possible to increase the size of the heating area to 50 μm with a constant size of the measuring area in the centre of the heating spot of about 10 μm. Experiments were carried out on metal and semiconductor samples. In the case of metal samples, a comparison of experimental data with modelling calculations showed that the recorded signal corresponds to the temporal dependences of thermal deformations. In the case of semiconductors, a more complex combination of thermo-optical effects was identified. The applicability of the proposed scheme for the characterisation of near-surface properties on microsites of the sample by the relaxation rate of a pulsed thermal disturbance was demonstrated.

For the presented samples, differences in the rate of signal change are clearly visible depending on their thermal diffusivity and the size of the heating area. If the gap between the end of the fibre and the surface of the sample has been fixed, then the size of the heating area will be close to a constant value. As a result, if we analyse the rate of signal change in the course of cooling stage, we can draw conclusions on the differences of the thermal diffusivity of several samples. It is also possible to evaluate the change in the local values of thermal diffusivity on the surface of an inhomogeneous sample.

The ongoing work will develop in the direction of isolating contributions from surface displacement and changes in the optical characteristics of semiconductor samples due to differences in the rate of their manifestation in the interferometer signal.

Author Contributions: Conceptualisation, A.A.S. and P.V.S.; methodology, A.N.K. and A.A.S.; software V.I.G.; hardware A.N.K. and V.I.G.; validation, A.N.K., A.A.S. and P.V.S.; writing—original draft preparation, A.A.S., P.V.S. and A.N.K.; writing—review and editing, A.A.S. and P.V.S. All authors have read and agreed to the published version of the manuscript.

Funding: The investigation has been conducted at the expense of a grant by the Russian Science Foundation (project no. 22-29-00789).

Data Availability Statement: Data available on request due to organisation rules.

Conflicts of Interest: The authors declare no conflict of interest.

List of Symbols and Abbreviations

	Definition
Subscripts	
A	Aperture
p	Pressure
d	Duration
r	Radiation
T	Thermal
z	z-axis direction

Variables and functions	
α_T	Coefficient of linear thermal expansion
β	Variable of integration
δ	Delta function
η	Attenuation coefficient of the radiation intensity
λ	Wavelength of the probing radiation
ν	Poisson's ratio
π	π number
ρ_0	Density of the material
τ	Characteristic thermal time
τ_d	Duration of the radiation pulse
φ	Phase shift of the radiation reflected from the sample
ω	Radius of the Gaussian light beam
ω_0	Half diameter of the mode spot of the light guide
a	Thermal diffusivity
c_p	Specific heat capacity
E	Pulse energy
E_0	Absorbed pulse energy
F	Heating source function
f	Normalised dependence of the pulse amplitude on time
I	Gaussian power distribution
I_r	Radiation intensity in time at the EFPI output
I_0	Initial radiation intensity at the EFPI input
L	Interferometer gap size
L_0	Rayleigh beam length
N_A	Fibre core numerical aperture
n	Refractive index of the medium in the interferometer gap
R	Average reflection coefficient
r	Radial distance from heating centre
T	Temperature change
t	Time
t'	Variable of integration
Δt	Integration time
u_0	Quasi-one-dimensional mode amplitude
u_z	Normal thermal deformations along heated surface
W	Local surface heat source
Abbreviations	
DFB	Distributed feedback laser diode
EFPI	External Fabry–Pérot interferometer

References

1. Kennedy, D.; Xue, Y.; Mihaylova, M. Current and Future Applications of Surface Engineering. *Eng. J.* **2005**, *59*, 287–292.
2. Kannatey-Asibu, E. Laser surface modification. In *Principles of Laser Materials Processing*, 2nd ed.; Wiley: Hoboken, NJ, USA, 2023; pp. 568–616.
3. Kwok, C.T. Pulsed laser surface treatment of multilayer gold-nickel-copper (Au/Ni/Cu) coatings to improve the corrosion resistance of components in electronics. In *Laser Surface Modification of Alloys for Corrosion and Erosion Resistance*; Kwok, C.T., Ed.; Elsevier Science: Amsterdam, The Netherlands, 2012.
4. Orazi, L.; Rota, A.; Reggiani, B. Experimental investigation on a novel approach for laser surface hardening modelling. *Int. J. Mech. Mater. Eng.* **2021**, *16*, 2. [CrossRef]
5. Malinauskas, M.; Žukauskas, A.; Hasegawa, S.; Hayasaki, Y.; Mizeikis, V.; Buividas, R.; Juodkazis, S. Ultrafast laser processing of materials: From science to industry. *Light Sci. Appl.* **2016**, *5*, e16133. [CrossRef] [PubMed]
6. Shugaev, M.V.; He, M.; Levy, Y.; Mazzi, A.; Miotello, A.; Bulgakova, N.M.; Zhigilei, L.V. Laser-Induced Thermal Processes: Heat Transfer, Generation of Stresses, Melting and Solidification, Vaporization, and Phase Explosion. In *Handbook of Laser Micro—And Nano-Engineering*; Sugioka, K., Ed.; Springer: Cham, Switzerland, 2020; pp. 83–163. [CrossRef]
7. Starostin, A.A.; Shangin, V.V.; Lonchakov, A.T.; Kotov, A.N.; Bobin, S.B. Laser Pump Probe Fiber Optic Technique for Characterization of Near Surface Layers of Solids: Development and Application Prospects for Studying Semiconductors and Weyl Semimetals. *Annal. der Physik.* **2020**, *532*, 1900586–1900593. [CrossRef]

8. Lipnyagov, E.V.; Gurashkin, A.L.; Starostin, A.A.; Skripov, P.V. Going to Spontaneous Boiling-Up Onset. *J. Eng. Thermophys.* **2018**, *27*, 307–318. [CrossRef]
9. Kotov, A.N.; Gurashkin, A.L.; Starostin, A.A.; Skripov, P.V. Low-Energy Activation of Superheated N-pentane Booling-Up by Laser Pulse at the Fiber-Liquid Interface. *Interfacial Phenom. Heat Transf.* **2022**, *10*, 15–23. [CrossRef]
10. Zhu, J.; Wu, X.; Lattery, D.M.; Zheng, W.; Xi, W. The Ultrafast Laser Pump-Probe Technique for Thermal Characterization of Materials with Micro/nanostructures. *Nanoscale Microscale Thermophys. Eng.* **2017**, *21*, 177–198. [CrossRef]
11. Magunov, A.N. Laser Thermometry of Solids: State of the Art and Problems. *Meas. Tech.* **2002**, *45*, 173–181. [CrossRef]
12. Hovell, T.; Petzing, J.; Justham, L.; Kinnell, P. From Light to Displacement: A Design Framework for Optimising Spectral-Domain Low-Coherence Interferometric Sensors for In Situ Measurement. *Appl. Sci.* **2020**, *10*, 8590. [CrossRef]
13. Escola, F.Z.; Kunik, D.; Martinez, O.E.; Mingolo, N. Photothermal Microscopy. *Procedia Mater. Sci.* **2015**, *8*, 665–673. [CrossRef]
14. Proskurnin, M.A.; Khabibullin, V.R.; Usoltseva, L.O.; Vyrko, E.A.; Mikheev, I.V.; Volkov, D.S. Photothermal and optoacoustic spectroscopy: State of the art and prospects. *Phys.-Uspekhi* **2022**, *65*, 270–312. [CrossRef]
15. Capeloto, O.A.; Lukasievicz, G.V.B.; Zanuto, V.S.; Herculano, L.S.; Souza Filho, N.E.; Novatski, A.; Malacarne, L.C.; Bialkowski, S.E.; Baesso, M.L.; Astrath, N.G.C. Pulsed photothermal mirror technique: Characterization of opaque materials. *Appl. Opt.* **2014**, *53*, 7985–7991. [CrossRef] [PubMed]
16. Marcano, A.; Gwanmesia, G.; Workie, B. Photothermal Mirror Method for the Study of Thermal Diffusivity and Thermo-Elastic Properties of Opaque Solid Materials. *Int. J. Thermophys.* **2017**, *38*, 136. [CrossRef]
17. Sato, F.; Malacarne, L.C.; Pedreira, P.R.B.; Belancon, M.P.; Mendes, R.S.; Baesso, M.L.; Astrath, N.G.C.; Shen, J. Time-resolved thermal mirror method: A theoretical study. *J. Appl. Phys.* **2008**, *104*, 053520. [CrossRef]
18. Flizikowski, G.A.S.; Anghinoni, B.; Rohling, J.H.; Belançon, M.P.; Mendes, R.S.; Baesso, M.L.; Malacarne, L.C.; Požar, T.; Bialkowski, S.E.; Astrath, N.G.C. Influence of edge effects on laser-induced surface displacement of opaque materials by photothermal interferometry. *J. Appl. Phys.* **2020**, *128*, 044509. [CrossRef]
19. Vintsents, S.V.; Dmitriev, S.G.; Shagimuratov, O.G. Instantaneous profiles of quasistatic deformations and displacements of solid surfaces during local laser irradiation. *Phys. Solid State* **1996**, *38*, 552–557.
20. Vintsents, S.V.; Dmitriev, S.G.; Spiridonov, K.I. Quasi-one-dimensional thermal deformation and displacement of the surface of a solid in a pulsed laser beam. *Phys. Solid State* **1997**, *39*, 1985–1988. [CrossRef]
21. Huang, Y.W.; Tao, J.; Huang, X.G. Research Progress on F-P Interference—Based Fiber-Optic Sensors. *Sensors* **2016**, *16*, 1424. [CrossRef]
22. Liokumovich, L.; Markvart, A.; Ushakov, N. Utilization of extrinsic Fabry-Perot interferometers with spectral interferometric interrogation for microdisplacement measurement. *J. Electron. Sci. Technol.* **2020**, *18*, 100030. [CrossRef]
23. Jiang, Y.; Ding, W. Recent developments in fiber optic spectral white-light interferometry. *Photonic Sens.* **2011**, *1*, 62–71. [CrossRef]
24. Lee, B.H.; Kim, Y.H.; Park, K.S.; Eom, J.B.; Kim, M.J.; Rho, B.S.; Choi, H.Y. Interferometric Fiber Optic Sensors. *Sensors* **2012**, *12*, 2467–2486. [CrossRef] [PubMed]
25. Ushakov, N.; Liokumovich, L. Resolution limits of extrinsic Fabry—Perot interferometric displacement sensors utilizing wavelength scanning interrogation. *Appl. Opt.* **2014**, *53*, 5092–5099. [CrossRef] [PubMed]
26. Chin, K.K.; Sun, Y.; Feng, G.; Georgiou, G.E.; Guo, K.; Niver, E.; Roman, H.; Noe, K. Fabry-Perot diaphragm fiber-optic sensor. *Appl. Opt.* **2007**, *46*, 7614–7619. [CrossRef] [PubMed]
27. Liokumovich, L.B.; Ushakov, N.A.; Markvart, A.A.; Evdokimenko, E.Y. The spectral characteristic of a multilayer extrinsic fiber Fabry—Perot interferometer, St. Petersburg State Polytechnical University. *J. Phys. Mat.* **2022**, *15*, 129–146. [CrossRef]
28. Zhukovsky, K. Operational Approach and Solutions of Hyperbolic Heat Conduction Equations. *Axioms* **2016**, *5*, 28. [CrossRef]

axiomS

Article

Structural Damage Identification Using the First-Order Vibration-Mode-Based Frequency-Shift Flexibility Sensitivity Algorithm

Shanshan Cao [1], Qiuwei Yang [2,3,*] and Xi Peng [2,3]

1 Experimental College, The Open University of China, Beijing 100039, China; oconan@163.com
2 School of Civil and Transportation Engineering, Ningbo University of Technology, Ningbo 315211, China
3 Engineering Research Center of Industrial Construction in Civil Engineering of Zhejiang, Ningbo University of Technology, Ningbo 315211, China
* Correspondence: yangqiuwei79@gmail.com

Abstract: The diagnosis of structural damage usually belongs to a mathematical inverse problem. This work presents a novel frequency-shift flexibility sensitivity algorithm for structural damage assessment using only the first-order vibration mode to achieve the goal of successfully identifying structural damage with fewer modal parameters. The core idea of the proposed method is to make the first-order vibration mode contribute the most to a structural flexibility matrix through the frequency-shift operation. A high-precision flexibility matrix can be obtained after the frequency-shift operation, which only needs the first mode of structural free vibration. Through this special advantage, structural damage coefficients can be accurately calculated by the frequency-shift flexibility sensitivity equation. Thus, a reliable identification result can be obtained according to the values of the calculated damage coefficients. In some engineering applications, another advantage of the proposed method is that it does not require a complete finite element modeling process, as long as a few lower-frequency vibration modes of the intact structure are measured. A truss structure and a beam structure are used as two numerical examples to demonstrate the proposed approach. The results show that the proposed method has higher calculation accuracy than the ordinary flexibility sensitivity method by using only the first-order vibration mode. The proposed method can overcome possible misdiagnosis of the ordinary flexibility sensitivity method. It also has been shown that the proposed method may have the potential to identify minor damage in a structure. Using the experimental data of a steel frame structure, the effectiveness and reliability of the proposed method have been further verified. The proposed method provides a simple way for structural damage identification with only a few vibration modal data.

Keywords: damage diagnosis; frequency-shift flexibility; sensitivity analysis; vibration mode; damage parameter

MSC: 65M32

Citation: Cao, S.; Yang, Q.; Peng, X. Structural Damage Identification Using the First-Order Vibration-Mode-Based Frequency-Shift Flexibility Sensitivity Algorithm. *Axioms* **2023**, *12*, 551. https://doi.org/10.3390/axioms12060551

Academic Editors: Nhon Nguyen-Thanh and Leonid Plotnikov

Received: 23 April 2023
Revised: 28 May 2023
Accepted: 31 May 2023
Published: 2 June 2023

1. Introduction

During the service period, an engineering structure will inevitably be damaged due to the influence of environmental corrosion or a disaster load. Local damage in a structure may lead to a sudden collapse of the whole structure, thus causing serious loss of life or property. In view of this, it is very necessary to conduct timely damage diagnosis for a structure to avoid catastrophic consequences. Due to the large volume and numerous components of engineering structures, traditional non-destructive testing techniques such as ultrasound, radiographic testing, and penetration testing cannot complete defect diagnosis of large engineering structures. In the past few decades, methods for diagnosing structural damage using the response parameters of structures under static or dynamic loads have been continuously studied in depth. The theoretical basis for this type of method is that faults in structures can cause changes in structural static and vibration response parameters. In practice, the response data of structures can be measured through special testing equipment, and then their changes can be used to diagnose structural fault conditions. In view of this, many methods based on static or dynamic response parameters have been developed for structural damage identification [1–3] in recent years. These methods can be mathematically attributed to an inverse problem.

Among these methods, the flexibility-based approach is favored because of its simple operation and wide application [4]. It is known that a structural flexibility matrix can be obtained from both static data and the low-order vibration modes. Pandey and Biswas [5] used the flexibility matrix change to determine the damage location of the beam structure without constructing the finite element model (FEM). It was found that the diagonal element in the flexibility difference matrix can indicate the damage location. Jaishi and Ren [6] employed the flexibility difference as the objective function to modify the structural FEM for detecting structural damage. Catbas et al. [7] found that the dynamic test without a fixed reference measurement position can also be used to generate data for calculating the modal flexibility. Then, the displacement distribution can be obtained by using the modal flexibility for damage detection. Duan et al. [8] extended the damage location method based on flexibility to the case of environmental vibration with incomplete measured degrees of freedom (DOFs). Tomaszewska [9] discussed the damage detection method based on the structural dynamic flexibility of the building structure. In order to distinguish the true and false damage detection results, the absolute damage index was proposed to constrain the influence of the modal identification errors. Yang [10] proposed a new damage identification method based on structural flexible disassembly. The scheme has a unique advantage in that it can accurately calculate the stiffness damage parameters without any high-order sensitivity analysis or iteration. Maghsoodi et al. [11] proposed a simple method based on local flexibility to detect, locate, and quantify multiple cracks in Euler–Bernoulli multi-step beams. The main advantage of their method is that it can detect a number of unknown cracks.

Weng et al. [12] presented a new substructure method for structural damage detection by using the substructure dynamic flexibility matrix. The main advantage of their method is that the substructure characteristic parameters are more sensitive to local damage than the global characteristic parameters. Using Dempster–Shafer evidence theory, Grande and Imbimbo [13] proposed a multi-stage flexibility method for damage detection in the case of multiple damage locations and three-dimensional systems. Hosseinzadeh et al. [14] developed a damage detection method by introducing an effective objective function based on modal assurance criteria and modal flexibility. It was found that the proposed method can only use the data of the first few modes to accurately identify the damage even if the incomplete noise modal data are taken as the input data. Altunisik et al. [15] used modal curvature and modal flexibility methods to locate cracks in steel cantilever beams. The comparison shows that the modal flexibility method is effective in determining the crack location. Wickramasinghe et al. [16] developed the vertical damage index and transverse damage index based on the modal flexibility to detect and locate the damage of the main cable and hanger of a suspension bridge. The results confirm the applicability of the vertical damage index to accurately detect the damage in the actual suspension bridge by using only the first few modes. Sarmadi et al. [17] improved the sensitivity function of modal flexibility and proposed a new iterative regularization method to solve the ill-posed problem to locate and quantify the damage. It was found that their method is robust enough to solve the ill-posed problem of damage location and quantification under noise-free and noisy modal data. Ahmadi-Nedushan and Fathnejat [18] proposed a two-stage structural damage detection method based on modal flexibility and an improved teaching–learning optimization algorithm to reduce the influence of measurement noise.

Feng et al. [19] proposed a Bayesian model updating method with modal flexibility to find the most probable value of the model parameters for damage identification.

Bernagozzi et al. [20] proposed a data-driven standard for structural type classification, which can be used in the framework of modal flexibility-based damage identification methods. Yang and Peng [21] developed a highly efficient model reduction method for structural damage identification based on the reduced flexibility matrix. Dinh-Cong et al. [22] used the damage index based on modal flexibility sensitivity to detect damages in functionally graded beams. The results indicate that when the noise level added to the vibration mode data is less than 10%, the provided method can correctly locate the position of damaged components. Darshan et al. [23] developed the damage detection procedure based on strain energy and a flexibility matrix to detect single and multiple damages in plate structures. Quqa and Landi [24] proposed a damage identification method based on bridge flexibility curvature with sparse acceleration measurement. The damage index proposed by them is particularly sensitive to the damage location and can be successfully applied to the steel truss bridge with different damage patterns. Nick et al. [25] proposed a damage index based on modal flexibility and modal strain energy, and a two-stage multi-criteria damage detection method using an artificial neural network (ANN) to locate and quantify the damage of steel frames. The modal flexibility matrix was obtained by the first three bending vibration modes. Cuomo et al. [26] proposed a new baseline free method for real-time structural damage diagnosis during low-speed and high-speed collisions, which is based on the decomposition of propagation patterns caused by collision events. Aulakh et al. [27] developed the curvature and coordinated modal assurance criteria based on strain modal flexibility for structural damage monitoring. They found that strain modal flexibility is more sensitive to structural damage than displacement mode flexibility.

The limitation of the existing flexibility-based method is that several low-order vibration modes are needed to approximately obtain the dynamic flexibility matrix. This leads to a large workload of dynamic analysis and a high requirement for analysis accuracy. To overcome this limitation, this work presents a novel frequency-shift flexibility sensitivity method for structural damage detection by only using the first-order vibration mode. Based on the frequency-shift operation, the first-order vibration mode will account for the majority of the dynamic flexibility of the structure. As a result, the damage coefficients in structural FEM can be solved accurately through the frequency-shift flexibility sensitivity equation, which indicates the damage locations and extents. The innovation of this work mainly lies in two aspects. The first innovation is that the ordinary flexibility sensitivity method has been improved through the frequency-shift operation for achieving the goal of identifying structural damage with fewer modal parameters. The second innovation is that in some engineering applications, the tested modal data can be used to directly compute the frequency-shift flexibility without the need for a complete finite element modeling process. Two numerical examples and one experimental example are used to validate the presented frequency-shift flexibility sensitivity method. It is found that the proportion of the higher-order vibration modes in the frequency-shift flexibility is greatly reduced. For this reason, the damage coefficients of the structure can be calculated accurately by the frequency-shift flexibility sensitivity analysis with only a few modal data. It has been shown that the proposed method requires fewer modal parameters but has higher calculation accuracy than the ordinary flexibility sensitivity method. It may be a valuable new approach to structural damage identification in engineering practice.

2. Theoretical Development

In this section, the ordinary flexibility sensitivity method is first briefly reviewed, and then a new frequency-shift flexibility sensitivity algorithm is developed for structural damage detection.

Based on structural FEM, the free vibration eigen-parameters can be computed by solving the following generalized eigenvalue problem as:

$$K\phi_j = \lambda_j M\phi_j \tag{1}$$

where K and M are the stiffness and mass matrices of a structure with n DOFs, λ_j and φ_j are the j-th eigenvalue and eigenvector (λ_j and φ_j, also called the j-th eigen-pair), respectively. Similarly, the eigen-solutions of the damaged structure can be also obtained by

$$K_d\phi_{dj} = \lambda_{dj}M\phi_{dj} \tag{2}$$

$$K_d = K - \Delta K \tag{3}$$

$$\Delta K = \sum_{i=1}^{N} \varepsilon_i K_i \tag{4}$$

where K_d is the damaged stiffness matrix, λ_{dj} and ϕ_{dj} are the j-th eigenvalue and eigenvector of the damaged structure, ΔK is the stiffness reduction due to structural damage, ε_i and K_i are the damage coefficient and elementary stiffness matrix of the i-th element in FEM, and N is the total number of the elements in FEM. It is known that the flexibility matrix is the inverse of the stiffness matrix and can be obtained approximately by several low-order eigen-pairs as:

$$F = K^{-1} \approx \sum_{j=1}^{m} \frac{1}{\lambda_j}\phi_j\phi_j^T \tag{5}$$

$$F_d = K_d^{-1} \approx \sum_{j=1}^{m} \frac{1}{\lambda_{dj}}\phi_{dj}\phi_{dj}^T \tag{6}$$

where F and F_d are the flexibility matrices of the undamaged and damaged structures, and m is the number of the measured modes in the vibration testing. Subtracting (5) from (6), one obtains:

$$\Delta F = F_d - F = K_d^{-1} - K^{-1} \tag{7}$$

$$\Delta F \approx \sum_{j=1}^{m} \left(\frac{1}{\lambda_{dj}}\phi_{dj}\phi_{dj}^T - \frac{1}{\lambda_j}\phi_j\phi_j^T\right) \tag{8}$$

Equation (8) shows that the flexibility change ΔF can be approximately obtained by the measured lower eigen-parameters of the structure before and after damage. Substituting Equation (3) into (7), one obtains:

$$\Delta F = (K - \Delta K)^{-1} - K^{-1} \tag{9}$$

Using Neumann series expansion, Equation (9) can be further simplified as:

$$\Delta F = F \cdot \Delta K \cdot F + F \cdot \Delta K \cdot F \cdot \Delta K \cdot F + \cdots \tag{10}$$

Ignoring the higher-order items in Equation (10) yields:

$$\Delta F = F \cdot \Delta K \cdot F \tag{11}$$

Substituting Equation (4) into (11), one obtains:

$$\Delta F = \sum_{i=1}^{N} \varepsilon_i \Pi_i \tag{12}$$

$$\Pi_i = FK_iF \tag{13}$$

Equation (12) is called the ordinary flexibility sensitivity equation and will be used to compute the unknown damage coefficients ε_i ($i = 1 \sim N$) for damage identification. The matrix Π_i is called the elementary flexibility sensitivity of the i-th element in FEM. The operation steps of the above flexibility sensitivity method are summarized as follows:

(1) Establish the FEM of the intact structure to obtain the stiffness and mass matrices K and M.
(2) Conduct dynamic analysis on the structure and measure the lower-order eigen-pairs of the intact and damaged structures.
(3) Compute the flexibility change ΔF by Equation (8) and compute the elementary flexibility sensitivity matrix Π_i by Equation (13).
(4) Compute the damage coefficients ε_i $(i = 1 \sim N)$ by solving Equation (12). Finally, the damage locations and extents in the structure can be determined according to the values of ε_i $(i = 1 \sim N)$.

The advantage of the flexibility sensitivity method is that the calculation formula of elementary flexibility sensitivity (i.e., Equation (13)) is very simple, especially compared with the calculation formula of the eigenvector sensitivity [28]. The limitation of the ordinary flexibility sensitivity method is that several low-order eigen-pairs are still needed when applying Equation (8) to calculate the flexibility change ΔF. This leads to a large workload of dynamic analysis and a high requirement for analysis accuracy. However, only the first mode of structural vibration, namely the fundamental frequency and mode shape, is usually measured in practical engineering. This limits the successful application of the traditional flexibility sensitivity method in actual engineering structural damage identification. The method of successfully identifying structural damage using only the fundamental frequency and mode of structural vibration will be very popular in engineering practice.

To overcome this limitation, a novel frequency-shift flexibility sensitivity method is proposed in this work to compute the damage coefficients only using the first-order eigen-pairs. The key idea of the proposed method is to make the first eigenvalue ($\overline{\lambda_1}$) of the new system very close to zero through the frequency-shift operation. This will cause the reciprocal of the first eigenvalue to be particularly large and far exceed the reciprocals of other eigenvalues. As a result, the flexibility matrix of the new system can be accurately calculated by only the first-order eigen-pairs. The specific formulas of the proposed method are derived as follows. From Equations (1) and (2), the generalized eigenvalue equations of the intact and damaged systems after frequency shift are expressed as:

$$\overline{K}\phi_j = \overline{\lambda_j} M\phi_j \quad (14)$$

$$\overline{K_d}\phi_{dj} = \overline{\lambda_{dj}} M\phi_{dj} \quad (15)$$

$$\overline{K} = K - \mu M \quad (16)$$

$$\overline{\lambda_j} = \lambda_j - \mu \quad (17)$$

$$\overline{K_d} = K_d - \mu M \quad (18)$$

$$\overline{\lambda_{dj}} = \lambda_{dj} - \mu \quad (19)$$

where μ denotes the frequency-shift distance, $\overline{K}$ and $\overline{K_d}$ denote the stiffness matrices of the intact and damaged systems after frequency shift, and $\overline{\lambda_j}$ and $\overline{\lambda_{dj}}$ denote the eigenvalues of the intact and damaged systems after frequency shift. The frequency-shift operation shown in Equations (16)–(19) has been shown to be an effective means for quickly calculating the eigenvalues of large structures. The related content can be referred to in [29,30]. Similar to Equations (5) and (6), the intact and damaged flexibility matrices of the systems after frequency shift can also be obtained approximately by several low-order eigen-pairs as:

$$\overline{F} = \overline{K}^{-1} \approx \sum_{j=1}^{m} \frac{1}{\overline{\lambda_j}} \phi_j \phi_j^T = \sum_{j=1}^{m} \frac{1}{\lambda_j - \mu} \phi_j \phi_j^T \tag{20}$$

$$\overline{F_d} = \overline{K_d}^{-1} \approx \sum_{j=1}^{m} \frac{1}{\overline{\lambda_{dj}}} \phi_{dj} \phi_{dj}^T = \sum_{j=1}^{m} \frac{1}{\lambda_{dj} - \mu} \phi_{dj} \phi_{dj}^T \tag{21}$$

where $\overline{F}$ and $\overline{F_d}$ are the intact and damaged flexibility matrices of the systems after frequency shift. As can be seen in Equation (20) or (21), the first-order eigen-pairs will contribute most to the flexibility matrix when the frequency-shift distance μ is close to λ_1 or λ_{d1}. The basis for determining the frequency-shift distance μ is that the contribution of the first mode to the flexibility matrix after the frequency-shift operation exceeds 95% or more, since the allowable data error level in the engineering field is usually around 5%. The vibration frequencies of actual engineering structures are all greater than zero and sorted in ascending order. In view of this, a simple criterion for determining the value of μ is that the ratio of the first to second frequency after the frequency-shift operation is less than 5% or more. In most cases, the frequency-shift distance μ can be taken as a number between 0.90 and 0.99 times the first eigenvalue, which can basically meet the above requirement. Without loss of generality, taking $\lambda_1 = 1$, $\lambda_2 = 5$, and $\lambda_3 = 10$ as an example, the reciprocals of these eigenvalues are $\frac{1}{\lambda_1} = 1$, $\frac{1}{\lambda_2} = 0.2$, and $\frac{1}{\lambda_3} = 0.1$. The ratio of the reciprocal of the first-order eigenvalue to the sum of all reciprocals is $\frac{1}{1+0.2+0.1} = 76.92\%$. Letting $\mu = 0.99\ \lambda_1 = 0.99$, the eigenvalues of the system after frequency shift are $\overline{\lambda_1} = 1 - 0.99 = 0.01$, $\overline{\lambda_2} = 5 - 0.99 = 4.01$, and $\overline{\lambda_3} = 10 - 0.99 = 9.01$, and the corresponding reciprocals are $\frac{1}{\overline{\lambda_1}} = 100$, $\frac{1}{\overline{\lambda_2}} = 0.25$, and $\frac{1}{\overline{\lambda_3}} = 0.11$. The ratio of the reciprocal of the first-order eigenvalue to the sum of all reciprocals is $\frac{100}{100+0.25+0.11} = 99.64\%$. It is clear that the contribution of the first-order eigen-pairs to the flexibility matrix increased from 76.92% to 99.64% for the system before and after frequency shift. In other words, the contribution of the higher-order modes to the flexibility matrix can be greatly reduced by the frequency-shift operation. As a result, the frequency-shift flexibility change $\overline{\Delta F}$ can be accurately estimated by only the first-order eigen-pairs as:

$$\overline{\Delta F} = \frac{1}{\lambda_{d1} - \mu} \phi_{d1} \phi_{d1}^T - \frac{1}{\lambda_1 - \mu} \phi_1 \phi_1^T \tag{22}$$

On the other hand, the frequency-shift flexibility sensitivity equation can also be derived by using the similar process from Equation (9) to (13) as:

$$\overline{\Delta F} = \sum_{i=1}^{N} \varepsilon_i \overline{\Pi}_i \tag{23}$$

$$\overline{\Pi}_i = \overline{F} K_i \overline{F} \tag{24}$$

$$\overline{F} = \overline{K}^{-1} = (K - \mu M)^{-1} \tag{25}$$

Equation (23) is called the frequency-shift flexibility sensitivity equation and will be used to compute the unknown damage coefficients ε_i ($i = 1 \sim N$) for damage identification. The matrix $\overline{\Pi}_i$ is called the elementary frequency-shift flexibility sensitivity of the i-th element in FEM.

In the end, the operation steps of the proposed frequency-shift flexibility sensitivity method are summarized as follows:

(1) Establish the FEM of the intact structure to obtain the stiffness and mass matrices K and M.

(2) Conduct dynamic analysis on the structure and measure the first-order eigen-pairs of the intact and damaged structures.
(3) Compute the frequency-shift flexibility change $\overline{\Delta F}$ by Equation (22) and compute the elementary frequency-shift flexibility sensitivity matrix $\overline{\Pi}_i$ by Equation (24).
(4) Compute the damage coefficients ε_i $(i = 1 \sim N)$ by solving Equation (23). Finally, the damage locations and extents in the structure can be determined according to the values of ε_i $(i = 1 \sim N)$. To resist the adverse effects of data noise due to measurement error, the singular-value truncation algorithm [31,32] is used in the process of solving the linear Equation (23) for achieving stable computational results in engineering applications. The core idea of the singular-value truncation algorithm is to ignore small singular values to partially eliminate the impact of data noise on the calculation results. The main formulas of the singular-value truncation algorithm are briefly illustrated as follows. Firstly, Equation (23) can be rewritten as a system of linear equations as:

$$\eta = \Omega \cdot \alpha \tag{26}$$

$$\alpha = (\varepsilon_1, \varepsilon_2, \cdots, \varepsilon_N)^T \tag{27}$$

where η is a column vector derived from the matrix $\overline{\Delta F}$, and Ω is the corresponding coefficient matrix derived from the matrices $\overline{\Pi}_i$ $(i = 1 \sim N)$. Performing the singular-value decomposition of Ω yields:

$$\Omega = U \Lambda V^T \tag{28}$$

$$U = [u_1, u_2, \cdots] \tag{29}$$

$$V = [v_1, v_2, \cdots] \tag{30}$$

$$\Lambda = \begin{bmatrix} Z & 0 \\ 0 & 0 \end{bmatrix} \tag{31}$$

$$Z = diag(\sigma_1, \sigma_2, \cdots, \sigma_t) \tag{32}$$

where $\sigma_1, \sigma_2, \cdots, \sigma_t$ are the nonzero singular values of Ω with $\sigma_1 \geq \sigma_2 \geq \cdots \geq \sigma_t$. Based on Equations (26) and (28), the singular-value truncation solution of the damage coefficient vector α can be obtained by ignoring a few smaller singular values as:

$$\alpha_{svt} = (\sum_{x=1}^{z} \sigma_x^{-1} v_x u_x^T)\eta \tag{33}$$

where z is the number of remaining singular values. These remaining singular values σ_x $(x = 1 \sim z)$ all satisfy $\sigma_x / \sigma_{\max} \geq \zeta$ (ζ is a predefined threshold value; for example, $\zeta = 0.001$ is used in the next beam and frame structure examples).

Compared with the ordinary flexibility sensitivity method, the frequency-shift flexibility sensitivity approach greatly reduces the adverse effect of the higher-order modal truncation on damage identification. Theoretically, the proposed method can accurately calculate the structural damage parameters only by using the first-order eigen-parameters of structural free vibration. From Equations (24) and (25), one can find that the system matrices K and M of the undamaged FEM are used in the computation of the frequency-shift flexibility sensitivity. However, Equation (25) can be approximated by using Equation (20) with a few lower-frequency vibration modes.

Combining Equations (20) and (24), it can be found that the frequency-shift flexibility sensitivity equations can also be established by only using the tested vibration modes of the intact structure, rather than the system matrices K and M obtained from the FEM of the intact structure. This advantage makes it possible to conduct structural damage identification even without finite element modeling. In other words, another advantage of the proposed method is that it does not require a complete finite element modeling process, as long as a few lower-frequency vibration modes of the intact structure are measured. Therefore, the proposed method is simpler in operation compared to other existing sensitivity methods, as these existing sensitivity methods require the use of the system matrices K and M of the structural FEM. The case verification of this advantage is detailed in the following section of the experimental example.

3. Numerical Example

3.1. A Truss Structure

A 23-bar truss structure as shown in Figure 1 is used as the first numerical example to verify the frequency-shift flexibility sensitivity method. The main parameters of this steel truss structure are as follows: Young's modulus is 200 GPa, density is 7800 kg/m^3, length of each bar is 1 m, and cross-sectional area is 1.759×10^{-4} m^2. The vibration test data used were generated through the numerical finite element model of the undamaged and damaged systems. Note that the data noise was not considered in this example in order to purely investigate the improvement effect of the frequency-shift process on the solution accuracy. Three damage scenarios are simulated in this example. The first damage scenario assumes that the elastic modulus of bar elements 10 and 13 are both reduced by 5%. The second damage scenario assumes that the elastic modulus of bar elements 9, 10, and 11 are reduced by 10%, 5%, and 5%, respectively. The third damage scenario assumes that the elastic modulus of bar element 13 is reduced by 2%. Table 1 presents the first natural frequencies of the undamaged and damage scenarios. In engineering practice, fatigue and corrosion of materials will lead to a decrease in the elastic modulus. Note that the proposed method is also applicable to other types of damage such as cracks or notches, as shown in the next experimental example. Figures 2–4 present the calculation results of the damage coefficients by the proposed method and the ordinary flexibility sensitivity method, respectively. Only the first-order eigen-parameters are used in the calculation, and the frequency-shift distance μ is taken as $\mu = 0.9\lambda_{d1}$, $\mu = 0.95\lambda_{d1}$, and $\mu = 0.99\lambda_{d1}$, respectively. Tables 2–4 present the comparison between the calculated value and the actual value of the damage extent for these three damage scenarios. The values in brackets in these tables represent the relative error between the calculated values and the true values.

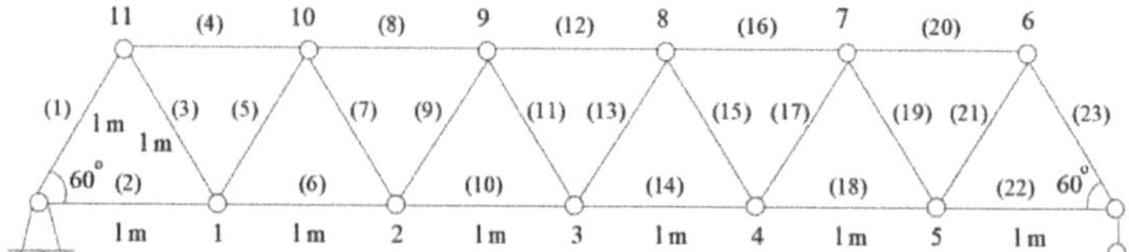

Figure 1. A 23-bar truss structure.

Table 1. The first natural frequencies of the undamaged and damage scenarios.

Case	Undamaged	Damage Scenario 1	Damage Scenario 2	Damage Scenario 3
Natural frequencies	56.8522	56.6298	56.6210	56.8516

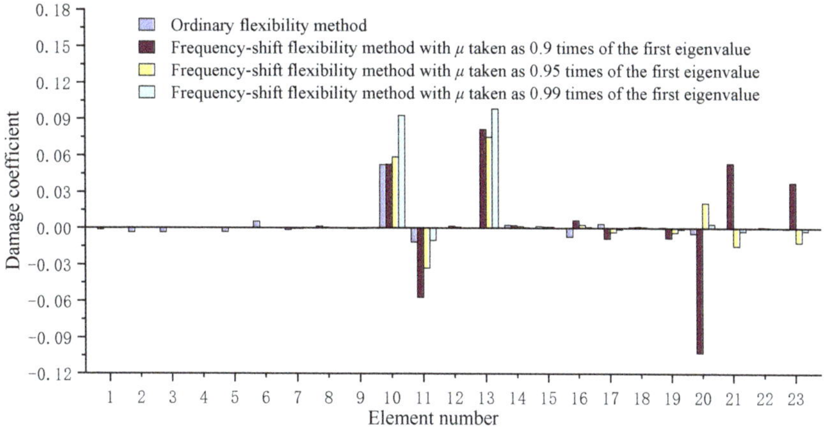

Figure 2. Calculation result comparison of the proposed method and the ordinary flexibility sensitivity method for the first damage scenario.

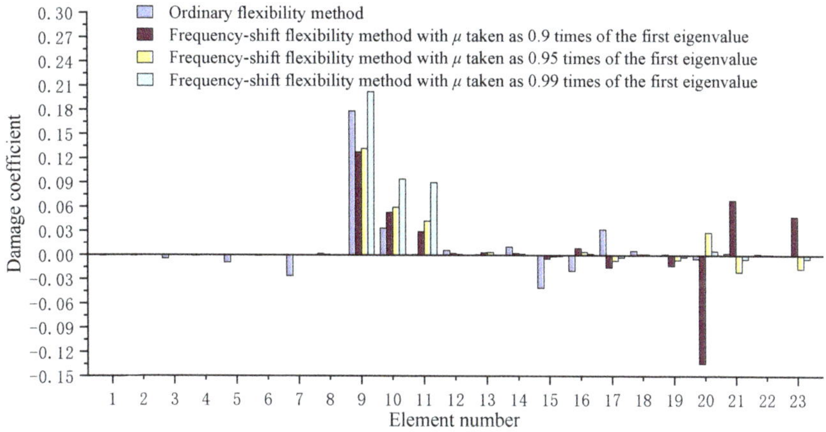

Figure 3. Calculation result comparison of the proposed method and the ordinary flexibility sensitivity method for the second damage scenario.

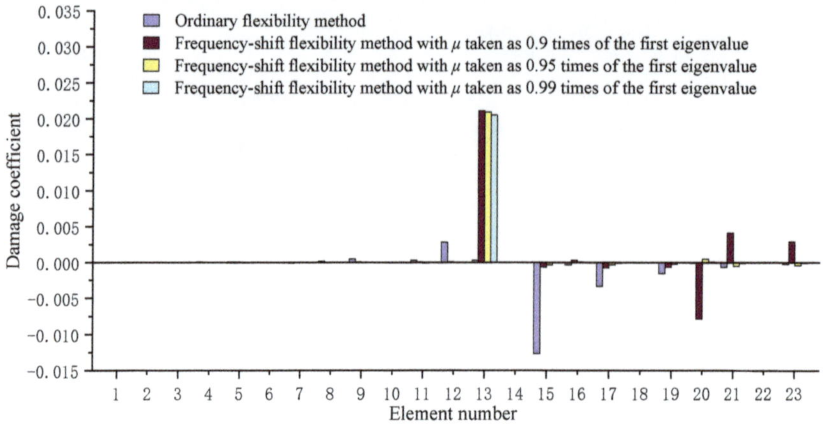

Figure 4. Calculation result comparison of the proposed method and the ordinary flexibility sensitivity method for the third damage scenario.

Table 2. Comparison between the calculated value and the actual value of the damage coefficient for the first damage scenario.

Damaged Element Number	True Value	Ordinary Flexibility Method	Frequency-Shift Flexibility Method		
			$\mu = 0.9\lambda_{d1}$	$\mu = 0.95\lambda_{d1}$	$\mu = 0.99\lambda_{d1}$
10	0.05	0.0524 (4.8%) *	0.0526 (4.9%)	0.0585 (14.5%)	0.0928 (46.1%)
13	0.05	−0.0001 (/)	0.0815 (38.7%)	0.0749 (33.2%)	0.0982 (49.1%)

* The values in brackets represent the relative error.

Table 3. Comparison between the calculated value and the actual value of the damage coefficient for the second damage scenario.

Damaged Element Number	True Value	Ordinary Flexibility Method	Frequency-Shift Flexibility Method		
			$\mu = 0.9\lambda_{d1}$	$\mu = 0.95\lambda_{d1}$	$\mu = 0.99\lambda_{d1}$
9	0.1	0.1787 (44.0%) *	0.1278 (21.8%)	0.1323 (24.4%)	0.2026 (50.6%)
10	0.05	0.0333 (33.4%)	0.0527 (5.1%)	0.0589 (15.1%)	0.0943 (46.9%)
11	0.05	0.0007 (98.6%)	0.029 (42%)	0.0419 (16.2%)	0.0901 (44.5%)

* The values in brackets represent the relative error.

Table 4. Comparison between the calculated value and the actual value of the damage coefficient for the third damage scenario.

Damaged Element Number	True Value	Ordinary Flexibility Method	Frequency-Shift Flexibility Method		
			$\mu = 0.9\lambda_{d1}$	$\mu = 0.95\lambda_{d1}$	$\mu = 0.99\lambda_{d1}$
13	0.02	0.0003 (98.5%) *	0.0211 (5.5%)	0.0209 (4.3%)	0.0205 (2.4%)

* The values in brackets represent the relative error.

For the first damage scenario, one can find from Figure 2 that the proposed method can successfully identify elements 10 and 13 as the damaged bars. However, the ordinary flexibility sensitivity method can only identify element 10 as the damaged bar, while element 13 cannot be identified. This indicates that the traditional flexibility sensitivity method may result in a missed diagnosis, while the proposed frequency-shift flexibility sensitivity method performs well. When the frequency-shift distance changes, Table 2 shows that the calculation error of the damage coefficient is relatively minimal when $\mu = 0.9\lambda_{d1}$. However, Figure 2 shows that the accuracy of damage localization is highest when $\mu = 0.99\lambda_{d1}$, since there is a misjudgment of damage location when $\mu = 0.9\lambda_{d1}$. When $\mu = 0.95\lambda_{d1}$, a good balance can be achieved between the damage localization and damage quantification of the frequency-shift flexibility method. For the second damage scenario, it can be seen from Figure 3 that the proposed method can successfully identify elements 9, 10, and 11 as the damaged bars. Again, the ordinary flexibility sensitivity method failed since element 11 cannot be identified by it. This once again demonstrates that the proposed frequency-shift flexibility sensitivity method is more reliable in calculating results than the traditional flexibility sensitivity method. When the frequency-shift distance changes, Table 3 shows that the calculation error of the damage coefficient is relatively minimal when $\mu = 0.95\lambda_{d1}$, and Figure 3 shows that the accuracy of damage localization is highest when $\mu = 0.99\lambda_{d1}$. For the third damage scenario (minor damage), one can find from Figure 4 that the proposed method more clearly indicates that element 13 is the damaged bar than the ordinary flexibility sensitivity method. It has been shown that the proposed method may have better ability to identify the minor damage than the ordinary flexibility sensitivity method. When the frequency-shift distance changes, Table 4 and Figure 4 show that the frequency-shift flexibility algorithm has the best accuracy in damage localization and quantification when $\mu = 0.99\lambda_{d1}$. Based on the above results, it can be concluded that the smaller the degree of damage, the larger the frequency-shift distance should be taken. In summary, the proposed frequency-shift flexibility algorithm is very efficient with only the first-order modal parameters. This new method can achieve the goal of improving the accuracy and reliability of the damage identification results through only the simple frequency-shift operation.

3.2. A Beam Structure

A beam structure with fixed ends (as shown in Figure 5) is used as the second numerical example to further verify the proposed method. The Young's modulus and density of this beam structure are 193 GPa and 7850 kg/m^3, respectively. The vibration test data used were generated through the numerical FEMs of the undamaged and damaged systems. In this example, a 5% random noise level is added to the first mode shape to simulate the measurement error as

$$\varphi'_{d1} = \varphi_{d1} \times [1 + \tau \cdot unifrnd(-1,1)] \tag{34}$$

where φ'_{d1} is the contaminated mode shape, τ is the noise level, and $unifrnd(-1,1)$ represents a random number located in the interval of $[-1,1]$.

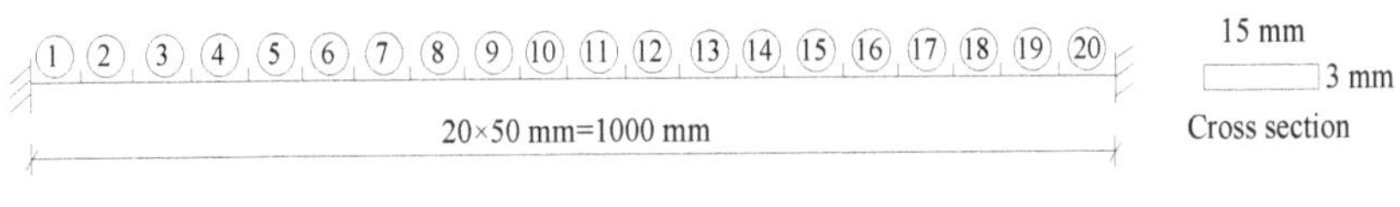

Figure 5. A beam structure.

Two fault scenarios are simulated in this example. The first fault scenario assumes that the elastic modulus of elements 3 and 10 are reduced by 10% and 15%, respectively. The second fault scenario assumes that the elastic modulus of elements 6, 10, and 14 are all reduced by 10%. As stated before, the singular-value truncation algorithm is used in this example to overcome the adverse effects of data noise. Figures 6 and 7 present the calculation results of the damage coefficients by the proposed method and the ordinary flexibility sensitivity method, respectively. Note that only the first-order eigen-parameters are used in the calculation, and the frequency-shift distance μ is taken as $\mu = 0.9\lambda_{d1}$. Tables 5 and 6 present the comparison between the calculated value and the actual value of the damage extent for the two damage scenarios. The values in brackets in these tables represent the relative error between the calculated values and the true values.

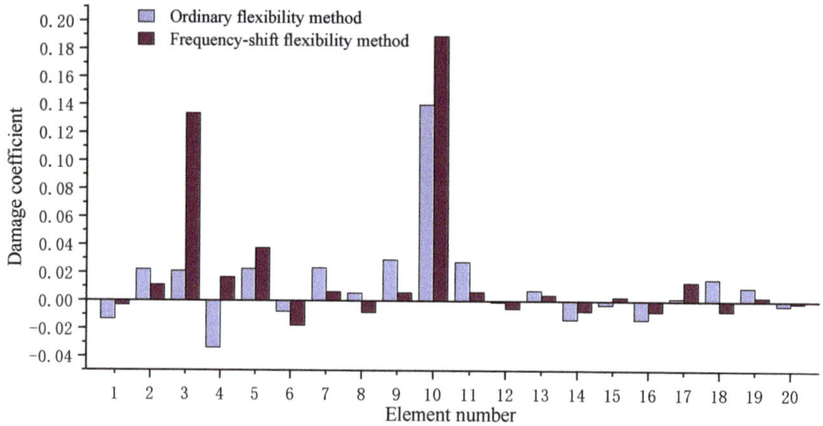

Figure 6. Calculation result comparison of the proposed method and the ordinary flexibility sensitivity method when elements 3 and 10 are damaged.

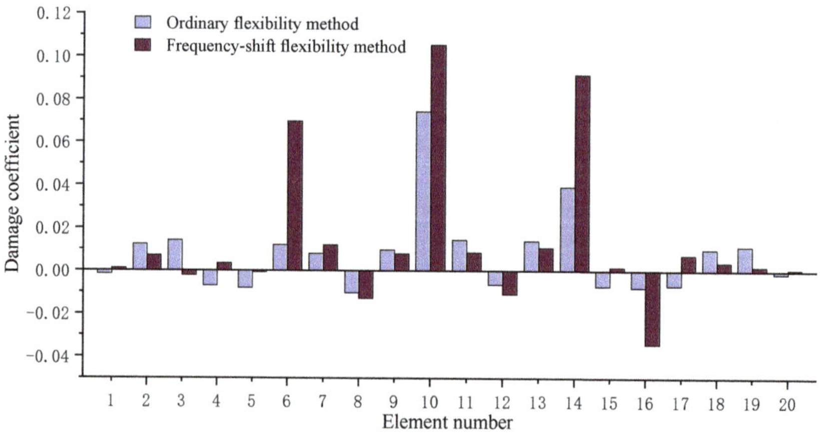

Figure 7. Calculation result comparison of the proposed method and the ordinary flexibility sensitivity method when elements 6, 10, and 14 are damaged.

Table 5. Comparison between the calculated value and the actual value of the damage coefficient when elements 3 and 10 are damaged.

Damaged Element Number	True Value	Ordinary Flexibility Method	Frequency-Shift Flexibility Method with $\mu = 0.9\lambda_{d1}$
3	0.1	0.021 (79.0%) *	0.1336 (25.1%)
10	0.15	0.1402 (6.5%)	0.1894 (26.3%)

* The values in brackets represent the relative error.

Table 6. Comparison between the calculated value and the actual value of the damage coefficient when elements 6, 10, and 14 are damaged.

Damaged Element Number	True Value	Ordinary Flexibility Method	Frequency-Shift Flexibility Method with $\mu = 0.9\lambda_{d1}$
6	0.1	0.0122 (87.8%) *	0.0696 (30.4%)
10	0.1	0.0745 (25.5%)	0.1055 (5.2%)
14	0.1	0.0393 (60.7%)	0.0916 (8.4%)

* The values in brackets represent the relative error.

For the first fault scenario, one can find from Figure 6 that the proposed method can successfully identify elements 3 and 10 as the fault locations. However, the ordinary flexibility sensitivity method can only identify element 10 as the fault location, while element 3 cannot be identified. This indicates that the traditional flexibility sensitivity method may result in misdiagnosis, while the proposed frequency-shift flexibility sensitivity method performs well. For the second fault scenario, it can be seen from Figure 7 that the proposed method can successfully identify elements 6, 10, and 14 as the damage locations. Again, the ordinary flexibility sensitivity method failed since element 6 cannot be identified by it. From Tables 5 and 6, one can find that the proposed frequency-shift flexibility method achieves better computational accuracy than the ordinary flexibility method. This once again demonstrates that the proposed method can achieve the goal of improving the accuracy and reliability of fault identification through only the simple frequency-shift operation.

4. Validation by the Experimental Data of a Steel Frame Structure

The presented algorithm is further verified by the experimental data obtained from a three-story steel frame structure from [33]. As shown in Figure 8a, the experimental structure consists of three steel plates and four rectangular columns. These steel plates and columns are welded to form a rigid shear system as shown in Figure 8b. Detailed descriptions of material physical parameters and testing processes are provided in [33]. The fault condition is simulated by cutting from 75 mm to 51.3 mm on the first floor as shown in Figure 9, with a corresponding damage severity of 11.6%. The first frequency and vibration mode of the undamaged system measured are $f_1 = 3.369$Hz and $\varphi_1 = (0.021108,\ 0.03922,\ 0.048427)^{\mathrm{T}}$. The second frequency and vibration mode of the undamaged system are $f_2 = 9.704$ and $\xi_2 = (0.048758, 0.02031, -0.03923)^T$. The third frequency and vibration mode of the undamaged system are $f_3 = 14.282$ and $\xi_3 = (0.037936, -0.04866, 0.022852)^T$. For the structure with damage, the measured first frequency and vibration mode are $f_{d1} = 3.259$Hz and $\varphi_{d1} = (0.022735,\ 0.039331,\ 0.047594)^{\mathrm{T}}$. As stated before, the frequency-shift flexibility sensitivity equations in this example are established by directly using the tested vibration modes of the intact structure, rather than the system matrices K and M obtained from the FEM of the intact structure. Therefore, this example does not require a complete finite element modeling process to perform damage identification. Figure 10 presents the calculation results of the damage coefficients by the proposed method with the frequency-shift distance $\mu = 0.9\lambda_{d1}$. From Figure 10, one can find that the proposed method can accurately identify the first floor as the damage location, and the calculated value of damage degree is 0.22. It has been shown that the proposed method can also complete the task of damage identification without the need for a complete finite element modeling process.

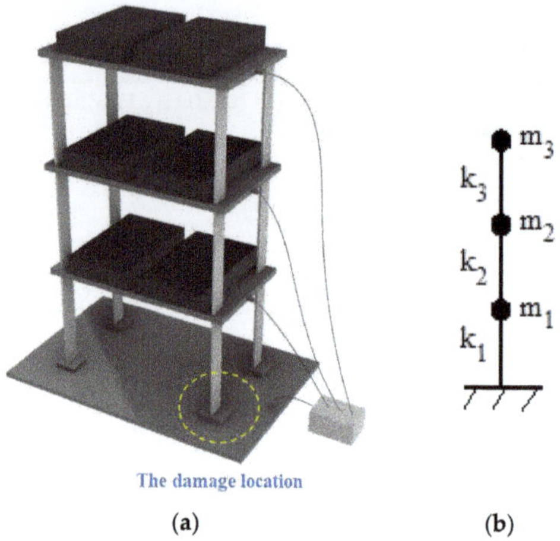

Figure 8. (**a**) A three-story experimental steel frame structure; (**b**) simplified shear system corresponding to the experimental structure.

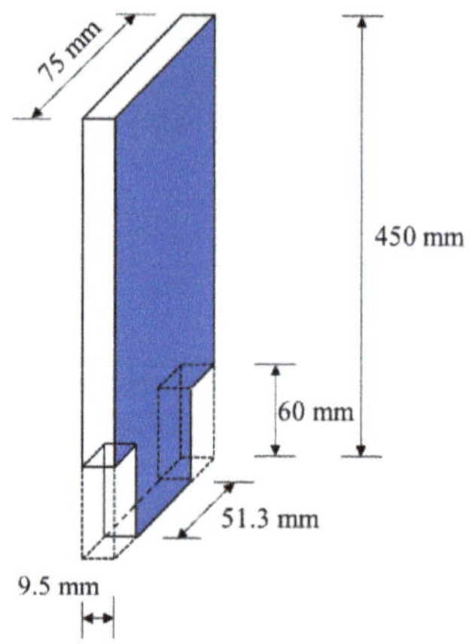

Figure 9. The column width is cut from 75 mm to 51.3 mm to simulate the damage.

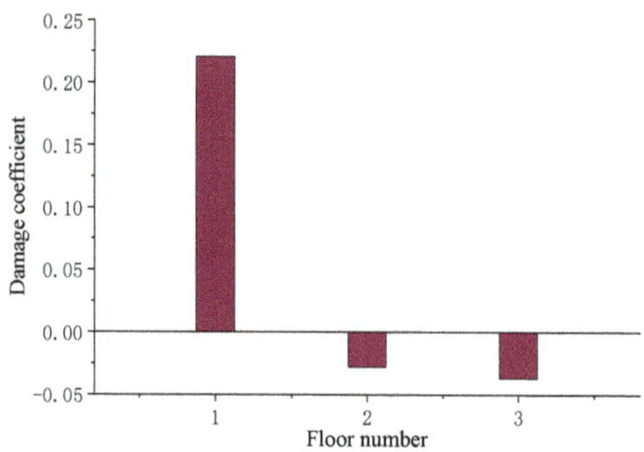

Figure 10. Calculation result of the proposed method when the first floor is damaged.

5. Conclusions

In this work, a new frequency-shift flexibility sensitivity technique was developed for structural damage evaluation using only the first-order mode of structural free vibration. With the help of the frequency-shift operation, the first-order vibration mode will contribute the most to the structural flexibility matrix. This results in a significant reduction in the adverse effect of the higher-order modal truncation on structural damage identification. Theoretically, the proposed method can accurately calculate the structural damage coefficients by using only the first-order modal parameters. As a result, the reliable identification results can be achieved according to the values of the calculated damage coefficients. Another advantage of the proposed method is that it does not require a complete finite element modeling process, as long as a few lower-frequency vibration modes of the intact structure are measured. Based on the calculation results of the numerical and experimental examples, it can be concluded that the proposed method requires fewer modal parameters but has higher calculation accuracy than the ordinary flexibility sensitivity method. For the multiple damages case, this new approach can overcome the possible missed diagnosis of the ordinary flexibility sensitivity method. For the minor damage case, the numerical results also showed that the proposed algorithm may have the potential to identify the minor damage in the structure. The proposed method provides a new way for structural damage identification with only the first-order modal parameters. Note that the proposed method is only applicable to linear structures, and research on damage identification of nonlinear structures will be conducted in the future.

Author Contributions: Conceptualization, S.C. and Q.Y.; methodology, S.C. and Q.Y.; software, X.P.; validation, S.C., Q.Y. and X.P. All authors have read and agreed to the published version of the manuscript.

Funding: This work was supported by the Zhejiang public welfare technology application research project (LGF22E080021), Ningbo natural science foundation project (202003N4169), Natural Science Foundation of China (11202138, 52008215), the Natural Science Foundation of Zhejiang Province, China (LQ20E080013), and the major special science and technology project (2019B10076) of "Ningbo science and technology innovation 2025".

Data Availability Statement: The data used to support the findings of this study are included in this study and also available from the corresponding author upon request.

Conflicts of Interest: The authors declare no conflict of interest.

References

1. Torzoni, M.; Manzoni, A.; Mariani, S. A multi-fidelity surrogate model for structural health monitoring exploiting model order reduction and artificial neural networks. *Mech. Syst. Signal Process.* **2023**, *197*, 110376. [CrossRef]
2. Peng, X.; Yang, Q.W. Damage detection in beam-like structures using static shear energy redistribution. *Front. Struct. Civ. Eng.* **2022**, *16*, 1552–1564. [CrossRef]
3. Peng, X.; Tian, C.; Yang, Q. Structural Damage Identification Using the Optimal Achievable Displacement Variation. *Materials* **2022**, *15*, 8440. [CrossRef] [PubMed]
4. Yang, Q.W.; Liu, J.K. Structural damage identification with flexibility changed: A review. *J. Vib. Shock* **2011**, *30*, 147–153.
5. Pandey, A.K.; Biswas, M. Damage detection in structures using changes in flexibility. *J. Sound Vib.* **1994**, *169*, 3–17. [CrossRef]
6. Jaishi, B.; Ren, W.X. Damage detection by finite element model updating using modal flexibility residual. *J. Sound Vib.* **2006**, *290*, 369–387. [CrossRef]
7. Catbas, F.N.; Brown, D.L.; Aktan, A.E. Use of modal flexibility for damage detection and condition assessment: Case studies and demonstrations on large structures. *J. Struct. Eng.* **2006**, *132*, 1699–1712. [CrossRef]
8. Duan, Z.; Yan, G.; Ou, J.; Spencer, B.F. Damage detection in ambient vibration using proportional flexibility matrix with incomplete measured DOFs. *Struct. Control Health Monit.* **2007**, *14*, 186–196. [CrossRef]
9. Tomaszewska, A. Influence of statistical errors on damage detection based on structural flexibility and mode shape curvature. *Comput. Struct.* **2010**, *88*, 154–164. [CrossRef]
10. Yang, Q.W. A new damage identification method based on structural flexibility disassembly. *J. Vib. Control* **2011**, *17*, 1000–1008. [CrossRef]
11. Maghsoodi, A.; Ghadami, A.; Mirdamadi, H.R. Multiple-crack damage detection in multi-step beams by a novel local flexibility-based damage index. *J. Sound Vib.* **2013**, *332*, 294–305. [CrossRef]

12. Weng, S.; Zhu, H.P.; Xia, Y.; Mao, L. Damage detection using the eigenparameter decomposition of substructural flexibility matrix. *Mech. Syst. Signal Process.* **2013**, *34*, 19–38. [CrossRef]
13. Grande, E.; Imbimbo, M. A multi-stage approach for damage detection in structural systems based on flexibility. *Mech. Syst. Signal Process.* **2016**, *76*, 455–475. [CrossRef]
14. Hosseinzadeh, A.Z.; Amiri, G.G.; Razzaghi, S.S.; Koo, K.Y.; Sung, S.H. Structural damage detection using sparse sensors installation by optimization procedure based on the modal flexibility matrix. *J. Sound Vib.* **2016**, *381*, 65–82. [CrossRef]
15. Altunışık, A.C.; Okur, F.Y.; Karaca, S.; Kahya, V. Vibration-based damage detection in beam structures with multiple cracks: Modal curvature vs. modal flexibility methods. *Nondestruct. Test. Eval.* **2019**, *34*, 33–53. [CrossRef]
16. Wickramasinghe, W.R.; Thambiratnam, D.P.; Chan, T.H. Damage detection in a suspension bridge using modal flexibility method. *Eng. Fail. Anal.* **2020**, *107*, 104194. [CrossRef]
17. Sarmadi, H.; Entezami, A.; Ghalehnovi, M. On model-based damage detection by an enhanced sensitivity function of modal flexibility and LSMR-Tikhonov method under incomplete noisy modal data. *Eng. Comput.* **2022**, *38*, 111–127. [CrossRef]
18. Ahmadi-Nedushan, B.; Fathnejat, H. A modified teaching–learning optimization algorithm for structural damage detection using a novel damage index based on modal flexibility and strain energy under environmental variations. *Eng. Comput.* **2022**, *38*, 847–874. [CrossRef]
19. Feng, Z.; Wang, W.; Zhang, J. Probabilistic Structural Model Updating with Modal Flexibility Using a Modified Firefly Algorithm. *Materials* **2022**, *15*, 8630. [CrossRef]
20. Bernagozzi, G.; Quqa, S.; Landi, L.; Diotallevi, P. Structure-type classification and flexibility-based detection of earthquake-induced damage in full-scale RC buildings. *J. Civ. Struct. Health Monit.* **2022**, *12*, 1443–1468. [CrossRef]
21. Yang, Q.W.; Peng, X. A highly efficient method for structural model reduction. *Int. J. Numer. Methods Eng.* **2023**, *124*, 513–533. [CrossRef]
22. Dinh-Cong, D.; Nguyen-Huynh, P.; Nguyen, S.; Nguyen-Thoi, T. Damage Identification of Functionally Graded Beams using Modal Flexibility Sensitivity-based Damage Index. *Period. Polytech. Civ. Eng.* **2023**, *67*, 272–281. [CrossRef]
23. Darshan, B.U.; Siddesha, H.; Rajanna, T. Structural Damage Detection for Plates Using Flexibility Based Strain Energy Method. *Lect. Notes Civ. Eng.* **2023**, *256*, 285–300.
24. Quqa, S.; Landi, L. Integrating flexibility-based curvature with quasi-static features induced by traffic loads for high-resolution damage localization in bridges. *Mech. Syst. Signal Process.* **2023**, *186*, 109907. [CrossRef]
25. Nick, H.; Ashrafpoor, A.; Aziminejad, A. Damage identification in steel frames using dual-criteria vibration-based damage detection method and artificial neural network. *Structures* **2023**, *51*, 1833–1851. [CrossRef]
26. Cuomo, S.; Boccaccio, M.; Meo, M. Damage identification during an impact event using the Hilbert-Huang transform of decomposed propagation modes. *Mech. Syst. Signal Process.* **2023**, *191*, 110126. [CrossRef]
27. Aulakh, D.S.; Bhalla, S. Piezo sensor based multiple damage detection under output only structural identification using strain modal flexibility. *Mech. Syst. Signal Process.* **2023**, *194*, 110272. [CrossRef]
28. He, S.; Shi, Y.; Jonsson, E.; Martins, J. Eigenvalue problem derivatives computation for a complex matrix using the adjoint method. *Mech. Syst. Signal Process.* **2023**, *185*, 109717. [CrossRef]
29. Zheng, S.; Ni, W.; Wang, W. Combined method for calculating eigenvector derivatives with repeated eigenvalues. *AIAA J.* **1998**, *36*, 428–431. [CrossRef]
30. Xu, T.; Guo, G.; Zhang, H. Vibration reanalysis using frequency-shift combined approximations. *Struct. Multidiscip. Optim.* **2011**, *44*, 235–246. [CrossRef]
31. Ren, W.X. A singular value decomposition based truncation algorithm in solving the structural damage equations. *Acta Mech. Solida Sin.* **2005**, *18*, 181–188.
32. Yang, B.; Huang, H.; Ma, H.; Zhou, L.; Du, Q. Calculation of microwave heating temperature distribution based on SVD truncation. *J. Microw. Power Electromagn. Energy* **2022**, *56*, 238–258. [CrossRef]
33. Li, L. Numerical and Experimental Studies of Damage Detection for Shear Buildings. Ph.D. Thesis, Huazhong University of Science and Technology, Wuhan, China, 2005.

axioms

MDPI

Article

An Energy-Efficient Optimal Operation Control Strategy for High-Speed Trains via a Symmetric Alternating Direction Method of Multipliers

Shan Ma [1], Feng Ma [2,*] and Chaoyu Tang [1]

[1] School of Automation, Central South University, Changsha 410083, China
[2] High-Tech Institute of Xi'an, Xi'an 710025, China
* Correspondence: fengma@outlook.com

Abstract: Train operation control is of great importance in reducing train operation energy consumption and improving railway operation efficiency. This paper investigates the design of optimal control inputs for multiple trains on a single railway line with several stations. Firstly, a distributed optimal control problem for multiple train operation is formulated to reduce the energy consumption and improve the punctuality of trains. Then, we propose an efficient algorithm based on the framework of the symmetric alternating direction method of multipliers to solve this optimization problem. Finally, numerical simulations show that the method can obtain the optimal train control sequence in fewer iterative steps compared to the alternating direction multiplier method, thus illustrating the effectiveness of the algorithm.

Keywords: high-speed train; distributed optimal control; discrete time; symmetric alternating direction method of multipliers

MSC: 90C30; 49M27

Citation: Ma, S.; Ma, F.; Tang, C. An Energy-Efficient Optimal Operation Control Strategy for High-Speed Trains via a Symmetric Alternating Direction Method of Multipliers. *Axioms* **2023**, *12*, 489. https://doi.org/10.3390/axioms12050489

Academic Editors: Leonid Plotnikov and Gustavo Olague

Received: 27 March 2023
Revised: 14 May 2023
Accepted: 15 May 2023
Published: 18 May 2023

1. Introduction

For a high-speed railway system, it is important to design operation control strategies for each train such that the trains can operate according to the scheduled timetable. Since the 1960s, the train operation control problem has received a lot of attention, and various train control strategies have been proposed [1–9]. In particular, Li et al. [10] investigated the robust train operation controller design problem using the framework of linear matrix inequalities. In [11,12], Li et al. extended the single train control problem to the multiple train movement control problem. By using LaSalle's invariance principle, a coordinated control strategy has been proposed for multiple train operations on a railway line [11,12]. The above works on train operation control are based on a feedback control approach. On the other hand, a large number of optimal control schemes have been proposed by addressing the train operation control problem as an optimization problem. Optimal control is a branch of numerical optimization, which deals with finding the control sequence of a plant in a period of time such that the objective function is optimized. Lin et al. [13] studied the design of single- and double-integrator operation feedback controllers for multiple trains operating on a railway line, and employed a convex optimization method to obtain the optimal control gains. Yan et al. [14] proposed a distributed cooperative optimal control algorithm for multiple high-speed train trajectory planning. Wang et al. [15] investigated the optimal trajectory planning problem for trains under operation constraints, and formulated it as a mixed-integer linear programming (MILP) problem. Since the train operation control problem is often a large-scale optimization problem, the most important issue is to find an efficient algorithm to obtain the optimal control inputs.

As an algorithm developed on the basis of the augmented Lagrange algorithm, the alternating direction method of multipliers (ADMM) aims to combine the decomposability of dual ascent with the superior convergence of the method of multipliers, and alternately minimize the decision variables [16]. A lot of studies have been performed to investigate the applications of the ADMM. For example, Fu et al. [17] designed optimal feedback gains via the ADMM, which can obey the sparsity constraints of controllers as well as optimizing the system performance. Li et al. [18] studied the distributed optimal control of multiple high-speed train movements by using the algorithm of ADMM with the objective of tracking the desired speed and position trajectories for each train. As an extension of ADMM, the symmetric alternating direction method of multipliers (SADMM) has been studied in 2014 [19–22]. SADMM is often used for the convex optimization problem with linear constraints and a separable objective function. This method has a better convergence rate compared with ADMM, though it requires additional assumptions to ensure its convergence [20,23,24]. In fact, SADMM has the potential to be used in various fields, including the train operation control problem.

In this paper, we consider the optimal control problem of multiple high-speed train movements on a single railway line with several stations. Different from the problem considered in [18], we consider a railway line consisting of several stations, and assume that the departure time of each train from every station is not earlier than the scheduled time in the timetable. Furthermore, the optimization model in [18] focuses on minimizing the deviation of the actual train operation from the desired operation, while in this paper, we treat the actual operation of each train as an optimization variable under the necessary safety and punctuality constraints. By so doing, the modeling error caused by the mismatch between the actual and nominal operations could be avoided. Furthermore, we use the symmetric alternating direction method of multipliers (SADMM) to solve the optimization problem, which usually outperforms the the alternating direction method of multipliers (ADMM), as used in [18].

This paper is structured as follows. In Section 2, we present the continuous-time dynamics of high-speed trains and some operation constraints. In Section 3, the dynamics of high-speed trains is discretized and the train operation control problem is formulated. In Section 4, SADMM is introduced to solve the problem. In Section 5, numerical simulations are performed to illustrate the effectiveness of the proposed method. Section 6 concludes the paper.

2. Problem Statement

2.1. Train Dynamics

The dynamical equation of a high-speed train i is modelled as

$$\begin{cases} \dot{x}_i(t) = v_i(t), \\ m_i\dot{v}_i(t) = F_i(t) - f, \quad i = 1, 2, \ldots, M, \end{cases} \tag{1}$$

where $x_i(t)$ and $v_i(t)$ represent the position and velocity of train i at time t, respectively. M is the total number of trains. m_i is the mass of train i and $F_i(t)$ is the control force of train i. f denotes the resistance, which includes ramp resistance, curve resistance, tunnel resistance and aerodynamic resistance, etc. For simplicity, we assume f is a constant.

2.2. Operation Constraints

In practice, train i cannot depart from station j before the scheduled departure time $t_{i,j,out}$. This constraint is expressed as

$$x_i(t_{i,j,out}) \leq l_j, \quad i = 1, 2, \ldots, M, \quad j = 1, 2, \ldots, J, \tag{2}$$

where l_j denotes the position of station j and $x_i(t_{i,j,out})$ represents the actual position of train i at time $t_{i,j,out}$. J is the number of stations.

The speed constraint of train i is expressed as

$$0 \leq v_i(t) \leq v_{max}, \quad i = 1, 2, \ldots, M, \tag{3}$$

where v_{max} represents the maximum speed of the trains.

The control force constraint is expressed as

$$F_{min} \leq F_i(t) \leq F_{max}, \quad i = 1, 2, \ldots, M, \tag{4}$$

where F_{min} and F_{max} represent the minimum and maximum allowed control force, respectively.

In train operations, a train has to keep a minimum safe distance from the preceding train, which is determined by the reaction time and the braking performance of the train. By Newton's second law, the minimum safe distance constraint is expressed as

$$x_{i-1}(t) - x_i(t) \geq v_i(t)d_s + \frac{v_i^2(t)}{2a_{max}}, \quad i = 2, \ldots, M, \tag{5}$$

where d_s is the reaction time to start braking and a_{max} is the maximum deceleration of a train. Constraint (5) is a nonlinear inequality because of the term $v_i^2(t)$. In practice, for simplicity, we usually replace constraint (5) with a linear inequality constraint

$$x_{i-1}(t) - x_i(t) \geq v_i(t)d_s + \frac{v_{max}v_i(t)}{2a_{max}}, \quad i = 2, \ldots, M. \tag{6}$$

2.3. Optimization Objective

The objective is formulated as follows

$$\Psi = \min \sum_{i=1}^{M} \sum_{j=1}^{J} (a_i(x_i(t_{i,j,in}) - l_j)^2 + b_i(v_i^2(t_{i,j,in}))) + \sum_{i=1}^{M} c_i \int_{t=t_0}^{t_l} F_i^2(t)dt, \tag{7}$$

where t_0 denotes the time that the first train begins to operate and t_l denotes the time that the last train finishes operating. a_i, b_i, and c_i are positive penalty factors. $x_i(t_{i,j,in})$ and $v_i(t_{i,j,in})$ represent the actual position and the actual speed of train i at the scheduled arrival time $t_{i,j,in}$ to station j, respectively. Note that we assume that the length of each station is small compared to the segment between stations, such that it can be treated as zero. The first term in (7) penalizes deviations of x_i from station j at the scheduled arrival time $t_{i,j,in}$. The second term in (7) penalizes large values of the velocity v_i at the scheduled arrival time $t_{i,j,in}$, which should be zero in the ideal case. These two terms are used to promote the punctuality of train i arriving at station j. The third term in (7) is included to generate an energy-efficient optimal trajectory.

3. Discrete-Time Optimal Control Problem

For numerical calculation purposes, the above continuous-time optimization problem will be transformed into a discrete-time form. Suppose d is the sampling period. Then, Equation (1) can be transformed as

$$\begin{cases} x_i(k+1) - x_i(k) = v_i(k)d + \frac{d^2}{2m_i}(F_i(k) - f), \\ v_i(k+1) - v_i(k) = \dfrac{d}{m_i}(F_i(k) - f), \quad i = 1, 2, \ldots, M, \end{cases} \tag{8}$$

and constraints (2)–(6) can be, respectively, transformed as

$$x_i(k_{i,j,out}) \le l_j, \quad i = 1,2,\ldots,M, \quad j = 1,2,\ldots,J, \tag{9}$$

$$0 \le v_i(k) \le v_{max}, \quad i = 1,2,\ldots,M, \tag{10}$$

$$F_{min} \le F_i(k) \le F_{max}, \quad i = 1,2,\ldots,M, \tag{11}$$

$$x_{i-1}(k) - x_i(k) \ge v_i(k)d_s + \frac{v_{max}v_i(k)}{2a_{max}}, \quad i = 2,\ldots,M. \tag{12}$$

Furthermore, the objective function (7) can be transformed into a discrete-time form as follows:

$$\Psi = \min \sum_{i=1}^{M}\sum_{j=1}^{J}(a_i(x_i(k_{i,j,in}) - l_j)^2 + b_i(v_i^2(k_{i,j,in}))) + \sum_{i=1}^{M}\sum_{k=0}^{N-1} q_i F_i^2(k), \tag{13}$$

where $x_i(k_{i,j,in})$ and $v_i(k_{i,j,in})$ represent the actual position and the actual speed of train i at the scheduled arrival time $k_{i,j,in}$ to station j, respectively. $q_i = c_i d$ is a positive penalty factor and N represents the time horizon of the optimal control problem. Let $\boldsymbol{x}_i = [x_i(1), x_i(2), \ldots, x_i(N)]^T$ denote the position information of train i at all sampling times. Let $\boldsymbol{y}_i = \left[y_i^T(1), y_i^T(2), \ldots, y_i^T(N)\right]^T$, $i = 2,3,\ldots,M$, where $y_1(k) = x_1(k)$ and $y_i(k) = [x_{i-1}(k), x_i(k)]^T$, $i = 2,3,\ldots,M$. Then, we have

$$\boldsymbol{y}_i = E_i \boldsymbol{z}, \tag{14}$$

where $\boldsymbol{z} = [\boldsymbol{x}_1^T, \boldsymbol{x}_2^T, \ldots, \boldsymbol{x}_M^T]^T$ and E_i is a 0–1 matrix which can be expressed as

$$E_i = \begin{cases} \left[I_N, O_{N\times(M-1)N}\right], & i = 1, \\ \left[O_{2N\times(i-2)N}, H, O_{2N\times(M-i)N}\right], & i = 2,3,\ldots,M, \end{cases} \tag{15}$$

$$H = \left[\begin{array}{cccc|cccc} B_1 & O & \cdots & O & B_2 & O & \cdots & O \\ O & B_1 & \cdots & O & O & B_2 & \cdots & O \\ \vdots & \vdots & \ddots & \vdots & \vdots & \vdots & \ddots & \vdots \\ O & O & \cdots & B_1 & O & O & \cdots & B_2 \end{array}\right], \tag{16}$$

$$B_1 = \begin{bmatrix} 1 \\ 0 \end{bmatrix}, B_2 = \begin{bmatrix} 0 \\ 1 \end{bmatrix}. \tag{17}$$

We also have $x_i(k) = Y_i y_i(k)$, where $Y_i = 1$ if $i = 1$, and $Y_i = [0,1]$ if $i = 2,3,\ldots,M$. The problem (13), with the constraints (8)–(12), can be reformulated as

$$\Psi = \min \quad \sum_{i=1}^{M}\sum_{j=1}^{J}(a_i(Y_i y_i(k_{i,j,in}) - l_j)^2 + b_i(v_i^2(k_{i,j,in}))) + \sum_{i=1}^{M}\sum_{k=0}^{N-1} q_i F_i^2(k) \tag{18}$$

subject to

$$y_i = E_i z, \tag{19}$$

$$Y_i(y_i(k+1) - y_i(k)) = v_i(k)d + \frac{d^2}{2m_i}(F_i(k) - f), \tag{20}$$

$$v_i(k+1) = v_i(k) + \frac{1}{m_i}(F_i(k) - f)d, \tag{21}$$

$$Y_i y_i(k_{i,j,out}) \leq l_j, \tag{22}$$

$$0 \leq v_i(k) \leq v_{max}, \tag{23}$$

$$F_{min} \leq F_i(k) \leq F_{max}, \tag{24}$$

$$\begin{bmatrix} 1 & -1 \end{bmatrix} y_i(k) \geq v_i(k)d_s + \frac{v_i(k)v_{max}}{2a_{max}}, \; i = 2, \ldots, M. \tag{25}$$

To deal with the optimal control problem (18) via a symmetric alternating direction method of multipliers, we need to transform constraints (19)–(25) to linear matrix constraints. Defining $\xi_i(k) = \begin{bmatrix} y_i(k) \\ v_i(k) \end{bmatrix}$, from Equations (20) and (21), we obtain

$$C_i\xi_i(k+1) = G_i\xi_i(k) + D_iF_i(k) + P_i, k = 0, 1, \ldots, N-1, \tag{26}$$

where the matrices of C_i, G_i, D_i, P_i, and $\xi_i(k)$ are given by

$$C_i = \begin{cases} \begin{bmatrix} 1 & 0 \\ 0 & 1 \end{bmatrix}, & i = 1, \\ \begin{bmatrix} 0 & 1 & 0 \\ 0 & 0 & 1 \end{bmatrix}, & i = 2, 3, \ldots, M, \end{cases} \quad G_i = \begin{cases} \begin{bmatrix} 1 & d \\ 0 & 1 \end{bmatrix}, & i = 1, \\ \begin{bmatrix} 0 & 1 & d \\ 0 & 0 & 1 \end{bmatrix}, & i = 2, 3, \ldots, M, \end{cases} \tag{27}$$

$$D_i = \begin{bmatrix} \frac{d^2}{2m_i} \\ \frac{d}{m_i} \end{bmatrix}, P_i = \begin{bmatrix} -\frac{fd^2}{2m_i} \\ -\frac{fd}{m_i} \end{bmatrix},. \tag{28}$$

Here, $F_i(0), \ldots, F_i(N-1)$ and $\xi_i(1), \xi_i(2), \ldots, \xi_i(N)$ are the optimization variables of the problem, and the initial state $\xi_i(0)$ is given. Then, we define the overall optimization variable w_i as $w_i = \left[F_i(0), \ldots, F_i(N-1), \xi_i^T(1), \ldots, \xi_i^T(N)\right]^T$ and reformulate constraint (26) as $A_i w_i = \phi_i$, where

$$A_i = \begin{bmatrix} -D_i & O & O & \cdots & O & C_i & O & \cdots & O & O \\ O & -D_i & O & \cdots & O & -G_i & C_i & \cdots & O & O \\ O & O & -D_i & \cdots & O & O & -G_i & \ddots & O & O \\ \vdots & \vdots & \vdots & \ddots & \vdots & \vdots & \vdots & \ddots & \ddots & \vdots \\ O & O & O & \cdots & -D_i & O & O & \cdots & -G_i & C_i \end{bmatrix}, \tag{29}$$

$$\phi_i = \begin{bmatrix} P_i + G_i\xi_i(0) \\ P_i \\ P_i \\ \vdots \\ P_i \end{bmatrix}. \tag{30}$$

It can be seen that $A_1 \in \mathbb{R}^{2N\times 3N}$, $A_i \in \mathbb{R}^{2N\times 4N}, i = 2,3,\ldots,M, \phi_i \in \mathbb{R}^{2N}$. Inequality (25) can be transformed as $\Upsilon w_i \geq 0$, where

$$\Upsilon = \begin{bmatrix} O & O & \cdots & O & Z & O & \cdots & O \\ O & O & \cdots & O & O & Z & \cdots & O \\ \vdots & \vdots & \ddots & \vdots & \vdots & \vdots & \ddots & \vdots \\ O & O & \cdots & O & O & O & \cdots & Z \end{bmatrix} \in \mathbb{R}^{N\times 4N}, \tag{31}$$

$$Z = \left[1, -1, -d_s - \frac{v_{max}}{2a_{max}}\right]. \tag{32}$$

Constraints (22) and (23) are, respectively, equivalent to $O \leq \xi_i(k) \leq L_{i,j}$, where $L_{i,j} = \begin{cases} \left[l_j, v_{max}\right]^T, & i = 1 \\ \left[l_J, l_j, v_{max}\right]^T, i = 2,\ldots,M \end{cases}$ and $(k_{i,j-1,out}+1) \leq k \leq k_{i,j,out}$. Next, let $\underline{U}_{i,j}$ and $\overline{U}_{i,j}$ denote the lower bound and upper bound of the variable $\xi_i(k)$ for $(k_{i,j-1,out} + 1) \leq k \leq k_{i,j,out}$, respectively. Here, $\overline{U}_{i,j} = [L_{i,j},\ldots,L_{i,j}]^T$, $\underline{U}_{i,j} = [O_i,\ldots,O_i]^T$, $\overline{U}_{1,j} \in \mathbb{R}^{2\kappa_{i,j}}$, $\underline{U}_{1,j} \in \mathbb{R}^{2\kappa_{i,j}}$, $\overline{U}_{i,j} \in \mathbb{R}^{3\kappa_{i,j}}$, $\underline{U}_{i,j} \in \mathbb{R}^{3\kappa_{i,j}}$, $i = 2,\ldots,M$, $\kappa_{i,j} = k_{i,j,out} - k_{i,j-1,out}$, $\sum_{j=1}^{J} \kappa_{i,j} = N$. Then, constraints (22)–(24) can be reformulated into a box constraint of w_i, expressed as $\underline{W_i} \leq w_i \leq \overline{W_i}$, where $\overline{W_i} = \left[F_{max},\ldots,F_{max},\overline{U}_{i,1},\overline{U}_{i,2},\ldots,\overline{U}_{i,J}\right]$, $\underline{W_i} = \left[F_{min},\ldots,F_{min},\underline{U}_{i,1},\underline{U}_{i,2},\ldots,\underline{U}_{i,J}\right]$.

By using w_i instead of the variables (F_i, y_i, v_i) in the objective function, the optimal problem (18) is reformulated as

$$\min \quad \Psi = \sum_{i=1}^{M} (w_i - p_i)^T Q_i (w_i - p_i) \tag{33}$$

$$\text{subject to} \quad K_i w_i = E_i z, \tag{34}$$

$$A_i w_i = \phi_i, \tag{35}$$

$$\Upsilon w_i \geq 0, \quad i = 2,\ldots,M, \tag{36}$$

$$\underline{W_i} \leq w_i \leq \overline{W_i}, \tag{37}$$

where

$$Q_i = \begin{bmatrix} R_i & O & O & \cdots & O \\ O & Q_{i1} & O & \cdots & O \\ O & O & Q_{i2} & \cdots & O \\ \vdots & \vdots & \vdots & \ddots & \vdots \\ O & O & O & \cdots & Q_{iJ} \end{bmatrix} \in \begin{cases} \mathbb{R}^{3N\times 3N}, & i=1 \\ \mathbb{R}^{4N\times 4N}, & i\neq 1 \end{cases} \tag{38}$$

$$R_i = \begin{bmatrix} q_i & 0 & \cdots & 0 \\ 0 & q_i & \cdots & 0 \\ \vdots & \vdots & \ddots & \vdots \\ 0 & 0 & \cdots & q_i \end{bmatrix} \in \mathbb{R}^{N\times N}, Q_{ij} = \begin{bmatrix} O_{\alpha_i} & O & O \\ O & \mathcal{J}_i & O \\ O & O & O_{\beta_{ij}} \end{bmatrix}, \tag{39}$$

$$\mathcal{J}_i = \begin{cases} \begin{bmatrix} a_i & 0 \\ 0 & b_i \end{bmatrix}, & i=1, \\ \begin{bmatrix} 0 & 0 & 0 \\ 0 & a_i & 0 \\ 0 & 0 & b_i \end{bmatrix}, & i\neq 1, \end{cases} \quad \beta_{ij} = \begin{cases} 2(k_{i,j,out}-k_{i,j,in}), & i=1, \\ 3(k_{i,j,out}-k_{i,j,in}), & i\neq 1, \end{cases} \tag{40}$$

$$K_i = \left[\begin{array}{ccc|cccc} O & \cdots & O & V_i & O & \cdots & O \\ O & \cdots & O & O & V_i & \cdots & O \\ \vdots & \ddots & \vdots & \vdots & \vdots & \ddots & \vdots \\ O & \cdots & O & O & O & \cdots & V_i \end{array}\right] \in \begin{cases} \mathbb{R}^{2N\times 3N}, & i=1, \\ \mathbb{R}^{2N\times 4N}, & i\neq 1, \end{cases} \tag{41}$$

$$V_i = \begin{cases} \begin{bmatrix} 1 & 0 \\ 0 & 1 \end{bmatrix}, & i=1, \\ \begin{bmatrix} 1 & 0 & 0 \\ 0 & 1 & 0 \end{bmatrix}, & i\neq 1, \end{cases} \tag{42}$$

$$p_i = \begin{cases} \left[O_{1\times N}, l_1, O_{k_{1,out}-k_{1,in}}\right], i=1,\ldots,M-1, \\ (K_i w_i^{k+1})_{2\ell}, \qquad i=M, \end{cases} \tag{43}$$

The optimization problem (33) could be further formulated as

$$\min \quad \sum_{i=1}^{M} (w_i - p_i)^T Q_i (w_i - p_i) \tag{44}$$

$$\text{subject to} \quad K_i w_i = E_i z, \tag{45}$$

$$w_i \in \mathcal{D}_i, \tag{46}$$

where $\mathcal{D}_i$ denotes constraints (35)–(37).

4. Symmetric Alternating Direction Method of Multipliers

4.1. The Algorithm Framework for the Control Problem

Consider the constrained optimization problem (44)–(46). The augmented Lagrangian associated with the equation constraint is given by

$$\mathcal{L}_\rho(w_i, z, \lambda_i) = \sum_{i=1}^{M} \left[f_i(w_i) + \lambda_i^T (K_i w_i - E_i z) + \frac{\rho}{2}\|K_i w_i - E_i z\|^2\right], \tag{47}$$

where $f_i(w_i) = (w_i - p_i)^T Q_i (w_i - p_i)$. Using the method in [19], the scaled form of SADMM for this problem is

$$w_i^{k+1} = \arg\min_{w_i} (w_i^k)^T (Q_i + \frac{\rho}{2} K_i^T K_i) w_i^k - (2p_i^T Q_i + \rho (z^k)^T K_i - (\lambda^k)^T K_i) w_i^k, \tag{48}$$

$$\lambda_i^{k+\frac{1}{2}} = \lambda_i^k + \rho (K_i w_i^{k+1} - E_i z^k), \tag{49}$$

$$z^{k+1} = \arg\min_z \sum_{i=1}^{M} (-(\lambda_i^{k+\frac{1}{2}})^T E_i z^k + \frac{\rho}{2} \| K_i w_i^{k+1} - E_i z^k \|^2), \tag{50}$$

$$\lambda_i^{k+1} = \lambda_i^{k+\frac{1}{2}} + \rho (K_i w_i^{k+1} - E_i z^{k+1}). \tag{51}$$

SADMM consists of a w_i-minimization step (48), a z-minimization step (50), and dual variable update steps (49) and (51). The dual variable update step (51) uses a step size equal to the augmented Lagrangian parameter ρ, which ensures dual feasibility in each SADMM iteration.

4.1.1. w_i-Minimization Step

The w_i-minimization step (48) solves a quadratic program subject to linear constraints (46). The interior-point approach performs well on this type of problem [25].

4.1.2. z-Minimization Step

A necessary and sufficient condition for z_{opt}^k to be the optimal value of (50) is

$$\frac{\partial \mathcal{L}_\rho}{\partial z_{\text{opt}}^k} = 0, \tag{52}$$

which can be expressed as

$$\sum_{i=1}^{M} E_i^T (\lambda_i^{k+\frac{1}{2}} + \rho (K_i w_i^{k+1} - E_i z_{\text{opt}}^k)) = 0. \tag{53}$$

Let $z_{i,\ell}$ denote the $((i-1)N + \ell)$-th component of the vector z, where $i = 1, 2, \cdots, M$ and $\ell = 1, 2, \cdots, N$. We have

$$z_{i,\ell}^{k+1} = \begin{cases} (E_i z_{\text{opt}}^k)_\ell = (E_{i+1} z_{\text{opt}}^k)_{2\ell-1}, & i = 1, \\ (E_i z_{\text{opt}}^k)_{2\ell} = (E_{i+1} z_{\text{opt}}^k)_{2\ell-1}, & i = 2, \ldots, M-1 \\ (E_i z_{\text{opt}}^k)_{2\ell}, & i = M, \end{cases} \tag{54}$$

Combining (50) and (53), we have

$$z_{i,\ell}^{k+1} = \begin{cases} \frac{1}{2\rho} \bar{\xi}_{i,\ell}^{k+\frac{1}{2}} + \frac{1}{2} \bar{\omega}_{i,\ell}^{k+1}, & i = 1, \ldots, M-1, \\ \frac{1}{\rho} \bar{\xi}_{i,\ell}^{k+\frac{1}{2}} + \bar{\omega}_{i,\ell}^{k+1}, & i = M, \end{cases} \tag{55}$$

where

$$\bar{\xi}_{i,\ell}^{k+\frac{1}{2}} = \begin{cases} (\lambda_i^{k+\frac{1}{2}})_\ell + (\lambda_{i+1}^{k+\frac{1}{2}})_{2\ell-1}, & i = 1, \\ (\lambda_i^{k+\frac{1}{2}})_{2\ell} + (\lambda_{i+1}^{k+\frac{1}{2}})_{2\ell-1}, & i = 2, \ldots, M-1, \\ (\lambda_i^{k+\frac{1}{2}})_{2\ell}, & i = M, \end{cases} \tag{56}$$

$$\bar{\omega}_{i,\ell}^{k+1} = \begin{cases} (K_i w_i^{k+1})_\ell + (K_{i+1} w_{i+1}^{k+1})_{2\ell-1}, & i = 1, \\ (K_i w_i^{k+1})_{2\ell} + (K_{i+1} w_{i+1}^{k+1})_{2\ell-1}, & i = 2, \ldots, M-1, \\ (K_i w_i^{k+1})_{2\ell}, & i = M. \end{cases} \tag{57}$$

$(\lambda_i^{k+\frac{1}{2}})_{2\ell}$ denotes the 2ℓ-th component of the vector $\lambda_i^{k+\frac{1}{2}}$.

Furthermore, the dual variable update step (51), which contains $z_{i,\ell}^{k+1}$, could be expressed as

$$(\lambda_i^{k+1})_{2\ell} = (\lambda_i^{k+\frac{1}{2}})_{2\ell} + \rho((K_i w_i^{k+1})_{2\ell} - z_{i,\ell}^{k+1}). \tag{58}$$

We also have

$$(\lambda_{i+1}^{k+1})_{2\ell-1} = (\lambda_{i+1}^{k+\frac{1}{2}})_{2\ell-1} + \rho((K_{i+1} w_{i+1}^{k+1})_{2\ell-1} - z_{i,\ell}^{k+1}). \tag{59}$$

where $i = 2, \ldots, M-1$. By adding Equations (58) and (59), we have

$$\bar{\xi}_{i,\ell}^{k+1} = \begin{cases} \bar{\xi}_{i,\ell}^{k+\frac{1}{2}} + \rho\bar{\omega}_{i,\ell}^{k+1} - 2\rho z_{i,\ell}^{k+1}, \ i = 1, \ldots, M-1, \\ \bar{\xi}_{i,\ell}^{k+\frac{1}{2}} + \rho\bar{\omega}_{i,\ell}^{k+1} - \rho z_{i,\ell}^{k+1}, \quad i = M, \end{cases} \tag{60}$$

where

$$\bar{\xi}_{i,\ell}^{k+1} = \begin{cases} (\lambda_i^{k+1})_\ell + (\lambda_{i+1}^{k+1})_{2\ell-1}, \ i = 1 \\ (\lambda_i^{k+1})_{2\ell} + (\lambda_{i+1}^{k+1})_{2\ell-1}, i = 2, \ldots, M-1, \\ (\lambda_i^{k+1})_{2\ell}, \qquad\qquad i = M, \end{cases} \tag{61}$$

Substituting Equation (55) into Equation (60), we can find $\bar{\xi}_{i,\ell}^{k+1} = 0$, i.e., the sum of the dual variable entries that correspond to any given global index i, ℓ of variable z is zero. Thus, in the next iteration, the dual variable update step could be written as

$$\bar{\xi}_{i,\ell}^{(k+1)+\frac{1}{2}} = \begin{cases} \rho\bar{\omega}_{i,\ell}^{k+1} - 2\rho z_{i,\ell}^{k+1}, \ i = 1, \ldots, M-1, \\ \rho\bar{\omega}_{i,\ell}^{k+1} - \rho z_{i,\ell}^{k+1}, \quad i = M, \end{cases} \tag{62}$$

Substituting (62) into Equation (55) of the next iteration, we have

$$z_{i,\ell}^{(k+1)+1} = \bar{\omega}_{i,\ell}^{k+1} - z_{i,\ell}^{k+1}. \tag{63}$$

Furthermore, we have

$$z_i^{k+1} = \begin{cases} T_1 K_i w_{i+1}^{k+1} + T_2 K_i w_i^{k+1} - z_i^k, i = 1, 2, \ldots, M-1, \\ 2T_2 K_i w_M^{k+1} - z_i^k, \qquad\qquad i = M, \end{cases} \tag{64}$$

where

$$z_i^{k+1} = \left[z_{i,1}^{k+1}, z_{i,2}^{k+1}, \ldots, z_{i,N}^{k+1} \right]^T, \tag{65}$$

$$T_1 = \begin{bmatrix} B_1^T & O & \cdots & O \\ O & B_1^T & \cdots & O \\ \vdots & \vdots & \ddots & \vdots \\ O & O & \cdots & B_1^T \end{bmatrix} \in \mathbb{R}^{N \times 2N}, \tag{66}$$

$$T_2 = \begin{bmatrix} B_2^T & O & \cdots & O \\ O & B_2^T & \cdots & O \\ \vdots & \vdots & \ddots & \vdots \\ O & O & \cdots & B_2^T \end{bmatrix} \in \mathbb{R}^{N \times 2N}. \tag{67}$$

4.2. Convergence of the SADMM and Stopping Criterion

A necessary and sufficient condition for (w_i^*, z^*, λ^*) to be the convergent point of the solution sequence (w_i^k, z^k, λ^k) is

$$0 \in \partial f(w_i^*) + K_i^T \lambda_i^*, \tag{68}$$

$$0 \in -E_i^T \lambda_i^*, \tag{69}$$

$$K_i w_i^* - E_i z^* = 0. \tag{70}$$

If (w_i^*, z^*) satisfy the optimal conditions (68)–(70), then the algorithm of SADMM converges to an optimal point of the problem (18) [20].

A practical termination criterion for SADMM is that the primal and dual residuals must be smaller than the values ϵ^{pri} and ϵ^{dual}, respectively. That is

$$\|r_i^k\|_2 \le \epsilon^{pri} \text{ and } \|s^k\|_2 \le \epsilon^{dual}, \tag{71}$$

where r_i^k is the primal residual and s^k is the dual residual at iteration k, defined as follows:

$$\|r_i^k\|_2 = \frac{1}{\rho}\|\lambda_i^k - \lambda_i^{k-1}\|_2, \tag{72}$$

$$\|s^k\|_2 = MN\rho^2\|z^{k-1} - z^k\|_2, \tag{73}$$

$\epsilon^{pri} = \sqrt{MN}\epsilon^{abs} + \epsilon^{rel} max\{\|y^k\|_2, \|z^k\|_2\}$ and $\epsilon^{dual} = \sqrt{MN}\epsilon^{abs} + \epsilon^{rel}\|\lambda^k\|_2$. The value ϵ^{abs} is an absolute tolerance and ϵ^{rel} is a relative tolerance. They may be chosen as $\epsilon^{rel} = 10^{-3}$ or 10^{-4} [16]. The proposed SADMM algorithm for optimal control problem (44) is given in Algorithm 1. The dual variable λ_i-updates and the w_i-updates can be carried out for each i. Algorithm 1 decomposes a large optimal control problem into several smaller optimal control problems that can be computed in parallel, thus could improve the overall computation performance.

Algorithm 1 Proposed SADMM for Problem (44)–(46)

1: Initialize $\lambda = 0$, $z = 0$ and $\rho = \frac{1}{2}$;
2: **repeat**
3: $w_i^{k+1} := \arg\min_{w_i} w_i^{kT}(Q_i + \frac{\rho}{2}K_i^T K_i)w_i^k - (2p_i^T Q_i + \rho z^T K_i - \lambda^{kT} K_i)w_i^k$,
4: $\lambda_i^{k+\frac{1}{2}} = \lambda_i^k + \rho(K_i w_i^{k+1} - E_i z^k)$.
5: $z_i^{k+1} = \begin{cases} T_1 K_i w_{i+1}^{k+1} + T_2 K_i w_{i+1}^{k+1} - z_i^k, & i = 1, 2, \dots, M-1, \\ 2T_2 K_i w_M^{k+1} - z_i^k, & i = M, \end{cases}$
6: $\lambda_i^{k+1} = \lambda_i^{k+\frac{1}{2}} + \rho(K_i w_i^{k+1} - E_i z^{k+1})$.
7: **until** $\|r_i^k\|_2 \le \epsilon^{pri}$ *and* $\|s^k\|_2 \le \epsilon^{dual}$.

5. Numerical Simulations

In this section, we give a numerical experiment to illustrate the efficiency of our proposed algorithm. Our experiments are all executed on a computer with an Intel(R) Core (TM) i5-11300H processor (Intel Corporation, Santa Clara, CA, USA)CPU 3.10 GHz and 16 GB memory. The source code is available from the GitHub repository on 14 May 2023 (https://github.com/ShanMa1/operation-control-of-trains.git).

The railway line in our experiment includes six stations and five trains. The speed limit of the trains is 300 km/h (83.3 m/s). The five trains are numbered G1001, G1003, G1005, G1007, and G1009. We assume that the distance between two adjacent stations is 135 km, the operation time of the trains between two adjacent stations is 30 min, and the headway buffer between two adjacent trains is 5 min. The planned timetable is shown in Table 1 and the train parameters are listed in Table 2. The weights a_i, b_i, and q_i in the experiment are chosen as 10^7, 10^7, and 10, respectively. By using the proposed algorithm, our aim is to generate the optimal operation trajectories of trains while guaranteeing the safety and punctuality of trains.

Table 1. Scheduled timetable.

Train \ Station	State	S1	S2	S3	S4	S5	S6
G1001	arrive	8:00	8:28	8:58	9:28	9:58	10:28
	depart	8:00	8:30	9:00	9:30	10:00	10:30
G1003	arrive	-	8:33	9:03	9:33	10:03	10:33
	depart	8:05	8:35	9:05	9:35	10:05	10:35
G1005	arrive	-	8:38	9:08	9:38	10:08	10:38
	depart	8:10	8:40	9:10	9:40	10:10	10:40
G1007	arrive	-	8:43	9:13	9:43	9:13	10:43
	depart	8:15	8:45	9:15	9:45	10:15	10:45
G1009	arrive	-	8:48	9:18	9:48	10:18	10:48
	depart	8:20	8:50	9:20	9:50	10:20	10:50

Table 2. Parameters of high-speed trains [18].

Parameters	Value	Unit
The weight of trains, $m_i, i = 1, 2, \ldots, 5$	450	ton
Maximum acceleration, $a_{i,max}$	0.56	N/kg
Maximum deceleration, $a_{i,min}$	0.8	N/kg
Maximum control force, F_{max}	500	kN
Minimum control force, F_{min}	−110	kN
Resistance force, f	−110	kN
Sampled time period, d	60	s

By solving the optimal train operation control problem via SADMM, the optimal operation trajectory of each train can be obtained as shown in Figure 1. Figure 2 shows the optimal time-distance-speed profiles for 5 trains. Figure 3 shows the accumulated energy consumptions for the five trains under the optimal trajectory. In this figure, the blue lines denote the time-speed profiles, and the yellow lines denote the real-time energy consumption profiles.

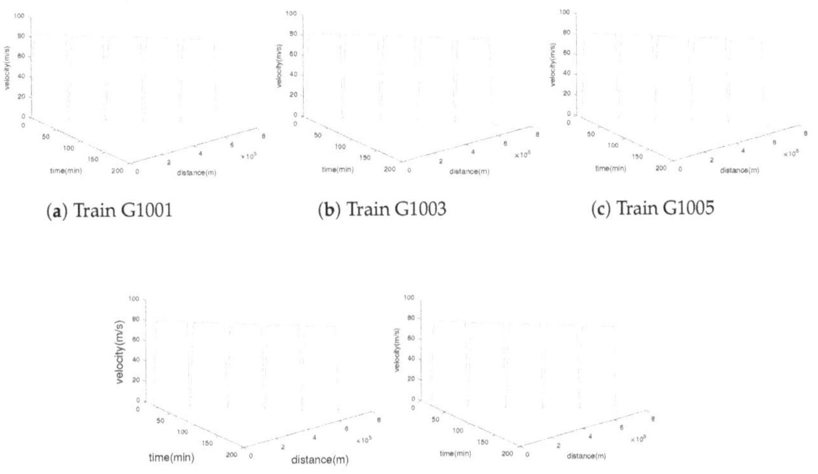

(**a**) Train G1001 (**b**) Train G1003 (**c**) Train G1005

(**d**) Train G1007 (**e**) Train G1009

Figure 1. Optimal trajectory for five high-speed trains.

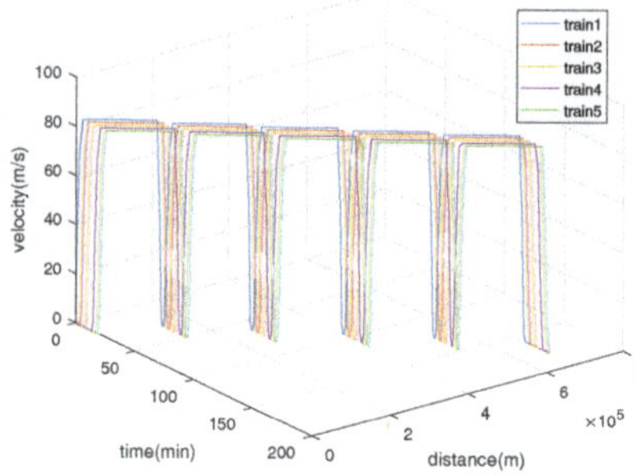

Figure 2. The optimal time–distance–speed profiles for five high-speed trains.

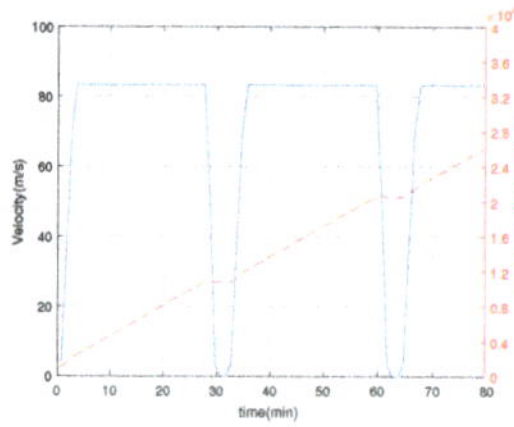

(**a**) Train G1001

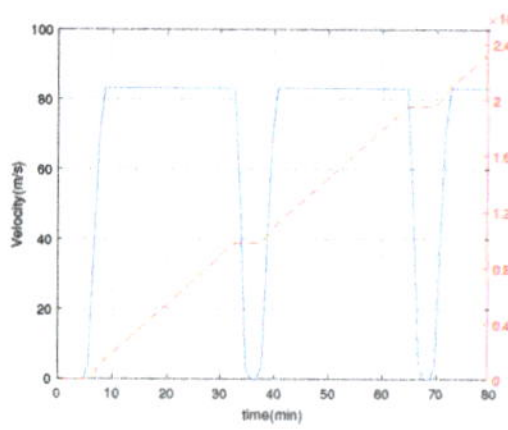

(**b**) Train G1003

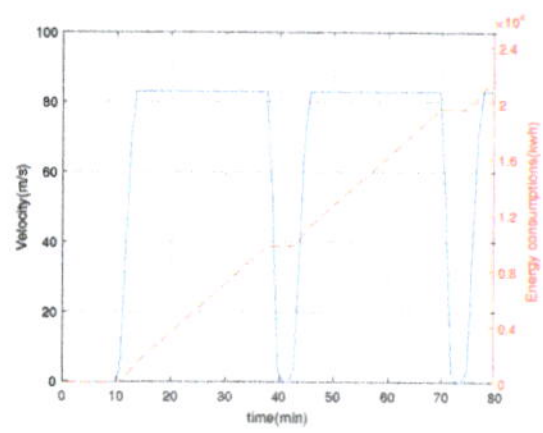

(**c**) Train G1005

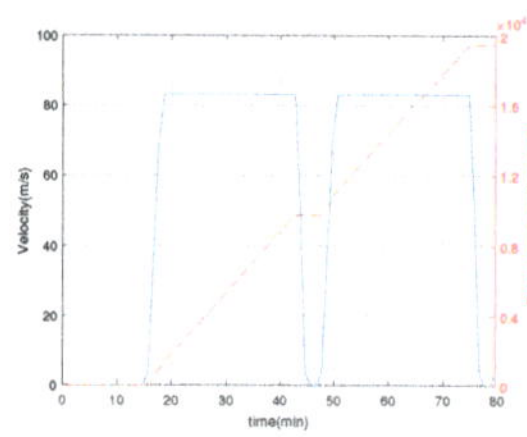

(**d**) Train G1007

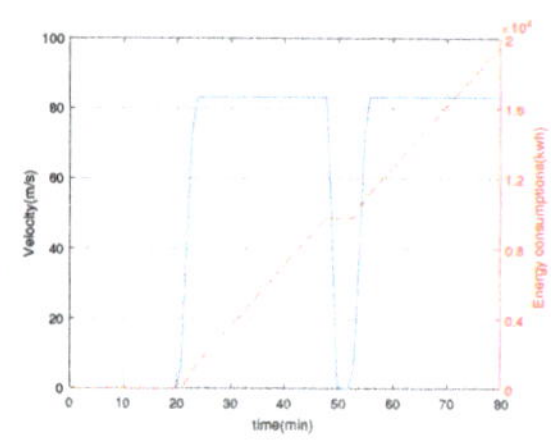

(**e**) Train G1009

Figure 3. Energy consumption for five high-speed trains.

Next, we consider the case that an emergency occurs, such that the first train receives a sudden speed limitation commend between stations $S2$ and $S3$. The speed of the first train is limited to 30 m/s. The duration of the emergency is assumed to be 15 min. By using our proposed algorithm, the optimal distance-time profile is obtained, as shown in Figure 4. In this figure, the dotted line represents the train operation profile without speed limitation, while the black line and red line denote the actual operation profile of the first train and the second train under the emergency, respectively. Since the speed of train $G1001$ is limited to 30 m/s, train $G1003$ has to slow down to keep a safe headway between train $G1001$. In this case, the minimum headway between train $G1001$ and train $G1003$ is 3 km. When the state of emergency is lifted, the speed of train $G1001$ will increase to achieve punctuality. In Figure 4, we can also find that the solution calculated by our proposed algorithm indicates that the trains could keep at least a minimum safe headway under the emergency.

Finally, we compare the computation effectiveness between SADMM and ADMM. The convergences of ADMM and SADMM are given as shown in Figure 5. Figure 6 shows the primal residual versus iterations. In this figure, we can see that SADMM can solve the optimal problem in 4 iterations, while ADMM can solve the optimal problem in 41 iterations. This means that SADMM converges faster than ADMM in our distributed control problem.

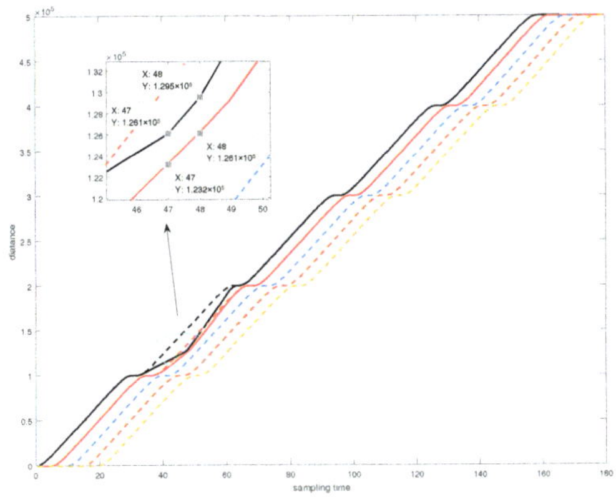

Figure 4. Rescheduling solutions of first train and second train.

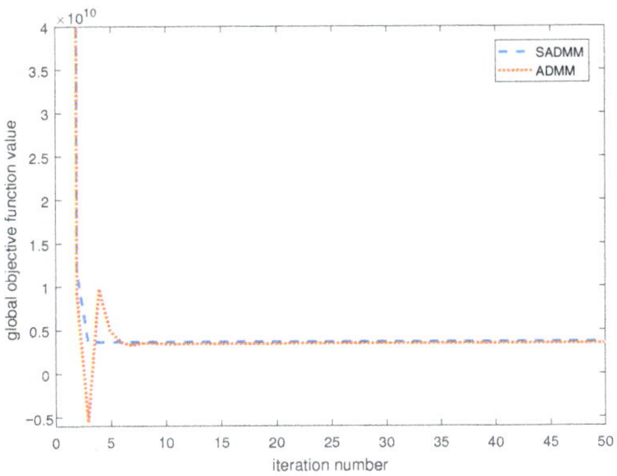

Figure 5. Convergence curves of ADMM and SADMM.

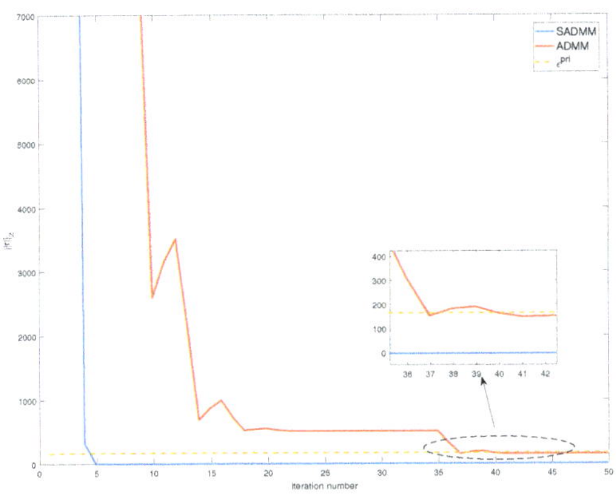

Figure 6. Primal residual versus iterations of ADMM and SADMM.

6. Conclusions

In this paper, for the dynamics of multiple trains with headway constraints and punctuality constraints, a distributed optimal control problem has been formulated to obtain the energy-efficient optimal train operation trajectories. The problem has been transformed into an optimization problem with several constraints. Then, we have proposed an efficient algorithm based on the framework of the symmetric alternating direction method of multipliers (SADMM) to solve this optimization problem. SADMM includes solving two convex optimization problems: the w-minimization problem and the z-minimization problem. The w-minimization problem could be solved by using the interior-point method, and the z-minimization problem could be solved via an analytical formula. Numerical simulations show that SADMM can obtain the optimal train control sequence in fewer iterative steps compared to the alternating direction multiplier method, thus illustrating the effectiveness of the algorithm. The results developed in this paper may have potential applications in the operational control of trains.

In particular, the results may find potential applications in future automatic train operation systems, where trains operate automatically and no driver is needed. In this case, the control inputs are generated according to algorithms, to ensure the punctuality and safety of trains.

Author Contributions: Formal analysis, S.M.; funding acquisition, F.M.; methodology, F.M.; writing—original draft, C.T. and S.M.; writing—review and editing, S.M. and F.M. All authors have read and agreed to the published version of the manuscript.

Funding: This research was funded by the NSF of Shaanxi Province grant 2020JQ-485.

Data Availability Statement: The source code is available from the GitHub repository on 14 May 2023 (https://github.com/ShanMa1/operation-control-of-trains.git).

Conflicts of Interest: The authors declare no conflict of interest.

References

1. Ichikawa, K. Application of optimization theory for bounded state variable problems to the operation of train. *Bull. JSME* **1968**, *11*, 857–865. [CrossRef]
2. Howlett, P. Optimal strategies for the control of a train. *Automatica* **1996**, *32*, 519–532. [CrossRef]
3. Guerra, T.-M.; Aguiar, B.; Berdjag, D.; Demaya, B. Robust estimation for nonlinear continuous-discrete systems with missing outputs: Application to automatic train control. *IEEE Trans. Control. Syst. Technol.* **2022**, *30*, 1304–1310. [CrossRef]
4. Havaei, P.; Sandidzadeh, M.A. Intelligent-PID controller design for speed track in automatic train operation system with heuristic algorithms. *J. Rail Transp. Plan. Manag.* **2022**, *22*, 100321. [CrossRef]
5. Ichikawa, S.; Miyatake, M. Energy efficient train trajectory in the railway system with moving block signaling scheme. *IEEJ J. Ind. Appl.* **2019**, *8*, 586–591. [CrossRef]
6. Sato, K.; Kato, H.; Fukushima, T. Development of SiC applied traction system for next-generation Shinkansen high-speed trains. *IEEJ J. Ind. Appl.* **2018**, *9*, 453–459. [CrossRef]
7. Nallaperuma, S.; Fletcher, D.; Harrison, R. Optimal control and energy storage for DC electric train systems using evolutionary algorithms. *Railw. Eng. Sci.* **2021**, *29*, 327–335. [CrossRef]
8. Khmelnitsky, E. On an optimal control problem of train operation. *IEEE Trans. Autom. Control* **2000**, *45*, 1257–1266. [CrossRef]
9. Bai, W.; Lin, Z.; Dong, H. Coordinated control in the presence of actuator saturation for multiple high-speed trains in the moving block signaling system mode. *IEEE Trans. Veh. Technol.* **2020**, *69*, 8054–8064. [CrossRef]
10. Li, S.; Yang, L.; Li, K.; Gao, Z. Robust sampled-data cruise control scheduling of high speed train. *Transp. Res. Part C Emerg. Technol.* **2014**, *46*, 274–283. [CrossRef]
11. Li, S.; Yang, L.; Gao, Z. Coordinated cruise control for high-speed train movements based on a multi-agent model. *Transp. Res. Part C Emerg. Technol.* **2015**, *56*, 281–292. [CrossRef]
12. Li, S.; Yang, L.; Li, K.; Gao, Z. Adaptive coordinated control of multiple high-speed trains with input saturation. *Nonlinear Dyn.* **2016**, *83*, 2157–2169. [CrossRef]
13. Lin, F.; Fardad, M.; Jovanovic, M.R. Optimal control of vehicular formations with nearest neighbor interactions. *IEEE Trans. Autom. Control* **2012**, *57*, 2203–2218. [CrossRef]
14. Yan, X.; Cai, B.; Ning, B.; ShangGuan, W. Online distributed cooperative model predictive control of energy-saving trajectory planning for multiple high-speed train movements. *Transp. Res. Part C Emerg. Technol.* **2016**, *69*, 60–78. [CrossRef]
15. Wang, Y.; De Schutter, B.; van den Boom, T.J.; Ning, B. Optimal trajectory planning for trains-a pseudospectral method and a mixed integer linear programming approach. *Transp. Res. Part C Emerg. Technol.* **2013**, *29*, 97–114. [CrossRef]
16. Boyd, S.; Parikh, N.; Eric, C.; Peleato, B.; Eckstein, J. Distributed optimization and statistical learning via the alternating direction method of multipliers. *Found. Trends Mach. Learn.* **2011**, *3*, 1–122. [CrossRef]
17. Lin, F.; Fardad, M.; Jovanović, M.R. Design of optimal sparse feedback gains via the alternating direction method of multipliers. *IEEE Trans. Autom. Control* **2013**, *58*, 2426–2431. [CrossRef]

18. Li, S.; Yang, L.; Gao, Z. Distributed optimal control for multiple high-speed train movement: An alternating direction method of multipliers. *Automatica* **2020**, *112*, 108646–108654. [CrossRef]
19. He, B.; Liu, H.; Wang, Z.; Yuan, X. A strictly contractive peaceman-rachford splitting method for convex programming. *SIAM J. Optim.* **2014**, *24*, 1011–1040. [CrossRef]
20. He, B.; Ma, F.; Yuan, X. Convergence study on the symmetric version of ADMM with larger step sizes. *SIAM J. Imaging Sci.* **2016**, *9*, 1467–1501. [CrossRef]
21. He, B.; Ma, F.; Yuan, X. Optimally linearizing the alternating direction method of multipliers for convex programming. *Comput. Optim. Appl.* **2020**, *75*, 361–388. [CrossRef]
22. Gao, B.; Ma, F. Symmetric alternating direction method with indefinite proximal regularization for linearly constrained convex optimization. *J. Optim. Theory Appl.* **2018**, *176*, 178–204. [CrossRef]
23. Jiao, Y.; Jin, Q.; Lu, X.; Wang, W. Alternating direction method of multipliers for linear inverse problems. *SIAM J. Numer. Anal.* **2016**, *54*, 2114–2137. [CrossRef]
24. Yang, L.; Luo, J.; Xu, Y.; Zhang, Z.; Dong, Z. A distributed dual consensus ADMM based on partition for dc-dopf with carbon emission trading. *IEEE Trans. Ind. Inform.* **2019**, *16*, 1858–1872. [CrossRef]
25. Nocedal, J.; Wright, S.J.; Mikosch, T.V.; Resnick, S.I.; Robinson, S.M. *Numerical Optimization*; Springer: Berlin/Heidelberg, Germany, 1999.

axioms
MDPI

Article

An Analytic Solution for 2D Heat Conduction Problems with General Dirichlet Boundary Conditions

Heng-Pin Hsu [1], Te-Wen Tu [2] and Jer-Rong Chang [1,*]

[1] Department of Aircraft Engineering, Air Force Institute of Technology, 1 Associate Jyulun Road, Gang-Shan District, Kaoshiung City 820, Taiwan
[2] Department of Mechanical Engineering, Air Force Institute of Technology, 1 Associate Jyulun Road, Gang-Shan District, Kaoshiung City 820, Taiwan
* Correspondence: jerrong.chang@gmail.com; Tel.: +886-7-6256040

Abstract: This paper proposed a closed-form solution for the 2D transient heat conduction in a rectangular cross-section of an infinite bar with the general Dirichlet boundary conditions. The boundary conditions at the four edges of the rectangular region are specified as the general case of space–time dependence. First, the physical system is decomposed into two one-dimensional subsystems, each of which can be solved by combining the proposed shifting function method with the eigenfunction expansion theorem. Therefore, through the superposition of the solutions of the two subsystems, the complete solution in the form of series can be obtained. Two numerical examples are used to investigate the analytic solution of the 2D heat conduction problems with space–time-dependent boundary conditions. The considered space–time-dependent functions are separable in the space–time domain for convenience. The space-dependent function is specified as a sine function and/or a parabolic function, and the time-dependent function is specified as an exponential function and/or a cosine function. In order to verify the correctness of the proposed method, the case of the space-dependent sinusoidal function and time-dependent exponential function is studied, and the consistency between the derived solution and the literature solution is verified. The parameter influence of the time-dependent function of the boundary conditions on the temperature variation is also investigated, and the time-dependent function includes harmonic type and exponential type.

Keywords: analytic solution; 2D heat conduction; space–time-dependent dependent; Dirichlet boundary conditions; shifting function method

MSC: 35K05; 80M99

check for updates

Citation: Hsu, H.-P.; Tu, T.-W.; Chang, J.-R. An Analytic Solution for 2D Heat Conduction Problems with General Dirichlet Boundary Conditions. *Axioms* **2023**, *12*, 416. https://doi.org/10.3390/axioms12050416

Academic Editors: Leonid Plotnikov and Giovanni Mascali

Received: 2 March 2023
Revised: 8 April 2023
Accepted: 20 April 2023
Published: 24 April 2023

1. Introduction

The application of heat conduction problems with time-dependent boundary conditions can be broadly applied in a wide range of engineering fields, such as time-varying heating on walls or plate panels, laser heating on solids, and the design of mechanical parts (such as those in turbines and engines [1,2]). In general, the types of time-dependent boundary conditions at the boundary surface include (1) the first type: specified temperature distribution (Dirichlet boundary condition); (2) the second type: specified heat flux distribution; and (3) the third type: convective heat exchange with the environment at a specified temperature. There are many methods to solve these three types of problems, such as pure numerical method, approximate method, and exact method. The literature review focuses on the study of 1D and 2D transient heat conduction problems with various time-dependent boundary conditions, as shown below.

For one-dimensional heat conduction problems with different kinds of time-dependent boundary conditions, these problems cannot be solved directly by the variable separation method due to the nonhomogeneity of boundary conditions. In the early 1970s, Ivanov and Salomatov [3,4] and Postol'Nik [5] were the first to transform the governing differential equation of the linear one-dimensional system into a nonlinear equation by introducing new variables. After neglecting the nonlinear terms, they obtained an approximated solution that they said was valid for the system with the Biot number less than 0.25. At the same time, Kozlov [6] used the Laplace transformation technique to study the problem with the Biot function in a rational combination of sine, cosine, polynomial and exponential functions. Although exact series solutions can be obtained for a given transformation system, great difficulties arise in the inversion of the transformation function, which is often not so straightforward. In addition, various approximation methods such as the iterative perturbation method [7], the eigenfunction expansion method [8], and the Lie point symmetry analysis method [9] have been used to study such heat conduction problems. Later in 2010, Lee and colleagues [10–12] proposed an integration-free solution method, which is an extension of the shifting function method developed in their previous research [13], to derive an analytic closed solution for the heat conduction with time-dependent boundary conditions of the second and the third types. Using the same method, they [14–16] had successfully performed one-dimensional inverse estimation of the heat treatment problem with unknown time-dependent boundary conditions of various types.

For the two-dimensional heat conduction problems with time-dependent boundary conditions, a considerable amount of work can be found in the literature on the development of exact, approximate, and numerical methods. In some advanced heat conduction books [17–19], some classical techniques such as Laplace transform, Duhamel's theorem, and Green's function have been proposed to solve them. Applying the Laplace transform starts with finding the solution of a 2D problem with nonhomogeneous boundary conditions in the transformed domain. Taking the inverse Laplace transform from the complex domain always has difficulties. The typical surveys included Zhu [20], Zhu et al. [21], and Sutradhar et al. [22] who dealt with the time derivative term in the diffusion equation by using the Laplace transform techniques. On the other hand, using Duhamel's theorem [17], an auxiliary 2D problem with associated nonhomogeneous boundary conditions must first be solved. Therefore, the result will be obtained by differentiating under the integration. Similarly, Green's function solution method [17] requires the derivation of the associated Green's function, which satisfies a differential equation with a delta function and homogeneous boundary conditions. To obtain the general solution, the associated Green's function must be directionally differentiated and integrated over the space and time domains. In addition, some numerical techniques, such as finite difference method and boundary element method, have also been used to solve 2D heat conduction problems with time-dependent boundary conditions. Bulgakov et al. [23] used the finite difference method to advance the solution in the time domain with the numerical schemes based on the boundary element method, while Walker [24] applied the diffusion fundamental solution combined with the time integration to solve the diffusion equation. Later, Chen et al. [25] applied the method of fundamental solutions for diffusion equations by using the modified Helmholtz fundamental solution. Burgess and Mahajerin [26] used the fundamental collocation method to solve the problems of arbitrary shapes subjected to arbitrary initial conditions and mixed time-dependent boundary conditions. The time-dependent fundamental solutions for diffusion equations were directly used by Young et al. [27] to obtain the solution as a linear combination of the fundamental solution of the diffusion operator. On the other hand, Cole and Yen [28] involved the method of Green's function to obtain fast-converging expressions for the temperature and heat flux in a rectangular plate. Beck et al. [29] have developed the transient temperatures of the plates under time-varying heating conditions to an integer power at a surface. Lei et al. [30] presented a space–time generalized finite difference method (GFDM) to solve the transient heat conduction problem by integrating direct space–time discretization techniques into the meshless GFDM. Alam et al. [31] proposed a novel generalized (G′/G) extension technique for two nonlinear evolution equations: the (2+1) dimensional Konopelchenko–Dubrovsky (KD) equation and the (2+1) dimensional Kadomtsev–Petviashvili (KP) equations and obtained some new precise answers. The secured answers include a particular variety of solitary wave solutions. Islam et al. [32]

applied a modified (G′/G) expansion method to seek new calculations of the Zakharov–Kuznetsov (ZK) equations developed in electrical engineering. Illustrated by 3D and contour plots, the mathematical results clearly demonstrate the complete honesty and high performance of the proposed algorithm. Krishnan et al. [33] proposed a technique based on eigenfunction expansion for solving 1D phase change heat transfer problems with time-dependent temperature or heat flux boundary conditions. By using Duhamel's theorem, Belekar et al. [34] derived an analytic solution for the transient axisymmetric temperature distribution in a cylindrical geometry with time-dependent boundary conditions.

To the best of the authors' knowledge, there is no literature formulating an analytic solution for the 2D heat conduction problems with the general Dirichlet boundary conditions specifying space–time-dependent dependent boundary conditions at the four edges of rectangular region. This paper develops a simplified exact solution method for the transient heat conduction in a rectangular cross-section of an infinite bar with space–time-dependent dependent boundary conditions using the shifting function method proposed by Lee and colleagues [10–16]. The study focuses on the 2D heat conduction problems with the general Dirichlet boundary conditions. For the two-dimensional problem, the original two-dimensional system is separated into two independent one-dimensional subsystems. The boundary conditions of the subsystems can then be changed from nonhomogeneous to homogeneous using the shifting function method, and an analytic solution can be derived using the eigenfunction expansion theorem. The solutions obtained from the two separate subsystems are combined to construct the solution of the original two-dimensional system. Finally, a numerical example is given, and the correctness of the obtained solution is verified via comparison with the literature [27]. Other case studies illustrate the feasibility of this approach.

The contributions of this paper are as follows:

(1) Lee and colleagues [10–16] used the shifting function method to derive an analytic solution for the heat conduction with time-dependent boundary conditions. They also performed an inverse estimation of a heat treatment problem with unknown time-dependent boundary conditions. However, their research is limited to the scope of one-dimensional heat conduction problems. The greatest contribution of this work is the first investigation of the analytic solution to 2D heat conduction problems with the general Dirichlet boundary conditions by using the proposed method, combining the shifting function method with the expansion theorem method. The applicability of the present method is in solving the heat conduction problems of a rectangular cross-section of an infinite rod with specified space–time-dependent dependent boundary conditions at the four edges of the rectangular region;

(2) Some advanced heat conduction books [17–19] proposed some classical techniques such as the Laplace transform, Duhamel's theorem, and Green's function to solve the heat conduction problem. However, they are limited to the integration situation during the solution process. The correctness of the solution in this study is verified by comparing it with the results of Young et al. [27]. To the best of the authors' knowledge, the other cases in this paper have never been presented in past studies. Although the number of series expansion terms determines the accuracy of the solution, the case study shows that the proposed method has good convergence to the solution using series expansion and can quickly reach a convergence value. The influence of the parameters of the time-dependent boundary function on the temperature variation is also studied.

2. Mathematical Modeling

Consider the transient heat conduction for a rectangular cross-section in an infinite bar with the space–time-dependent Dirichlet boundary conditions on its four sides and no heat generation in the medium. Figure 1 shows the geometry, the boundary conditions and initial condition of a rectangular cross-section in an infinite bar. The governing equation, boundary conditions and initial condition of the problem are as follows:

$$k\left[\frac{\partial^2 T(x,y,t)}{\partial x^2}+\frac{\partial^2 T(x,y,t)}{\partial y^2}\right]=\rho c\frac{\partial T(x,y,t)}{\partial t} \quad \text{in } 0<x<L_x,\ 0<y<L_y,\ t>0, \tag{1}$$

$$T(0,y,t)=f_1(y,t) \quad \text{at } x=0,\quad 0\le y\le L_y, \tag{2}$$

$$T(L_x,y,t)=f_2(y,t) \quad \text{at } x=L_x,\quad 0\le y\le L_y, \tag{3}$$

$$T(x,0,t)=f_3(x,t) \quad \text{at } y=0,\quad 0\le x\le L_x, \tag{4}$$

$$T(x,L_y,t)=f_4(x,t) \quad \text{at } y=L_y,\quad 0\le x\le L_x, \tag{5}$$

$$T(x,y,0)=T_0(x,y) \quad \text{at } t=0,\quad 0\le x\le L_x,\quad 0\le y\le L_y \tag{6}$$

where $T(x,y,t)$ denotes the temperature function, x and y are the two-dimensional space variables, L_x and L_y are the thicknesses of the rectangular region at x and y directions, respectively, and t is the time variable. In addition, k is the thermal conductivity, ρ is the mass density, and c is the specific heat. It is noted that $f_i(y,t)$ $i=1$, 2 and $f_i(x,t)$ $i=3$, 4 denote the general case of space–time-dependent temperatures prescribed along the surfaces at the left and right ends and bottom and top ends, respectively. Furthermore, considering the matching of the boundary conditions with the initial conditions, one has

$$f_1(y,0)=T_0(0,y),\quad f_2(y,0)=T_0(L_x,y),\quad f_3(x,0)=T_0(x,0),\quad f_4(x,0)=T_0(x,L_y). \tag{7}$$

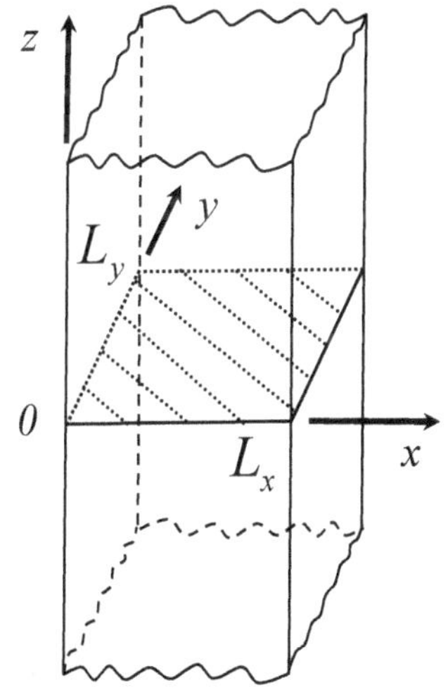

(a)

Figure 1. *Cont.*

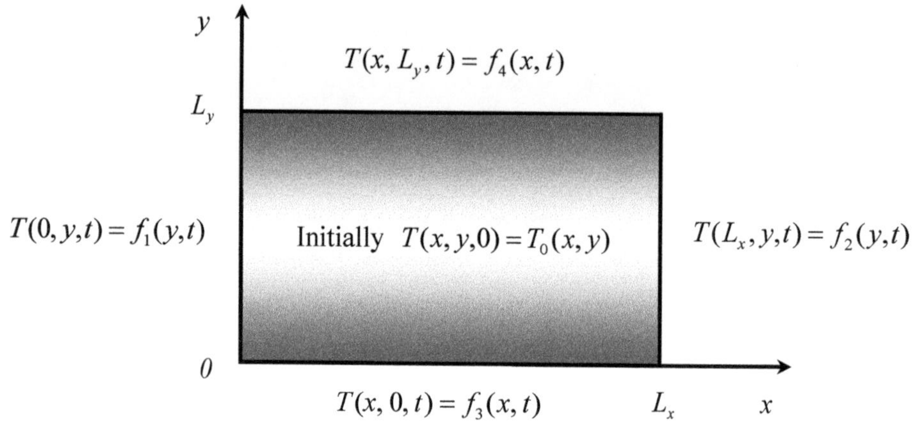

(**b**)

Figure 1. The 2D heat conduction in a rectangular cross-section of an infinite bar with general Dirichlet boundary conditions. (**a**) An infinite bar with a rectangular cross-section. (**b**) The 2D heat transfer system in a rectangular region (cross-section).

3. The Solution Methodology

A dimensionless form of the 2D heat conduction system is first derived and split into two subsystems, each of which can be solved as a 1D problem. By properly introducing the shifting functions, the second-order governing differential equation with space–time-dependent boundary conditions are transformed into the differential equation with homogeneous boundary conditions.

3.1. The Dimensionless Form of Physical System

The dimensionless parameters are defined as follows:

$$\theta(X,Y,\tau) = \frac{T(x,y,t)}{T_r},\ \tau = \frac{\alpha t}{L_y^2},\ X = \frac{x}{L_x},\ Y = \frac{y}{L_y},\ L_r = \frac{L_y}{L_x},\ F_1(Y,\tau) = \frac{f_1(y,t)}{T_r},$$

$$F_2(Y,\tau) = \frac{f_2(y,t)}{T_r},\ F_3(X,\tau) = \frac{f_3(x,t)}{T_r},\ F_4(X,\tau) = \frac{f_4(x,t)}{T_r},\ \theta_0(X,Y) = \frac{T_0(x,y)}{T_r}. \quad (8)$$

The dimensionless form of the boundary-initial value problem is derived as follows:

$$\left[L_r^2 \frac{\partial^2\theta(X,Y,\tau)}{\partial X^2} + \frac{\partial^2\theta(X,Y,\tau)}{\partial Y^2}\right] = \frac{\partial\theta(X,Y,\tau)}{\partial\tau} \quad \text{in } 0 < X < 1,\ 0 < Y < 1,\ \tau > 0, \quad (9)$$

$$\theta(0,Y,\tau) = F_1(Y,\tau) \quad \text{at } X = 0, \quad 0 \le Y \le 1, \quad (10)$$

$$\theta(1,Y,\tau) = F_2(Y,\tau) \quad \text{at } X = 1, \quad 0 \le Y \le 1, \quad (11)$$

$$\theta(X,0,\tau) = F_3(X,\tau) \quad \text{at } Y = 0, \quad 0 \le X \le 1, \quad (12)$$

$$\theta(X,1,\tau) = F_4(X,\tau) \quad \text{at } Y = 1, \quad 0 \le X \le 1, \quad (13)$$

$$\theta(X,Y,0) = \theta_0(X,Y) \quad \text{at } \tau = 0, \quad 0 \le X \le 1,\ 0 \le Y \le 1, \tag{14}$$

$$F_1(Y,0) = \theta_0(0,Y),\ \ F_2(Y,0) = \theta_0(1,Y),\ \ F_3(X,0) = \theta_0(X,0),\ \ F_4(X,0) = \theta_0(X,1) \tag{15}$$

where the parameter $\alpha = \frac{k}{\rho c}$ in Equation (8) represents the thermal diffusivity and T_r is the reference temperature.

3.2. Principle of Superposition

Due to the linear property of the boundary value problem, the physical system can be divided into two subsystems, A and B along the X and Y directions by using the superposition principle, as shown in Figure 2; $\theta(X,Y,\tau)$ is spilt into two parts as follows:

$$\theta(X,Y,\tau) = \theta_a(X,Y,\tau) + \theta_b(X,Y,\tau). \tag{16}$$

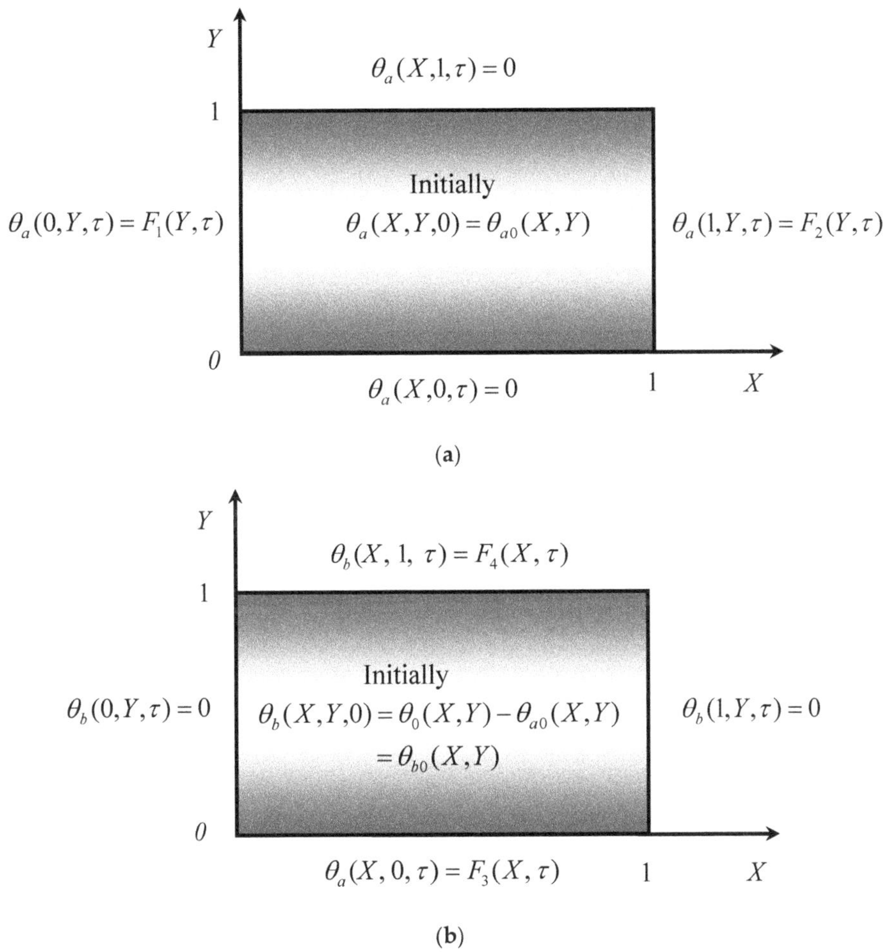

Figure 2. The two subsystems of the two-dimensional heat conduction system with general Dirichlet boundary conditions. (**a**) For subsystem A. (**b**) For subsystem B.

For the subsystem A, the governing equation, boundary conditions and initial condition for the heat conduction problem are

$$\left[L_r^2\frac{\partial^2\theta_a(X,Y,\tau)}{\partial X^2}+\frac{\partial^2\theta_a(X,Y,\tau)}{\partial Y^2}\right]=\frac{\partial\theta_a(X,Y,\tau)}{\partial\tau}\quad \text{in } 0<X<1,\ 0<Y<1,\ \ \tau>0, \tag{17}$$

$$\theta_a(0,Y,\tau)=F_1(Y,\tau)\quad \text{at } X=0,\quad 0\le Y\le 1, \tag{18}$$

$$\theta_a(1,Y,\tau)=F_2(Y,\tau)\quad \text{at } X=1,\quad 0\le Y\le 1, \tag{19}$$

$$\theta_a(X,0,\tau)=0\quad \text{at } Y=0,\quad 0\le X\le 1, \tag{20}$$

$$\theta_a(X,1,\tau)=0\quad \text{at } Y=1,\quad 0\le X\le 1, \tag{21}$$

$$\theta_a(X,Y,0)=\theta_{a0}(X,Y)\quad \text{at } \tau=0,\quad 0\le X\le 1,\ 0\le Y\le 1. \tag{22}$$

Likewise, for the subsystem B, the governing equation, boundary conditions and initial condition for the heat conduction problem are

$$\left[L_r^2\frac{\partial^2\theta_b(X,Y,\tau)}{\partial X^2}+\frac{\partial^2\theta_b(X,Y,\tau)}{\partial Y^2}\right]=\frac{\partial\theta_b(X,Y,\tau)}{\partial\tau}\quad \text{in } 0<X<1,\ 0<Y<1,\ \ \tau>0, \tag{23}$$

$$\theta_b(0,Y,\tau)=0\quad \text{at } X=0,\quad 0\le Y\le 1, \tag{24}$$

$$\theta_b(1,Y,\tau)=0\quad \text{at } X=1,\quad 0\le Y\le 1, \tag{25}$$

$$\theta_b(X,0,\tau)=F_3(X,\tau)\quad \text{at } Y=0,\quad 0\le X\le 1, \tag{26}$$

$$\theta_b(X,1,\tau)=F_4(X,\tau)\quad \text{at } Y=1,\quad 0\le X\le 1, \tag{27}$$

$$\theta_b(X,Y,0)=\theta_0(X,Y)-\theta_{a0}(X,Y)=\theta_{b0}(X,Y)\quad \text{at } \tau=0,\ \ 0\le X\le 1,\ 0\le Y\le 1. \tag{28}$$

For the two similar subsystems, first solve the subsystem A, and then solve the subsystem B, listed in the Appendix A for brevity.

3.3. Reduced to One-Dimensional Problem

Considering the two homogeneous boundary conditions at the opposite edges of the rectangular region, namely, $Y=0$ and $Y=1$, one can reasonably assume that the temperature $\theta_a(X,Y,\tau)$ and dimensionless quantities $F_i(Y,\tau)$ $(i=1,2)$ defined in Equations (18) and (19) are

$$\theta_a(X,Y,\tau)=\sum_{m=1}^{\infty}\left[\theta_m(X,\tau)\sin(m\pi Y)\right], \tag{29}$$

$$F_i(Y,\tau)=\sum_{m=1}^{\infty}\left[\overline{F}_{i,m}(\tau)\sin(m\pi Y)\right],\quad (i=1,2) \tag{30}$$

where $\overline{F}_{i,m}(\tau)$ $(i=1,2)$ is defined as

$$\overline{F}_{i,m}(\tau)=2\int_0^1 F_i(Y,\tau)\sin(m\pi Y)dY,\ \ (i=1,2). \tag{31}$$

Thus, $\theta_m(X,\tau)$ in Equation (29) is determined by satisfying the boundary conditions on both sides $X = 0$ and $X = 1$ (Equations (18) and (19)) and the governing equation. After substituting Equations (29) and (30) back into Equations (17)–(19), we can obtain the following results

$$\frac{\partial \theta_m(X,\tau)}{\partial \tau} - L_r^2 \frac{\partial^2 \theta_m(X,\tau)}{\partial X^2} + m^2\pi^2\theta_m(X,\tau) = 0 \quad \text{in } 0 < X < 1, \quad \tau > 0, \tag{32}$$

$$\theta_m(0,\tau) = \overline{F}_{1,m}(\tau) \quad \text{at } X = 0, \tag{33}$$

$$\theta_m(1,\tau) = \overline{F}_{2,m}(\tau) \quad \text{at } X = 1, \tag{34}$$

$$\theta_m(X,0) = 2\int_0^1 \theta_{a0}(X,Y)\sin(m\pi Y)dY \quad \text{at } \tau = 0. \tag{35}$$

3.4. The Shifting Function Method

3.4.1. Change of Variable

To solve the second-order partial differential equation (Equation (32)) with nonhomogeneous boundary conditions (Equations (33) and (34)), the shifting function method developed by Lee and colleagues [10] is extended by employing the following transformation equation

$$\theta_m(X,\tau) = \overline{\theta}_m(X,\tau) + \sum_{i=1}^{2} [g_{i,m}(X)\overline{F}_{i,m}(\tau)]. \tag{36}$$

where $\overline{\theta}_m(X,\tau)$ is a transformed function and $g_{i,m}(X)$ $(i = 1, 2)$ represents the two shift functions that need to be specified.

Substituting Equation (36) into Equations (32)–(34) can obtain

$$\begin{gathered}\dot{\overline{\theta}}_m(X,\tau) + \sum_{i=1}^{2} [g_{i,m}(X)\dot{\overline{F}}_{i,m}(\tau)] - L_r^2\{\overline{\theta}''{}_m(X,\tau) + \sum_{i=1}^{2} [g''_{i,m}(X)\overline{F}_{i,m}(\tau)]\} \\ +m^2\pi^2\{\overline{\theta}_m(X,\tau) + \sum_{i=1}^{2} [g_{i,m}(X)\overline{F}_{i,m}(\tau)]\} = 0\end{gathered} \tag{37}$$

where the double primes are used to represent the twice differentiation with respect to X and the dots represent the differentiation with respect to τ, respectively.

The associated boundary conditions become

$$\overline{\theta}_m(0,\tau) + \sum_{i=1}^{2} [g_{i,m}(0)\overline{F}_{i,m}(\tau)] = \overline{F}_{1,m}(\tau), \tag{38}$$

$$\overline{\theta}_m(1,\tau) + \sum_{i=1}^{2} [g_{i,m}(1)\overline{F}_{i,m}(\tau)] = \overline{F}_{2,m}(\tau). \tag{39}$$

3.4.2. The Shifting Functions

For the convenience of analysis, the shifting functions are specially selected so that they satisfy the following differential equations and boundary conditions

$$g''_{i,m}(X) = 0, \quad i = 1, 2, \quad 0 < X < 1, \tag{40}$$

$$g_{i,m}(0) = \delta_{i1}, \quad g_{i,m}(1) = \delta_{i2} \tag{41}$$

where δ_{ij} is the Kronecker delta. Therefore, two shifting functions can be easily determined as

$$g_{1,m}(X) = 1 - X, \quad g_{2,m}(X) = X \tag{42}$$

After substituting Equations (40)–(42) back into Equations (37)–(39) yields a differential equation for $\overline{\theta}_m(X,\tau)$ as below:

$$\dot{\overline{\theta}}_m(X,\tau) - L_r^2\overline{\theta}''{}_m(X,\tau) + m^2\pi^2\overline{\theta}_m(X,\tau) = \overline{G}_m(X,\tau) \tag{43}$$

and the homogeneous boundary conditions become

$$\overline{\theta}_m(0,\tau) = 0, \qquad \overline{\theta}_m(1,\tau) = 0 \tag{44}$$

where $\overline{G}_m(X,\tau)$ in Equation (43) is defined as

$$\overline{G}_m(X,\tau) = -\sum_{i=1}^{2}\{g_{i,m}(X)[\dot{\overline{F}}_{i,m}(\tau) + m^2\pi^2\overline{F}_{i,m}(\tau)]\}. \tag{45}$$

Moreover, the initial condition can be transformed as

$$\overline{\theta}_m(X,0) = 2\int_0^1 \theta_{a0}(X,Y)\sin(m\pi Y)dY - \sum_{i=1}^{2}[g_{i,m}(X)\overline{F}_{i,m}(0)]. \tag{46}$$

3.4.3. The Eigenfunction Expansion Theorem

The solution $\overline{\theta}_m(X,\tau)$ specified by Equations (43) and (44) can be expressed by applying the method of separation variable as

$$\overline{\theta}_m(X,\tau) = \sum_{n=1}^{\infty}[\overline{\theta}_{mn}(X)T_{mna}(\tau)] \tag{47}$$

where the space variable $\overline{\theta}_{mn}(X)$ satisfies the following Sturm–Liouville eigenvalue problem

$$\overline{\theta}''_{mn}(X) + \omega_n^2\overline{\theta}_{mn}(X) = 0, \qquad 0 < X < 1, \tag{48}$$

$$\overline{\theta}_{mn}(0) = 0 \quad \text{at} \;\; X = 0, \tag{49}$$

$$\overline{\theta}_{mn}(1) = 0 \quad \text{at} \;\; X = 1. \tag{50}$$

It is noted that the eigenfunctions $\overline{\theta}_{mn}(X)$ $(n = 1, 2, 3, \cdots)$ and the corresponding eigenvalues are

$$\overline{\theta}_{mn}(X) = \sin\omega_n X, \quad \omega_n = n\pi, \quad (n = 1, 2, 3, \cdots) \tag{51}$$

In addition, the eigenfunctions form an orthogonal set in the interval [0, 1] as

$$\int_0^1 \overline{\theta}_{mi}(X)\,\overline{\theta}_{mj}(X)dX = \begin{cases} 0 & \text{for } i \neq j, \\ \frac{1}{2} & \text{for } i = j. \end{cases} \tag{52}$$

Substituting Equation (47) into Equation (43), multiplying it by $\overline{\theta}_{mn}(X)$, and integrating from 0 to 1, one will obtain the following differential equation

$$\dot{T}_{mna}(\tau) + \lambda_{mna}^2 T_{mna}(\tau) = \gamma_{mna}(\tau) \tag{53}$$

where λ_{mna} and $\gamma_{mna}(\tau)$ are given as

$$\lambda_{mna} = \sqrt{m^2 + n^2L_r^2}\pi, \tag{54}$$

Therefore, one can obtain

$$T_{11a}(\tau) = \frac{4}{\pi}e^{-2\pi^2\tau} - \frac{2}{\pi}[\eta_1(\tau) + \eta_2(\tau)] + 2\pi\int_0^{\tau} e^{-2\pi^2(\tau-\phi)}[\eta_1(\phi) + \eta_2(\phi)]d\phi, \tag{77}$$

$$T_{11b}(\tau) = \frac{4}{\pi}e^{-2\pi^2\tau} - \frac{2}{\pi}[\eta_3(\tau) + \eta_4(\tau)] + 2\pi\int_0^{\tau} e^{-2\pi^2(\tau-\phi)}[\eta_3(\phi) + \eta_4(\phi)]d\phi. \tag{78}$$

(Case 1): Consider the time-dependent functions to be of exponential type, as follows:

$$\eta_i(\tau) = e^{-\pi^2\tau}, \quad (i = 1, 2, 3, 4). \tag{79}$$

From Equations (77) and (78) one obtains

$$T_{11a}(\tau) = T_{11b}(\tau) = 0. \tag{80}$$

The solution from the dimensionless form of Equation (72) becomes

$$\theta(X, Y, \tau) = \frac{[\sin(\pi X) + \sin(\pi Y)]e^{-\pi^2\tau}}{T_r}. \tag{81}$$

Substituting this back into dimensional form would be

$$T(x, y, t) = [\sin(\pi x) + \sin(\pi y)]e^{-\alpha\pi^2 t}. \tag{82}$$

It can be seen that the solution obtained in Equation (82) is exactly the same form as that given by Young et al. [25].

(Case 2): Consider the time-dependent functions to be of periodic type, as follows:

$$\eta_i(\tau) = \cos(\omega_i\tau), \quad (i = 1, 2, 3, 4). \tag{83}$$

One obtains

$$\gamma_{11a}(\tau) = -\frac{2}{\pi}[-\omega_1\sin(\omega_1\tau) - \omega_2\sin(\omega_2\tau) + \pi^2\cos(\omega_1\tau) + \pi^2\cos(\omega_2\tau)], \tag{84}$$

$$\gamma_{11b}(\tau) = -\frac{2}{\pi}[-\omega_3\sin(\omega_3\tau) - \omega_4\sin(\omega_4\tau) + \pi^2\cos(\omega_3\tau) + \pi^2\cos(\omega_4\tau)] \tag{85}$$

$$\begin{aligned} T_{11a}(\tau) = \; & \tfrac{2}{\pi}e^{-2\pi^2\tau}\left(\tfrac{2\pi^4+\omega_1^2}{4\pi^4+\omega_1^2} + \tfrac{2\pi^4+\omega_2^2}{4\pi^4+\omega_2^2}\right) - \tfrac{2}{\pi}\Big[\tfrac{(2\pi^4+\omega_1^2)\cos(\omega_1\pi)-\pi^2\omega_1\sin(\omega_1\pi)}{4\pi^4+\omega_1^2} \\ & + \tfrac{(2\pi^4+\omega_2^2)\cos(\omega_2\pi)-\pi^2\omega_2\sin(\omega_2\pi)}{4\pi^4+\omega_2^2}\Big], \end{aligned} \tag{86}$$

$$\begin{aligned} T_{11b}(\tau) = \; & \tfrac{2}{\pi}e^{-2\pi^2\tau}\left(\tfrac{2\pi^4+\omega_3^2}{4\pi^4+\omega_3^2} + \tfrac{2\pi^4+\omega_4^2}{4\pi^4+\omega_4^2}\right) - \tfrac{2}{\pi}\Big[\tfrac{(2\pi^4+\omega_3^2)\cos(\omega_3\pi)-\pi^2\omega_3\sin(\omega_3\pi)}{4\pi^4+\omega_3^2} \\ & + \tfrac{(2\pi^4+\omega_4^2)\cos(\omega_4\pi)-\pi^2\omega_4\sin(\omega_4\pi)}{4\pi^4+\omega_4^2}\Big]. \end{aligned} \tag{87}$$

Therefore, the exact solution in dimensionless form becomes

$$\begin{aligned} \theta(X, Y, \tau) = \; & [\sin(\pi X)T_{11a}(\tau) + (1 - X)\tfrac{\cos(\omega_1\tau)}{T_r} + X\tfrac{\cos(\omega_2\tau)}{T_r}]\sin(\pi Y) \\ & + [\sin(\pi Y)T_{11b}(\tau) + (1 - Y)\tfrac{\cos(\omega_3\tau)}{T_r} + Y\tfrac{\cos(\omega_4\tau)}{T_r}]\sin(\pi X). \end{aligned} \tag{88}$$

Three cases including $\omega_i = \pi (i = 1, 2, 3, 4)$, $\omega_i = 5 (i = 1, 2, 3, 4)$, and $\omega_i = 7$ $(i = 1, 2, 3, 4)$ will be considered in the numerical analysis. Figure 3 illustrates the temperature-time variation in the middle of a rectangular region with various parameter values of $\omega_i (i = 1, 2, 3, 4)$, which shows the oscillating behavior, as expected. In addition, the larger the parameter value of $\omega_i (i = 1, 2, 3, 4)$, the more frequent the temperature fluctuation.

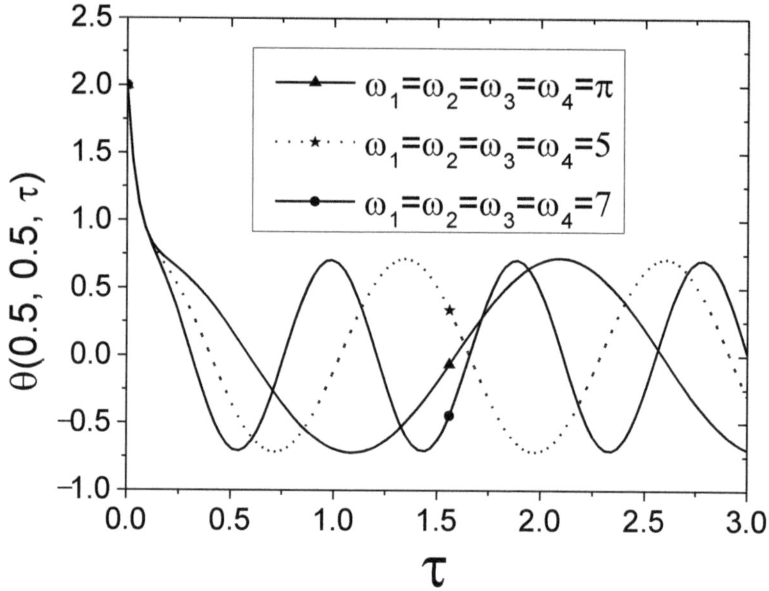

Figure 3. Temperature variation in the middle of the rectangular region with various parameters of harmonic-type time-dependent boundary conditions (Case 2 of Example 1).

4.2. The Space-Dependent Boundary Conditions of Parabolic Type

Example 2: *Consider a 2D transient heat conduction problem in a rectangular region ($L_x = L_y = L_r = 1$).*

The boundary and initial conditions are listed as follows:

$$T(0, y, t) = f_1(y, t) = (y - y^2)\eta_1(\alpha t) \quad \text{at } x = 0, \;\; 0 \le y \le 1, \tag{89}$$

$$T(1, y, t) = f_2(y, t) = (y - y^2)\eta_2(\alpha t) \quad \text{at } x = 1, \;\; 0 \le y \le 1, \tag{90}$$

$$T(x, 0, t) = f_3(x, t) = (x - x^2)\eta_3(\alpha t) \quad \text{at } y = 0, \;\; 0 \le x \le 1, \tag{91}$$

$$T(x, 1, t) = f_4(x, t) = (x - x^2)\eta_4(\alpha t) \quad \text{at } y = 1, \;\; 0 \le x \le 1, \tag{92}$$

$$T(x, y, 0) = (x - x^2) + (y - y^2) \quad \text{at } t = 0, \;\; 0 \le x \le 1, \;\; 0 \le y \le 1 \tag{93}$$

Using the dimensionless parameters generates

$$\theta(0, Y, \tau) = \frac{Y - Y^2}{T_r}\eta_1(\tau) \equiv F_1(Y, \tau) \quad \text{at } X = 0, \;\; 0 \le Y \le 1, \tag{94}$$

$$\theta(1, Y, \tau) = \frac{Y - Y^2}{T_r}\eta_2(\tau) \equiv F_2(Y, \tau) \quad \text{at } X = 1, \;\; 0 \le Y \le 1, \tag{95}$$

$$\theta(X,0,\tau) = \frac{X - X^2}{T_r}\eta_3(\tau) \equiv F_3(X,\tau) \quad \text{at } Y = 0, \quad 0 \le X \le 1, \tag{96}$$

$$\theta(X,1,\tau) = \frac{X - X^2}{T_r}\eta_4(\tau) \equiv F_4(X,\tau) \quad \text{at } Y = 0, \quad 0 \le X \le 1, \tag{97}$$

$$\theta(X,Y,0) = \frac{(X - X^2) + (Y - Y^2)}{T_r} \quad \text{at } \tau = 0, \quad 0 \le X \le 1, \quad 0 \le Y \le 1, \tag{98}$$

and separating $\theta(X,Y,0)$ into two parts yields

$$\theta_{a0}(X,Y,0) = \frac{Y - Y^2}{T_r}, \qquad \theta_{b0}(X,Y,0) = \frac{X - X^2}{T_r}. \tag{99}$$

Following the same solution procedure, the associated dimensionless quantity $\overline{F}_{i,m}(\tau)$ ($i = 1, 2, 3, 4$) becomes

$$\overline{F}_{i,m}(\tau) = \frac{4[1-(-1)^m]}{m^3\pi^3T_r}\eta_i(\tau), \quad i = 1, 2, 3, 4. \tag{100}$$

To determine $T_{mna}(\tau)$ and $T_{mnb}(\tau)$, one derives first

$$\lambda_{mna} = \lambda_{mnb} = \sqrt{m^2+n^2}\pi, \tag{101}$$

$$T_{mna}(0) = \frac{8[1-(-1)^m]}{m^3n\pi^4T_r}[1-(-1)^n - \eta_1(0) + (-1)^n\eta_2(0)], \tag{102}$$

$$T_{mnb}(0) = \frac{8[1-(-1)^m]}{m^3n\pi^4T_r}[1-(-1)^n - \eta_3(0) + (-1)^n\eta_4(0)], \tag{103}$$

$$\gamma_{mna}(\tau) = \frac{-8[1-(-1)^m]}{m^3n\pi^4T_r}\{\dot{\eta}_1(\tau) - (-1)^n\dot{\eta}_2(\tau) + m^2\pi^2[\eta_1(\tau) - (-1)^n\eta_2(\tau)]\}, \tag{104}$$

$$\gamma_{mnb}(\tau) = \frac{-8[1-(-1)^m]}{m^3n\pi^4T_r}\{\dot{\eta}_3(\tau) - (-1)^n\dot{\eta}_4(\tau) + m^2\pi^2[\eta_3(\tau) - (-1)^n\eta_4(\tau)]\}. \tag{105}$$

Accordingly, the solutions for $T_{mna}(\tau)$ and $T_{mnb}(\tau)$ are

$$\begin{aligned} T_{mna}(\tau) = {} & \tfrac{8[1-(-1)^m]}{m^3n\pi^4T_r}\{[1-(-1)^n]e^{-\lambda_{mna}^2\tau} - \eta_1(\tau) + (-1)^n\eta_2(\tau)] \\ & + n^2\pi^2\textstyle\int_0^\tau e^{-\lambda_{mna}^2(\tau-\phi)}[\eta_1(\phi) - (-1)^n\eta_2(\phi)]d\phi\}, \end{aligned} \tag{106}$$

$$\begin{aligned} T_{mnb}(\tau) = {} & \tfrac{8[1-(-1)^m]}{m^3n\pi^4T_r}\{[1-(-1)^n]e^{-\lambda_{mnb}^2\tau} - \eta_3(\tau) + (-1)^n\eta_4(\tau)] \\ & + n^2\pi^2\textstyle\int_0^\tau e^{-\lambda_{mnb}^2(\tau-\phi)}[\eta_3(\phi) - (-1)^n\eta_4(\phi)]d\phi\}. \end{aligned} \tag{107}$$

Therefore, the exact solution in dimensionless form is

$$\begin{aligned} \theta(X,Y,\tau) = {} & \sum_{m=1}^{\infty}\{\sum_{n=1}^{\infty}[\sin(n\pi X)T_{mna}(\tau)] + 4(1-X)\tfrac{[1-(-1)^m]}{m^3\pi^3T_r}\eta_1(\tau) + 4X\tfrac{[1-(-1)^m]}{m^3\pi^3T_r}\eta_2(\tau)\}\sin(m\pi Y) \\ & + \sum_{m=1}^{\infty}\{\sum_{n=1}^{\infty}[\sin(n\pi Y)T_{mnb}(\tau)] + 4(1-Y)\tfrac{[1-(-1)^m]}{m^3\pi^3T_r}\eta_3(\tau) + 4Y\tfrac{[1-(-1)^m]}{m^3\pi^3T_r}\eta_4(\tau)\}\sin(m\pi X). \end{aligned} \tag{108}$$

Considering the time-dependent term of exponential type as

$$\eta_i(\tau) = e^{-d_i\tau}, \qquad (i = 1, 2, 3, 4), \tag{109}$$

one has

$$T_{mna}(\tau) = \frac{8[1-(-1)^m]}{m^3 n\pi^4 T_r}\{[1-(-1)^n]e^{-\lambda_{mna}^2\tau} - e^{-d_1\tau} + (-1)^n e^{-d_2\tau}] + n^2\pi^2[\frac{e^{-d_1\tau} - e^{-\lambda_{mna}^2\tau}}{\lambda_{mna}^2 - d_1} - (-1)^n \frac{e^{-d_2\tau} - e^{-\lambda_{mna}^2\tau}}{\lambda_{mna}^2 - d_2}]\}, \quad (110)$$

$$T_{mnb}(\tau) = \frac{8[1-(-1)^m]}{m^3 n\pi^4 T_r}\{[1-(-1)^n]e^{-\lambda_{mnb}^2\tau} - e^{-d_3\tau} + (-1)^n e^{-d_4\tau}] + n^2\pi^2[\frac{e^{-d_3\tau} - e^{-\lambda_{mnb}^2\tau}}{\lambda_{mnb}^2 - d_3} - (-1)^n \frac{e^{-d_4\tau} - e^{-\lambda_{mnb}^2\tau}}{\lambda_{mnb}^2 - d_4}]\} \quad (111)$$

where d_i $(i = 1, 2, 3, 4)$.represents four arbitrary constants. Tables 1–3 shows the temperature variation of the midpoint of the rectangular region under the three kinds of exponential parameters of d_i $(i = 1, 2, 3, 4)$. It can be found that the solutions developed converge to convergence values as the number of series terms $(m = n)$ increases. The temperature at $0 \leq \tau \leq 1.2$ are the same between 10 and 20 terms expansion. The results converge when 10 terms expansion is used. By comparing the temperature at $0 \leq \tau \leq 1.2$ between 5 and 10 terms expansion in each table of Tables 1–3, one can see that when 5 terms expansion is used, the error of the solution evaluated is less than 1%. Therefore, 5 terms expansion $(m = n)$ of the series will be taken for the numerical analysis below. Figure 4 illustrates the temperature variation in the middle of the rectangular region with respect to time τ for three different kinds of d_i (i = 1, 2, 3, 4). It is seen from Figure 4 that the temperature curve of the set of $d_i = 1$ $(i = 1, 2, 3, 4)$ decays faster than the other two curves, and the trend of the temperature curves of three sets is the same.

Table 1. The temperature of the rectangular region at X = Y = 0.5 and at various times $[\eta_i(\tau) = e^{-\tau}$, $(i = 1, 2, 3, 4)]$.

τ	θ(X=0.5,Y=0.5,τ)				
	Number of Expansion Terms (m=n)				
	1	3	5	10	20
0	0.516	0.497	0.501	0.500	0.500
0.1	0.229	0.246	0.243	0.243	0.243
0.2	0.174	0.189	0.187	0.187	0.187
0.4	0.138	0.150	0.148	0.148	0.148
0.6	0.113	0.123	0.121	0.121	0.121
0.8	0.0921	0.100	0.0989	0.0994	0.0994
1.0	0.0754	0.0823	0.0810	0.0814	0.0814
1.2	0.0618	0.0674	0.0663	0.0666	0.0666

Table 2. The temperature of the rectangular region at X = Y = 0.5 and at various times $[\eta_i(\tau) = e^{-\tau}$, $(i = 1, 2)$; $\eta_i(\tau) = e^{-2\tau}$, $(i = 3, 4)]$.

τ	θ(X=0.5,Y=0.5,τ)				
	Number of Expansion Terms (m=n)				
	1	3	5	10	20
0	0.516	0.497	0.501	0.500	0.500
0.1	0.249	0.263	0.261	0.261	0.261
0.2	0.203	0.215	0.213	0.213	0.213
0.4	0.176	0.184	0.183	0.183	0.183
0.6	0.154	0.159	0.159	0.159	0.159
0.8	0.133	0.136	0.136	0.136	0.136
1.0	0.113	0.115	0.115	0.115	0.115
1.2	0.0954	0.0969	0.0967	0.0968	0.0968

Table 3. The temperature of the rectangular region at $X = Y = 0.5$ and at various times $[\eta_i(\tau) = e^{-i\tau}$, $(i = 1, 2, 3, 4)]$.

τ	$\theta(X=0.5,Y=0.5,\tau)$				
	Number of Expansion Terms (m=n)				
	1	3	5	10	20
0	0.516	0.497	0.501	0.500	0.500
0.1	0.285	0.295	0.293	0.293	0.293
0.2	0.251	0.257	0.256	0.256	0.256
0.4	0.229	0.230	0.230	0.230	0.230
0.6	0.202	0.201	0.202	0.202	0.202
0.8	0.174	0.172	0.172	0.172	0.172
1.0	0.147	0.145	0.146	0.146	0.146
1.2	0.123	0.121	0.122	0.122	0.122

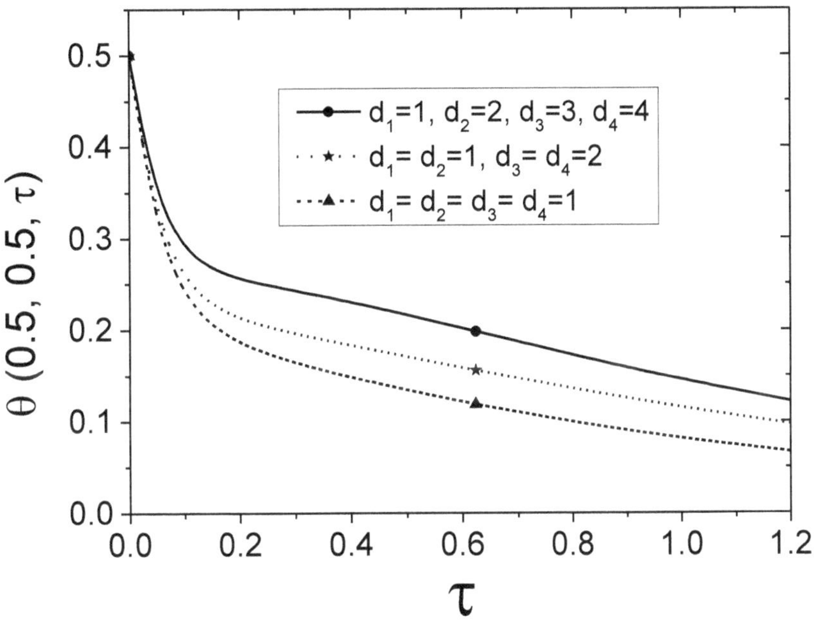

Figure 4. Temperature variation in the middle of the rectangular region with various parameters of exponential-type time-dependent boundary conditions (Example 2).

5. Conclusions

A closed form solution of the transient heat conduction in a rectangular cross-section in an infinite bar with the general space–time-dependent boundary conditions has been developed in terms of series expansion. The main advantages of the proposed solution method is that differentiation and/or integration of the Green's function is not required and the solution of the auxiliary 2D problem with associated nonhomogeneous boundary conditions is avoided. Two examples are given to illustrate the applicability of the method and the example of space-dependent boundary for periodic function is shown to be consistent with results in the literature.

The new findings of the present study are as follows:

(1) The proposed approach combining the shifting function method and the expansion theorem method can derive an analytic solution for the 2D heat conduction in a rectangular cross-section of an infinite bar with the general Dirichlet boundary condi-

tions specifying space–time-dependent boundary conditions at the four edges of the rectangular region;

(2) The series expansion derived from the proposed method has a good convergence to reach the convergence values. For space-dependent boundary with the parabolic-type case, one can take five terms of the series to obtain the series solutions within 1% error;

(3) When considering the time-dependent boundary of harmonic function, the fluctuation of the temperature variation increases as the frequency of the harmonic function increases. When considering the time-dependent boundary of exponential function, $e^{-d_i\tau}$, a smaller coefficient d_i will result in a lower and faster drop in temperature.

The analytic solution for the 2D heat conduction problems with general Dirichlet boundary conditions is obtained using the proposed method. However, the values of the boundary conditions and initial condition at the four corners of the rectangular region should be zero, which limits the applicability of this study. Transforming the temperature function before using the method proposed in this paper may overcome this limitation. A method that can be applied to the case for non-zero values at the four corners will be proposed in the near future.

Author Contributions: Conceptualization, T.-W.T. and J.-R.C.; formal analysis, H.-P.H.; investigation, J.-R.C.; methodology, H.-P.H., T.-W.T. and J.-R.C.; software, H.-P.H.; supervision, T.-W.T.; validation, J.-R.C.; writing—original draft, T.-W.T.; writing—review and editing, J.-R.C. All authors have read and agreed to the published version of the manuscript.

Funding: This research received no external funding. And the APC was funded by H.-P.H. and J.-R.C.

Data Availability Statement: Not applicable.

Conflicts of Interest: The authors declare no conflict of interest.

Nomenclature

A, B	two subsystems
c	specific heat (W · s/kg·°C)
$d_i (i = 1, 2, 3, 4)$	four arbitrary constants
$f_i(y, t)$, $i = 1, 2$	temperatures along the surface at the left end and the right end of the rectangular region
$f_i(x, t)$, $i = 3, 4$	temperatures along the surface at the bottom end and the top end of the rectangular region
$F_i(Y, \tau)$, $i = 1, 2$	dimensionless quantity defined in Equation (8)
$F_i(X, \tau)$, $i = 3, 4$	dimensionless quantity defined in Equation (8)
$\overline{F}_{i,m}(\tau)$, $i = 1, 2$	dimensionless quantity defined in Equation (31)
$\overline{F}_{i,m}(\tau)$, $i = 3, 4$	dimensionless quantity defined in Equation (A7)
$g_{i,m}(X), i = 1, 2$	shifting function
$g_{i,m}(Y), i = 3, 4$	shifting function
$\overline{G}_m(X, \tau)$	nonhomogeneous term in the differential equation of the transformed system defined in Equation (43)
k	thermal conductivity (W/m·°C)
L_r	aspect ratio, L_y/L_x defined in Equation (8)
L_x, L_y	thickness of the two-dimensional rectangular region at x- and y- directions (m)
$T(x, y, t)$	temperature function (°C)
$T_{mna}(\tau), T_{mnb}(\tau)$	dimensionless time variable of the transformed function defined in Equations (53) and (A22)
T_r	reference temperature (°C)
$T_0(x, y)$	initial temperature (°C)
t	time variable (s)
x	space variable in x-direction of a rectangular region (m)
X	dimensionless space variable in x-direction of a rectangular region

y	space variable in y-direction of a rectangular region (m)
Y	dimensionless space variable in y-direction of a rectangular region
α	thermal diffusivity (m^2/s)
ϕ	auxiliary integration variable
$\gamma_{mna}(\tau), \gamma_{mnb}(\tau)$	dimensionless quantity defined in Equations (55) and (A24)
$\eta_i(i = 1,2,3,4)$	time-dependent boundary condition
$\lambda_{mna}, \lambda_{mnb}$	n-th eigenvalues depend on ω_n defined in Equations (54) and (A23)
θ	dimensionless temperature
θ_0	dimensionless initial temperature
θ_a, θ_b	dimensionless temperatures for subsystems A and B
$\theta_m(X,\tau)$	generalized Fourier coefficient defined in Equation (29)
$\overline{\theta}_m(X,\tau)$	transformed function defined in Equation (36)
$\overline{\theta}_{mn}(X,\tau)$	n-th eigenfunction of the transformed function defined in Equation (47)
ρ	density (kg/m^3)
τ	dimensionless time
ω_n	n-th eigenvalue for Sturm–Liouville problem defined in Equation (48).
Subscripts	
$0, 1, 2, 3, 4, a, b, i, m, n, r$	described in the article

Appendix A. Analytic Solution of the Subsystem B

For the subsystem B, the boundary value problem is as follows:

$$\left[L_r^2 \frac{\partial^2\theta_b(X,Y,\tau)}{\partial X^2} + \frac{\partial^2\theta_b(X,Y,\tau)}{\partial Y^2}\right] = \frac{\partial\theta_b(X,Y,\tau)}{\partial\tau} \quad \text{in } 0 < X < 1,\ 0 < Y < 1, \quad \tau > 0, \tag{A1}$$

$$\theta_b(0,Y,\tau) = 0, \quad \theta_b(1,Y,\tau) = 0, \tag{A2}$$

$$\theta_b(X,0,\tau) = F_3(X,\tau), \quad \theta_b(X,1,\tau) = F_4(X,\tau), \tag{A3}$$

$$\theta_b(X,Y,0) = \theta_0(X,Y) - \theta_{a0}(X,Y) = \theta_{b0}(X,Y). \tag{A4}$$

Because the boundary conditions of the rectangular region at two opposite edges $X = 0$ and $X = 1$ are homogeneous, the temperature $\theta_b(X,Y,\tau)$ and the dimensionless quantities $F_3(X,\tau)$, $F_4(X,\tau)$ defined in Equation (A3), can be expressed as

$$\theta_b(X,Y,\tau) = \sum_{m=1}^{\infty} [\theta_m(Y,\tau)\sin(m\pi X)], \tag{A5}$$

$$F_i(X, \tau) = \sum_{m=1}^{\infty} \left[\overline{F}_{i,m}(\tau)\sin(m\pi X)\right], \quad (i = 3, 4) \tag{A6}$$

where m denotes a positive integer and $\overline{F}_{i,m}(\tau)$ $(i = 3, 4)$ is given as

$$\overline{F}_{i,m}(\tau) = 2\int_0^1 F_i(X,\tau)\sin(m\pi X)dX, \quad (i = 3, 4). \tag{A7}$$

Substituting Equations (A5) and (A6) back into Equations (A1)–(A4), one obtains

$$\frac{\partial\theta_m(Y, \tau)}{\partial\tau} - \frac{\partial^2\theta_m(Y, \tau)}{\partial Y^2} + m^2\pi^2L_r^2\theta_m(Y, \tau) = 0, \tag{A8}$$

$$\theta_m(0,\tau) = \overline{F}_{3,m}(\tau), \quad \theta_m(1,\tau) = \overline{F}_{4,m}(\tau), \tag{A9}$$

$$\theta_m(Y, 0) = 2\int_0^1 \theta_{b0}(X,Y)\sin(m\pi X)dX. \tag{A10}$$

To find the solution for the second-order differential Equation (A8) with nonhomogeneous boundary conditions (A9), one uses the shifting function method by taking

$$\theta_m(Y,\tau) = \overline{\theta}_m(Y,\tau) + \sum_{i=3}^{4} g_{i,m}(Y)\overline{F}_{i,m}(\tau) \tag{A11}$$

where $\overline{\theta}_m(Y,\tau)$ is the transformed function while $g_{i,m}(Y)$ $(i = 3, 4)$ indicates the shifting function to be specified.

Substituting Equation (A11) back into Equations (A8)–(A10), one obtains

$$\begin{aligned} &\dot{\overline{\theta}}_m(Y,\tau) + \sum_{i=3}^{4} g_{i,m}(Y)\dot{\overline{F}}_{i,m}(\tau) - \left[\overline{\theta}''_m(Y,\tau) + \sum_{i=3}^{4} g''_{i,m}(Y)\overline{F}_{i,m}(\tau)\right] \\ &+ m^2\pi^2 L_r^2[\overline{\theta}_m(Y,\tau) + \sum_{i=3}^{4} g_{i,m}(Y)\overline{F}_{i,m}(\tau)] = 0. \end{aligned} \tag{A12}$$

The associated boundary conditions become

$$\overline{\theta}_m(0,\tau) + g_{3,m}(0)\overline{F}_{3,m}(\tau) + g_{4,m}(0)\overline{F}_{4,m}(\tau) = \overline{F}_{3,m}(\tau), \tag{A13}$$

$$\overline{\theta}_m(1,\tau) + g_{3,m}(1)\overline{F}_{3,m}(\tau) + g_{4,m}(1)\overline{F}_{4,m}(\tau) = \overline{F}_{4,m}(\tau) \tag{A14}$$

As in the derivation process, the two shifting functions are determined as

$$g_{3,m}(Y) = 1 - Y, \quad g_{4,m}(Y) = Y. \tag{A15}$$

After substituting Equation (A15) into Equations (A12)–(A14), one has the differential equation for $\overline{\theta}_m(Y,\tau)$ as

$$\dot{\overline{\theta}}_m(Y,\tau) - \overline{\theta}''_m(Y,\tau) + m^2\pi^2 L_r^2\overline{\theta}_m(Y,\tau) = \overline{G}_m(Y,\tau), \tag{A16}$$

and the associated homogeneous boundary conditions as

$$\overline{\theta}_m(0,\tau) = 0, \quad \overline{\theta}_m(1,\tau) = 0. \tag{A17}$$

$\overline{G}_m(Y,\tau)$ is defined as

$$\overline{G}_m(Y,\tau) = -\sum_{i=3}^{4}\left[m^2\pi^2 L_r^2 g_{i,m}(Y)\overline{F}_{i,m}(\tau) + g_{i,m}(Y)\dot{\overline{F}}_{i,m}(\tau)\right]. \tag{A18}$$

Moreover, the initial condition is transformed to be

$$\overline{\theta}_m(Y,0) = 2\int_0^1 \theta_{b0}(X,Y)\sin(m\pi X)dX - \sum_{i=3}^{4}[g_{i,m}(Y)\overline{F}_{i,m}(0)]. \tag{A19}$$

The solution $\overline{\theta}_m(Y,\tau)$ specified by Equations (A16)–(A19) can be expressed in the form of eigenfunctions as

$$\overline{\theta}_m(Y,\tau) = \sum_{n=1}^{\infty} \overline{\theta}_{mn}(Y)T_{mnb}(\tau) \tag{A20}$$

where $\overline{\theta}_{mn}(Y)$ is

$$\overline{\theta}_{mn}(Y) = \sin n\pi Y. \tag{A21}$$

Substituting Equation (A20) into Equation (A16), multiplying it by $\overline{\theta}_{mn}(Y)$, and integrating from 0 to 1, one will obtain

$$\dot{T}_{mnb}(\tau) + \lambda_{mnb}^2 T_{mnb}(\tau) = \gamma_{mnb}(\tau) \tag{A22}$$

where λ_{mnb} and $\gamma_{mnb}(\tau)$ are

$$\lambda_{mnb} = \sqrt{m^2 L_r^2 + n^2 \pi}, \tag{A23}$$

$$\begin{aligned} \gamma_{mnb}(\tau) &= 2\int_0^1 \overline{\theta}_{mn}(Y)\overline{G}_m(Y,\tau)dY \\ &= \frac{-2}{n\pi}\left\{\left[\dot{\overline{F}}_{3,m}(\tau) - (-1)^n \dot{\overline{F}}_{4,m}(\tau)\right] + m^2\pi^2 L_r^2\left[\overline{F}_{3,m}(\tau) - (-1)^n \overline{F}_{4,m}(\tau)\right]\right\}. \end{aligned} \tag{A24}$$

Note that $T_{mnb}(0)$ can be determined from the initial condition of the transformed function defined in Equation (A19) as

$$T_{mnb}(0) = 4\int_0^1 \sin(n\pi Y)\int_0^1 \theta_{b0}(X,Y)\sin(n\pi X)dXdY - \frac{2}{n\pi}\left[\overline{F}_{3,m}(0) - (-1)^n \overline{F}_{4,m}(0)\right]. \tag{A25}$$

The general solution of Equation (A22) with the initial condition above is

$$T_{mnb}(\tau) = e^{-\lambda_{mnb}^2 \tau} T_{mnb}(0) + \int_0^\tau e^{-\lambda_{mnb}^2(\tau-\phi)}\gamma_{mnb}(\phi)d\phi. \tag{A26}$$

References

1. Perakis, N.; Haidna, O.J.; Ihme, M. Heat transfer augmentation by recombination reactions in turbulent reacting boundary layers at elevated pressures. *Int. J. Heat Mass Transf.* **2021**, *178*, 121628. [CrossRef]
2. Hassan, M.A.S.M.; Razlan, Z.M.; Bakar, S.A.; Rahman, A.A.; Rojan, M.A.; Wan, W.K.; Ibrahim, Z.; Ishak, A.A.; Ridzuan, M.J.M. Derivation and validation of heat transfer model for Spark-Ignition engine cylinder head. *Appl. Therm. Eng.* **2023**, *225*, 120240. [CrossRef]
3. Ivanov, V.; Salomatov, V. On the calculation of the temperature field in solids with variable heat-transfer coefficients. *J. Eng. Phys. Thermophys.* **1965**, *72*, 63–64. [CrossRef]
4. Ivanov, V.; Salomatov, V. Unsteady temperature field in solid bodies with variable heat transfer coefficient. *J. Eng. Phys. Thermophys.* **1966**, *11*, 151–152. [CrossRef]
5. Postol'Nik, Y.S. One-dimensional convective heating with a time-dependent heat-transfer coefficient. *J. Eng. Phys. Thermophys.* **1970**, *18*, 233–238. [CrossRef]
6. Kozlov, V. Solution of heat-conduction problem with variable heat-exchange coefficient. *J. Eng. Phys. Thermophys.* **1970**, *18*, 100–104. [CrossRef]
7. Holy, Z. Temperature and stresses in reactor fuel elements due to time-and space-dependent heat-transfer coefficients. *Nucl. Eng. Des.* **1972**, *18*, 145–197. [CrossRef]
8. Özişik, M.N.; Murray, R. On the solution of linear diffusion problems with variable boundary condition parameters. *J. Heat Transf.* **1974**, *96*, 48–51. [CrossRef]
9. Moitsheki, R.J. Transient heat diffusion with temperature-dependent conductivity and time-dependent heat transfer coefficient. *Math. Probl. Eng.* **2008**, *9*, 41–58. [CrossRef]
10. Chen, H.T.; Sun, S.L.; Huang, H.C.; Lee, S.Y. Analytic closed solution for the heat conduction with time dependent heat convection coefficient at one boundary. *Comput. Model. Eng. Sci.* **2010**, *59*, 107–126.
11. Lee, S.Y.; Huang, C.C. Analytic solutions for heat conduction in functionally graded circular hollow cylinders with time-dependent boundary conditions. *Math. Probl. Eng.* **2013**, *5*, 816385. [CrossRef]
12. Lee, S.Y.; Tu, T.W. Unsteady temperature field in slabs with different kinds of time-dependent boundary conditions. *Acta Mech.* **2015**, *226*, 3597–3609. [CrossRef]
13. Lee, S.Y.; Lin, S.M. Dynamic analysis of nonuniform beams with time-dependent elastic boundary conditions. *J. Appl. Mech.* **1996**, *63*, 474–478. [CrossRef]
14. Lee, S.Y.; Huang, T.W. A method for inverse analysis of laser surface heating with experimental data. *Int. J. Heat Mass Transf.* **2014**, *72*, 299–307. [CrossRef]
15. Lee, S.Y.; Huang, T.W. Inverse analysis of spray cooling on a hot surface with experimental data. *Int. J. Therm. Sci.* **2016**, *100*, 145–154. [CrossRef]
16. Lee, S.Y.; Yan, Q.Z. Inverse analysis of heat conduction problems with relatively long heat treatment. *Int. J. Heat Mass Transf.* **2017**, *105*, 401–410. [CrossRef]
17. Carslaw, H.; Jaeger, J. *Heat in Solids*, 2nd ed.; Clarendon Press: Oxford, UK, 1959.
18. Özışık, M.N. *Heat Conduction*; John Wiley & Sons: New York, USA, 1993.
19. Cole, K.D.; Beck, J.V.; Haji-Sheikh, A.; Litkouhi, B. *Heat Conduction Using Green's Functions*; Taylor & Francis: Boca Raton, FL, USA, 2010.
20. Zhu, S.P. Solving transient diffusion problems: Time-dependent fundamental solution approaches versus LTDRM approaches. *Eng. Anal. Bound. Elem.* **1998**, *21*, 87–90. [CrossRef]

21. Zhu, S.P.; Liu, H.W.; Lu, X.P. A combination of LTDRM and ATPS in solving diffusion problems. *Eng. Anal. Bound. Elem.* **1998**, *21*, 285–289. [CrossRef]
22. Sutradhar, A.; Paulino, G.H.; Gray, L. Transient heat conduction in homogeneous and non-homogeneous materials by the Laplace transform Galerkin boundary element method. *Eng. Anal. Bound. Elem.* **2002**, *26*, 119–132. [CrossRef]
23. Bulgakov, V.; Šarler, B.; Kuhn, G. Iterative solution of systems of equations in the dual reciprocity boundary element method for the diffusion equation. *Int. J. Numer. Methods Eng.* **1998**, *43*, 713–732. [CrossRef]
24. Walker, S. Diffusion problems using transient discrete source superposition. *Int. J. Numer. Methods Eng.* **1992**, *35*, 165–178. [CrossRef]
25. Chen, C.; Golberg, M.; Hon, Y. The method of fundamental solutions and quasi-Monte-Carlo method for diffusion equations. *Int. J. Numer. Methods Eng.* **1998**, *43*, 1421–1435. [CrossRef]
26. Burgess, G.; Mahajerin, E. Transient heat flow analysis using the fundamental collocation method. *Appl. Therm. Eng.* **2003**, *23*, 893–904. [CrossRef]
27. Young, D.; Tsai, C.; Murugesan, K.; Fan, C.; Chen, C. Time-dependent fundamental solutions for homogeneous diffusion problems. *Eng. Anal. Bound. Elem.* **2004**, *28*, 1463–1473. [CrossRef]
28. Cole, K.D.; Yen, D.H. Green's functions, temperature and heat flux in the rectangle. *Int. J. Heat Mass Transf.* **2001**, *44*, 3883–3894. [CrossRef]
29. Beck, J.V.; Wright, N.T.; Haji-Sheikh, A. Transient power variation in surface conditions in heat conduction for plates. *Int. J. Heat Mass Transf.* **2008**, *51*, 2553–2565. [CrossRef]
30. Lei, J.; Wang, Q.; Liu, X.; Gu, Y.; Fan, C.-M. A novel space-time generalized FDM for transient heat conduction problems. *Eng. Anal. Bound. Elem.* **2020**, *119*, 1–12. [CrossRef]
31. Alam, M.N.; Tunç, C. New solitary wave structures to the (2 + 1)-dimensional KD and KP equations with spatio-temporal dispersion. *J. King Saud Univ. Sci.* **2020**, *32*, 3400–3409. [CrossRef]
32. Islam, S.; Alam, M.N.; Fayz-Al-Asad, M.; Tunç, C. An analytical technique for solving new computational solutions of the modified Zakharov-Kuznetsov equation arising in electrical engineering. *J. Appl. Comput. Mech.* **2021**, *7*, 715–726.
33. Krishnan, G.; Parhizi, M.; Jain, A. Eigenfunction-based solution for solid-liquid phase change heat transfer problems with time-dependent boundary conditions. *Int. J. Heat Mass Transf.* **2022**, *189*, 122693. [CrossRef]
34. Belekar, V.V.; Murphy, E.J.; Subramaniam, S. Analytical solution to heat transfer in stationary wet granular mixtures with time-varying boundary conditions. *Int. Commun. Heat Mass Transf.* **2023**, *140*, 106500. [CrossRef]

axioms

Article

A Hermite Surface Triangle Modeling Method Considering High-Precision Fitting of 3D Printing Models

Ruichao Lian [1], Shikai Jing [2,*], Yang Chen [2] and Jiangxin Fan [2]

[1] Institute of Advanced Structure Technology, Beijing Institute of Technology, Beijing 100081, China; 7520210053@bit.edu.cn
[2] School of Mechanical Engineering, Beijing Institute of Technology, Beijing 100081, China; 3120205251@bit.edu.cn (Y.C.); fanjx0321@163.com (J.F.)
* Correspondence: jingshikai@bit.ed; Tel.: +86-13321166219

Abstract: Three-dimensional printing is a layer-by-layer stacking process. It can realize complex models that cannot be manufactured by traditional manufacturing technology. The most common model currently used for 3D printing is the STL model. It uses planar triangles to simplify the CAD model. This approach makes it difficult to fit complex surface shapes with high accuracy. The fitting result usually suffers from loss of local features of the model, poor fitting accuracy, or redundant data due to face piece subdivision, which will cause problems such as poor manufacturing accuracy or difficult data processing. To this end, this paper proposes a method for constructing Hermite surface models considering high-precision fitting of 3D printing models. The mapping relationship between different surface triangles and the same base triangle is established by analyzing the characteristics of Hermite surface triangles in AMF format files and using the radial variation property. By constructing a cubic surface model with general parameters and combining the vertex and tangent vector information, a cubic Hermite curve and surface triangle model are obtained. A sampling mapping point solution method is proposed, which transforms the volume integration problem between models into the summation problem of sampling point height difference. Considering the mean deviation and variance in multiple directions of the sampling points, a method for calculating and evaluating the model fitting error is constructed. Finally, the effectiveness of the proposed method is verified by rabbit and turbine.

Keywords: 3D printing; Hermite; surface model; high-precision fitting; error

MSC: 08C05; 14P99; 51P05

check for updates

Citation: Lian, R.; Jing, S.; Chen, Y.; Fan, J. A Hermite Surface Triangle Modeling Method Considering High-Precision Fitting of 3D Printing Models. *Axioms* **2023**, *12*, 370. https://doi.org/10.3390/axioms12040370

Academic Editor: Leonid Plotnikov

Received: 11 March 2023
Revised: 9 April 2023
Accepted: 10 April 2023
Published: 11 April 2023

1. Introduction

Three-dimensional printing is an advanced manufacturing technology that enables the "free fabrication" of complex structures quickly and efficiently with a simple device [1,2]. Compared with traditional manufacturing processes, it overcomes the limitations of complex configurations that are difficult to machine and reduces processing procedures, manufacturing cycle time, and manufacturing costs [3]. In recent years, 3D printing technology has been successfully applied in aerospace, automotive, and other areas [4–8]. Although 3D printing technology has made breakthroughs, due to its unique manufacturing process, the manufactured products usually have errors problems, which greatly restrict the widespread use of the technology [9].

There are three main sources of errors that exist in the 3D printing manufacturing process. The first error comes from the conversion between model formats, i.e., the process of converting a computer-aided design (CAD) model to a model in the format required for 3D printing; the second error comes from the layered slicing and path planning algorithm [10]; and the third error comes from the manufacturing accuracy of the device itself [11]. All of the above errors directly affect the molding accuracy of the final printed structure. However, compared to the latter two errors, if the first error cannot be effectively reduced, it will be difficult to manufacture a high-precision structure even if the subsequent process is highly accurate.

When performing 3D printing, we first need to build a digital model through modeling software. However, the model formats generated by different modeling software vary. They are not directly used to drive 3D printers. In order to generate Gcode that can "communicate" directly with the 3D printer, the model needs to be converted from different formats to the common STereoLithography (STL) format file for 3D printing. Currently, 3D printing models usually use planar triangles to form an envelope model to represent the CAD model, such as the more widely used STL model. It approximates the CAD model by setting the maximum chord height between the planar triangle and the surface of the model [12,13]. For this reason, when the CAD model has complex surfaces or high local accuracy, using planar triangles to simplify it will inevitably result in a loss of features and accuracy of the model [14]. In order to retain the features and accuracy of the CAD model as much as possible, the triangular facets of the overall model need to be continuously subdivided during the format conversion process. This will cause problems such as the too-large amount of model data or data redundancy, which inevitably increases the difficulty of model data processing [15]. Compared with the planar triangle model, the surface triangle model has higher degrees of freedom through parametric shape control. It can fit the surface and complex features of the model with high accuracy by using a relatively small number of face pieces, which effectively solves the problems caused by the simplification of the overall model by planar triangles. Several scholars have investigated the construction methods of surface triangles. Vlachos et al. [16] proposed a point-normal triangle in order to improve the visual quality in graphics rendering. Its main idea is to use a Bezier surface triangle (e.g., PN triangle) surface to replace each triangle in the original mesh. Compared with other Bezier triangles, PN triangles have lower degrees of freedom, and their shapes are influenced not only by the normal vectors but also by their different methods. Hamann et al. [17] constructs a C^0 continuous surface by using a triangular rational quadratic Bezier surface to approximate a cubic linear interpolation function profile. The construction of Bezier surfaces requires control point information, which is harder to obtain directly when performing model fitting. NURBS and B-sample surfaces using surface approximation control meshes all have a similar problem to Bezier surfaces in that it is difficult to construct a direct mathematical relationship between the surface model and the original model.

Unlike surfaces using control meshes, surface shape control based on boundary conditions is simpler and more intuitive, and easier to achieve stitching between surfaces. Márta et al. [18] proposed a new definition of a surface that uses three triangular surfaces instead of the original boundary curves on the triangular parameter domain to generate a triangular surface. This interpolation scheme has affine transformation invariance [19], while the connection between the resulting surface and its components is continuous along a common boundary curve, except for the vertices. This method involves a tremendous amount of data input and also contains the combined operation of three surfaces, which greatly increases the computational cost. In addition, Hagen [20] proposes an interpolation method based on the Hermite operator, which implements the interpolation of the boundary curvature of an arbitrary triangle.

In order to meet the growing demand for model formats for 3D printing, the American Society for Testing and Materials (ASTM) Special Advisory Panel has creatively proposed a surface triangle in the additive manufacturing file (AMF) format [21,22]. The surface triangle consists of cubic Hermite curves [23], but they only define the boundary curves of the surface triangle and do not define the Hermite surface triangle model completely. According to the authors' knowledge, there are few studies on Hermite surface triangles, but compared with other surface triangles, the definition and input quantity of Hermite surface triangles are relatively easy. It is not only suitable for 3D printing the required multi-surface sheet model, but also can make full use of the surface information contained in the original design model to achieve high accuracy fitting of the model. Therefore, how to construct a Hermite surface triangle model and make full use of the existing 3D printing manufacturing model information to fit the CAD model with high accuracy is a problem worthy of study.

Based on the above analysis, this paper proposes a Hermite surface triangle model construction method considering the high-precision fitting of 3D printing models. Affine transformation is used to establish the mapping relationship between multiple surface triangles and feature triangles. Then, the cubic surface model with general parameters is constructed and the cubic Hermite surface model is solved using the vertex and tangent vector information. Finally, the model fitting error calculation and comprehensive evaluation are realized by using the height difference between the model sampling points and the mapping points.

2. Hermite Surface Characterization

2.1. Definition of Hermite Curve in AMF

AMF is a format file that supports 3D printing, which contains the multi-color, multi-material, honeycomb structures and properties, etc. Its structure is similar to the STL file, which is a collection of several small spatial triangular facets. It is a model shell formed by combining triangular facets together after the triangular meshing of a 3D solid model. Each of its triangular face pieces consist of three vertices that obey the right-hand rule and whose corresponding normal vectors are directed outward. According to the AMF standard file, each edge of a surface triangle is a cubic Hermite curve, and the construction of each surface triangle depends on the Hermite curve of the boundary. Each Hermite curve is then determined by the position information of the triangle vertices recorded in the AMF file and the normal or tangential vector information. The surface triangles in the AMF file can be defined by the vertex and endpoint tangent vectors $\boldsymbol{t}_{ij}$ of the edges, as shown in Figure 1a. It can also be defined directly by the normal vectors $\boldsymbol{n}_i$ on the vertices, as shown in Figure 1b.

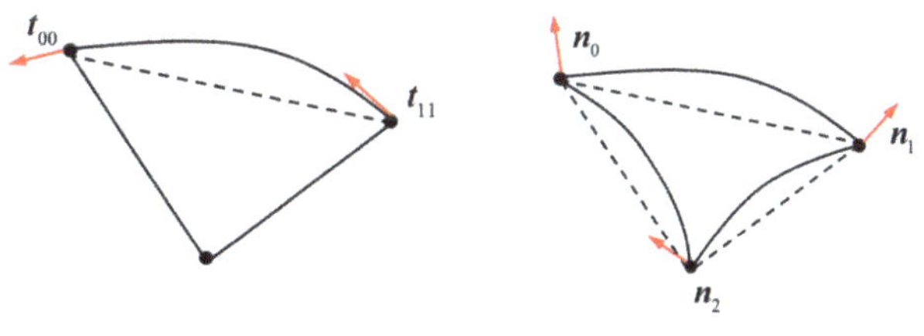

(a) (b)

Figure 1. Definition of AMF surface triangles: (**a**) tangent vector of the vertex; (**b**) normal vectors of the vertices.

In the surface triangle definition of the AMF file, the cubic Hermite curve consists of two endpoints and their tangent vectors, as shown in Figure 2.

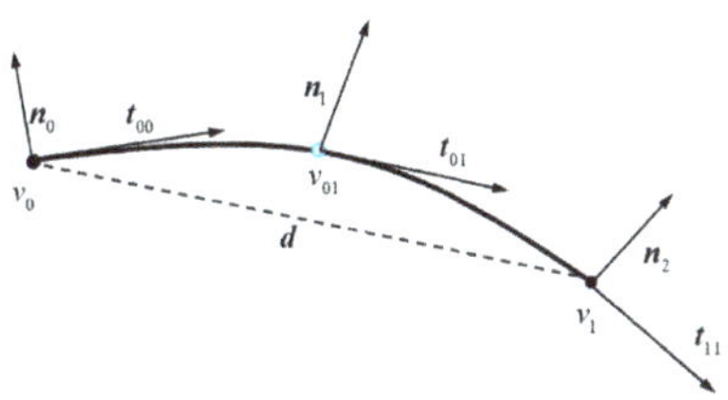

Figure 2. Hermite curve.

The mathematical expression of the cubic Hermite curve [24–26] can be described as:

$$h_e(s) = \left(2s^3 - 3s^2 + 1\right)v_0 + \left(s^3 - 2s^2 + s\right)t_0 + \left(-2s^3 + 3s^2\right)v_1 + \left(s^3 - s^2\right)t_1 \quad (1)$$

where v_0 and v_1 are the two endpoints of the curve. t_0 and t_1 are the two tangent vectors in the direction of the Hermite surface. The formula for calculating the tangent vector at any point on the curve can be formulated as:

$$t(s) = \left(6s^2 - 6s\right)v_0 + \left(3s^2 - 4s + 1\right)t_0 + \left(-6s^2 + 6s\right)v_1 + \left(3s^2 - 2s\right)t_1 \quad (2)$$

when a normal vector is defined at the endpoint, it can be converted to a tangent vector by Equation (3).

$$t_0 = |d_0| \frac{-(n_0 \times d_0) \times n_0}{|(n_0 \times d_0) \times n_0|} \quad (3)$$

where $d_0 = v_1 - v_0$. n_0 and n_1 are the normal vectors of the corresponding endpoints, respectively.

2.2. Determine Feature Triangle and Mapping Relationship

The Hermite surface triangle is constructed by means of cubic Hermite curves based on the vertices. The triangle is essentially parametric curves and each curve involves a parameter that takes values in the range [0, 1]. Therefore, its shape is theoretically related to the three parameters η, ξ and τ, and there is a certain correlation between η, ξ and τ, as shown in Figure 3.

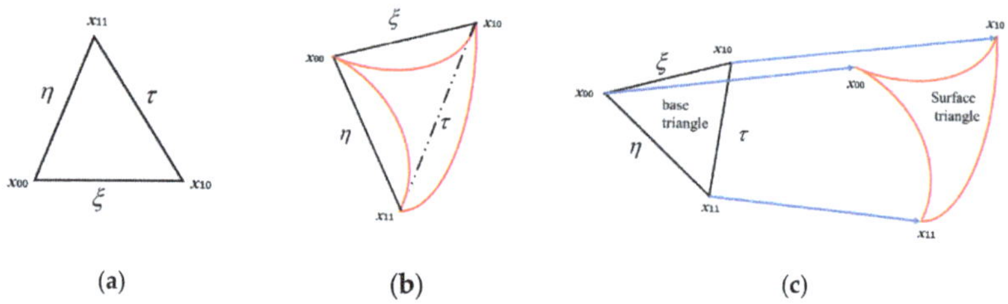

Figure 3. Curved triangle parameters: (**a**) parameters of each side; (**b**) curved triangle; (**c**) mapping of parameter fields.

The shape of the Hermite surface triangle is influenced by the location of its vertices, the direction of the tangent vectors, and the size of the tangent vectors. Each surface triangle has a different shape, so the shape of the base triangle formed by the three vertices of the surface triangle is also different. This would introduce too many parameters and increase the computational cost. To facilitate the calculation, the affine transformation is used in this subsection to simplify the calculation of the parameters in the surface triangle. Affine transformation is the process of transforming to another vector space by performing one linear transformation (multiplying by one matrix) and one translation (adding one vector) in the vector space. The basic idea of the proposed method is to use affine transformation to establish a mapping relationship between surface triangles, base triangles, and specified feature triangle. It can effectively normalize the complex problem and reduce the computational cost. More knowledge about affine transformations can be found in the literature [27].

According to the affine transformation property, all triangles can be obtained by the affine transformation of feature triangles. As shown in Figure 4, in the right-angle parameter domain, the feature triangle with right-angle characteristics and two right-angle sides of unit length is constructed with (0, 0), (1, 0), and (1, 1) as vertices, which satisfies the principle of regular shape and simple calculation. It effectively reduces the three parameter variables of the surface triangle to two. The feature triangle can be described as:

$$C = \{(u, v)|0 \le v \le u \le 1\} \quad (4)$$

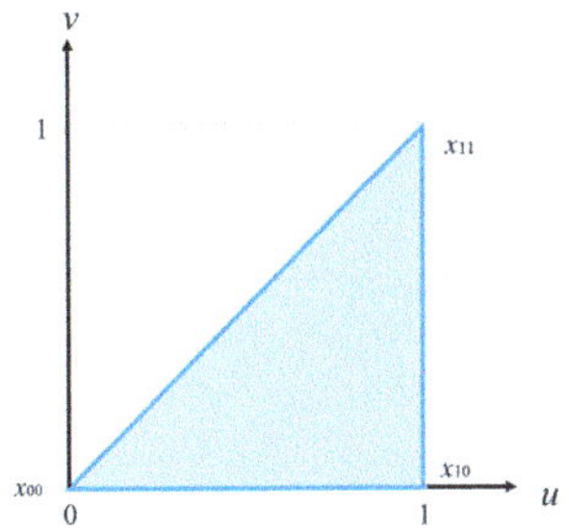

Figure 4. Feature triangle.

Based on the affine transformation property, any surface triangle and base triangle have a mapping relationship with the feature triangle C, i.e., their three vertices correspond to the three vertices of the base triangle, as shown in Figure 5.

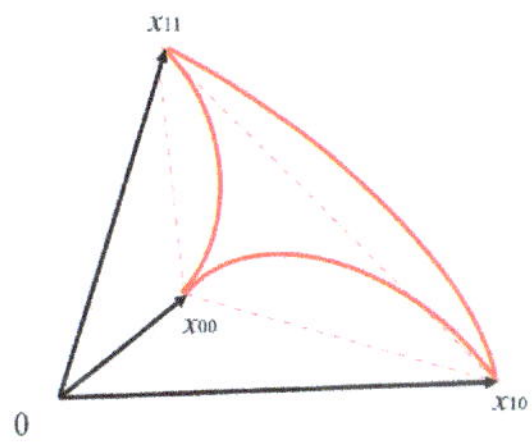

Figure 5. Vertex one-to-one correspondence.

The mapping relationship between the surface triangles and the feature triangles is described by $S(u, v)$, as shown in Figure 6. $S(u_i, v_i)$ is the corresponding point of any point (u_i, v_i) on the characteristic triangle in the surface triangle.

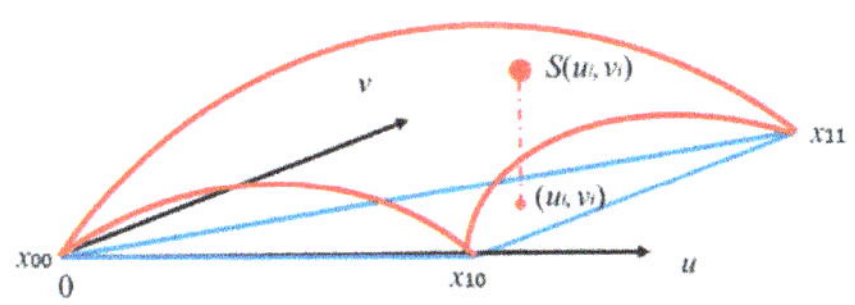

Figure 6. Mapping relationship between feature triangle and surface triangle.

The mapping relation is essentially a parametric formulation of surface triangles, as it satisfies the one-to-one correspondence between surface triangles and characteristic triangle vertices. Therefore, the mapping relation between them can be described as:

$$\begin{aligned} x_{00} &= S(0,0) \\ x_{10} &= S(1,0) \\ x_{11} &= S(1,1) \end{aligned} \quad (5)$$

The model is greatly simplified by Equation (5), which can effectively reduce the complexity of subsequent calculations. It is worth noting that for different surface triangles, the corresponding $S(u, v)$ is different.

3. Hermite Surface Triangle Model

In this subsection, based on the constructed feature triangles and mapping relations, the general parametric surface model is first established by the two parameters u and v. Then, the vertex coordinates and normal vectors of the Hermite curve are matched with the surface model to build the Hermite surface triangle model.

Let $S(u,v)$ be a general two-parameter surface. To ensure the smoothness of the surface, set it as a cubic equation. Then, its mathematical expression can be described as:

$$\begin{aligned} S(u,v) &= C_{00} + C_{10}u + C_{01}v + C_{20}u^2 + C_{11}uv + C_{02}v^2 \\ &+ C_{21}u^2v + C_{12}uv^2 + C_{30}u^3 + C_{03}v^3 \end{aligned} \tag{6}$$

where C_{ij} is the three-dimensional characteristic coefficient.

$$C_{ij} = \begin{bmatrix} C_{ij}^x \\ C_{ij}^y \\ C_{ij}^z \end{bmatrix}, i,j \in [0,3] \tag{7}$$

As shown in Figure 7, from the vertex coordinates and tangent vector information of the surface triangles, it is known that $v = 0$, $u = 1$ and $u = v$ are the three boundary curve parameters characterized by the vertices x_{00}, x_{10}, and x_{11} in the counterclockwise direction, respectively. Let $S(u,0)$, $S(1,v)$, and $S(u,u)$ be their corresponding curves, respectively.

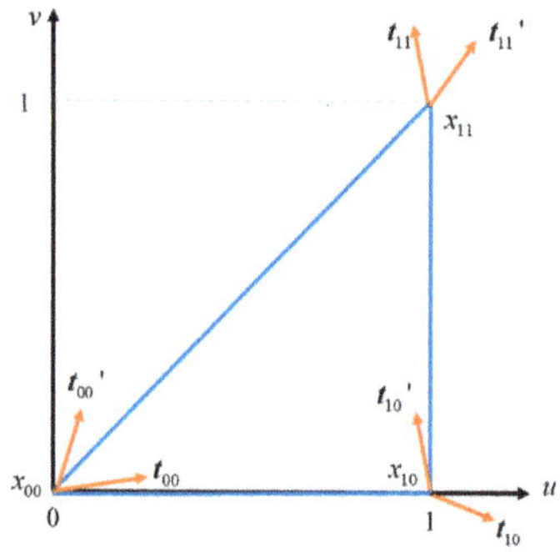

Figure 7. Coordinates and tangent vectors of the vertex of a curved triangle.

Substitute the characteristic parameters $v = 0$, $u = 1$, and $u = v$ into Equation (6). Then, the three boundary curves in the general parametric surface can be expressed as:

$$\begin{aligned} S(u,0) &= C_{00} + C_{10}u + C_{20}u^2 + C_{30}u^3 \\ S(1,v) &= C_{00} + C_{10} + C_{01}v + C_{20} + C_{11}v + C_{02}v^2 \\ &+ C_{21}v + C_{12}v^2 + C_{30} + C_{03}v^3 \\ S(u,u) &= C_{00} + C_{10}u + C_{01}u + C_{20}u^2 + C_{11}u^2 \\ &+ C_{02}u^2 + C_{21}u^3 + C_{12}u^3 + C_{30}u^3 + C_{03}u^3 \end{aligned} \tag{8}$$

The Hermite boundary curve corresponding to Equation (8) is obtained by substituting the vertex and tangent vector information in Figure 8 into Equation (1).

$$\begin{aligned} S(u,0) &= (2u^3 - 3u^2 + 1)x_{00} + (u^3 - 2u^2 + u)t_{00} + (-2u^3 + 3u^2)x_{10} + (u^3 - u^2)t_{10} \\ S(1,v) &= (2v^3 - 3v^2 + 1)x_{10} + (v^3 - 2v^2 + v)t_{10}' + (-2v^3 + 3v^2)x_{11} + (v^3 - v^2)t_{11} \\ S(u,u) &= (2u^3 - 3u^2 + 1)x_{00} + (u^3 - 2u^2 + u)t_{00}' + (-2u^3 + 3u^2)x_{11} + (u^3 - u^2)t_{11}' \end{aligned} \tag{9}$$

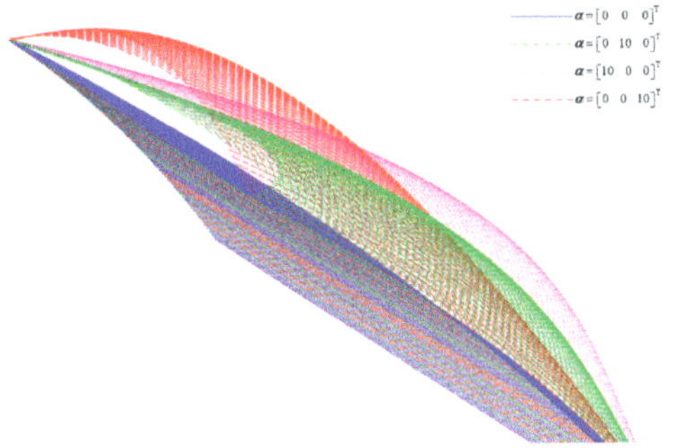

Figure 8. Hermite surfaces for different values of $\boldsymbol{a}$.

Equations (8) and (9) are combined to obtain Equation (10).

$$\begin{cases} C_{00} + C_{10}u + C_{20}u^2 + C_{30}u^3 = (2u^3 - 3u^2 + 1)x_{00} + (u^3 - 2u^2 + u)t_{00} \\ +(-2u^3 + 3u^2)x_{10} + (u^3 - u^2)t_{10} \\ C_{00} + C_{10} + C_{01}v + C_{20} + C_{11}v + C_{02}v^2 + C_{21}v + C_{12}v^2 + C_{30} + C_{03}v^3 \\ = (2v^3 - 3v^2 + 1)x_{10} + (v^3 - 2v^2 + v)t_{10}' + (-2v^3 + 3v^2)x_{11} + (v^3 - v^2)t_{11} \\ C_{00} + C_{10}u + C_{01}u + C_{20}u^2 + C_{11}u^2 + C_{02}u^2 + C_{21}u^3 + C_{12}u^3 + C_{30}u^3 + C_{03}u^3 \\ = (2u^3 - 3u^2 + 1)x_{00} + (u^3 - 2u^2 + u)t_{00}' + (-2u^3 + 3u^2)x_{11} + (u^3 - u^2)t_{11}' \end{cases} \tag{10}$$

The eigencoefficients of the Hermite surface model can be obtained by solving Equation (10).

$$\begin{aligned} &C_{00}{=}x_{00} \\ &C_{10}{=}t_{00} \\ &C_{01}{=}t_{00}' - t_{00} \\ &C_{20}{=} - 3x_{00} - 2t_{00} + 3x_{10} - t_{10} \\ &C_{11}{=}\boldsymbol{a} \\ &C_{02}{=} - 2t_{00}' + 3x_{11} - t_{11}' + 2t_{00} - 3x_{10} + t_{10} - \boldsymbol{a} \\ &C_{21}{=}t_{10}' - t_{00}' + t_{00} - \boldsymbol{a} \\ &C_{12}{=}2t_{00}' + t_{11}' - 2t_{10}' - 2t_{00} - t_{10} - t_{11} + \boldsymbol{a} \\ &C_{30}{=}2x_{00} + t_{00} - 2x_{10} + t_{10} \\ &C_{03}{=}2x_{10} + t_{10}' - 2x_{11} + t_{11} \end{aligned} \tag{11}$$

Then, Equation (6) can be written as Equation (12).

$$\boldsymbol{S}(u,v) = \boldsymbol{S}'(u,v) + f_a(u,v)\boldsymbol{a} \tag{12}$$

$$\begin{aligned} \boldsymbol{S}'(u,v) \; &= C_{00} + C_{10}u + C_{01}v + C_{20}u^2 + C_{02}'v^2 \\ &+C_{21}'u^2v + C_{12}'uv^2 + C_{30}u^3 + C_{03}v^3 \end{aligned} \tag{13}$$

$$\begin{aligned} &C_{21}'{=}t_{10}' - t_{00}' + t_{00} \\ &C_{02}'{=} - 2t_{00}' + 3x_{11} - t_{11}' + 2t_{00} - 3x_{10} + t_{10} \\ &C_{12}'{=}2t_{00}' + t_{11}' - 2t_{10}' - 2t_{00} - t_{10} - t_{11} \end{aligned} \tag{14}$$

where $f_a(u,v) = v(u-v)(1-u)$. $\boldsymbol{a} = [a_x \; a_y \; a_z]^{\mathrm{T}}$ is the set shape vector, which aims to improve the controllability of the surface model, and thus achieve the adjustment of the accuracy of fitting the local details of the CAD model. Figure 8 shows the change in the shape of the surface triangle when $\boldsymbol{a}$ is taken to different values. The focus of this paper is to construct a Hermite surface triangle for 3D printing with high-precision model fitting. A discussion of $\boldsymbol{a}$ will not be developed in this paper, and the specific details will be reflected in another paper. In order to verify the validity of the constructed surface model fit, we set $\boldsymbol{a} = \begin{bmatrix} 0 & 0 & 0 \end{bmatrix}^{\mathrm{T}}$ in the subsequent examples.

4. Error Calculation and Evaluation Method

In order to verify the effectiveness of the proposed method, the error calculation and comprehensive evaluation method between the fitted model and the CAD model are presented in this section. The fitting error is essentially the offset that exists between the fitted model and the original model surface. It is theoretically more accurate to use the inter-model volume error as the evaluation criterion for the offset of the model. However, since the distance $H(x, y)$ between model surfaces is related to both its location (x, y) and surface data, and the original model shape has uncertainty, this will make the fitting error difficult to be uniformly expressed by mathematical formulas. To this end, the fitting error is calculated in this paper by means of model sampling. The main idea is to transform the problem of integration of the volumes between the two models into a problem of summing the height differences at the sampling points.

In order to sample the data points of the original model reasonably and comprehensively, a method of calculating the sampling mapping points is proposed. The sampled data points are obtained by meshing the CAD model using *Hypermesh* software, as is shown Figure 9.

(a) (b) (c)

Figure 9. Sampling point acquisition for basketball model: (**a**) CAD model; (**b**) grid division; (**c**) sampling data points.

Figure 10 shows the schematic diagram of the model fitting error. $P_O(x_O, y_O, z_O)$ is any sampling point in the CAD model, and the fitting error at that point is the height difference between that point and the corresponding mapping point of the fitted model. The mapping point is the intersection point between a vertical line L_O made along the X/Y/Z direction past the $P_O(x_O, y_O, z_O)$ point and the fitted model, as shown in Figure 10. $P_H(x_O, y_O, z_H)$ and $P_S(x_O, y_O, z_S)$ are the mapping points of $P_O(x_O, y_O, z_O)$ on the Hermite surface model M_H and STL model M_S, respectively.

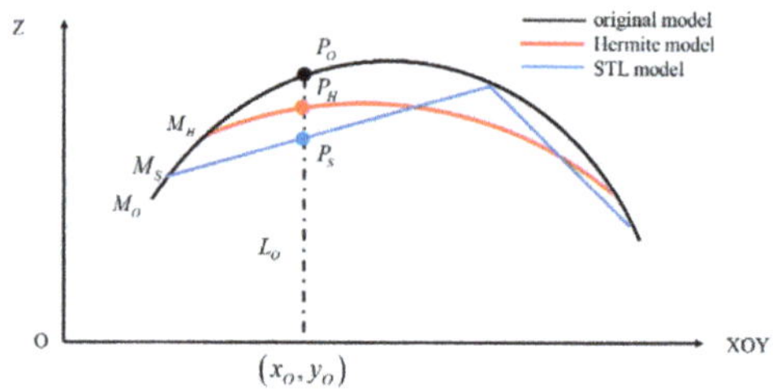

Figure 10. Local schematic of model fitting error.

The coordinates of the mapped point can be calculated from the information of the triangular surface slice where the vertical line intersects the fitted model. As shown in Figure 11, P_O' is the projection of the sampled points. $\overrightarrow{P_O'P_i'}$, $\overrightarrow{P_O'P_j'}$, and $\overrightarrow{P_O'P_k'}$ are the vectors constructed with the $\Delta P_i'P_j'P_k'$ vertices in a counterclockwise direction starting from P_O', respectively. $\overrightarrow{P_O'P_i'}$, $\overrightarrow{P_O'P_j'}$, and $\overrightarrow{P_O'P_k'}$ can be regarded as vectors rotating counterclockwise with P_O' as the center. When P_O' lies within $\Delta P_i'P_j'P_k'$, $\overrightarrow{P_O'P_i'} \times \overrightarrow{P_O'P_j'}$, $\overrightarrow{P_O'P_j'} \times \overrightarrow{P_O'P_k'}$, and $\overrightarrow{P_O'P_k'} \times \overrightarrow{P_O'P_i'}$ have the same sign, indicating that the vertical line L_O intersects the triangle $\Delta P_iP_jP_k$. Conversely, when P_O' is outside $\Delta P_i'P_j'P_k'$, the vertical line L_O does not intersect with triangle $\Delta P_iP_jP_k$.

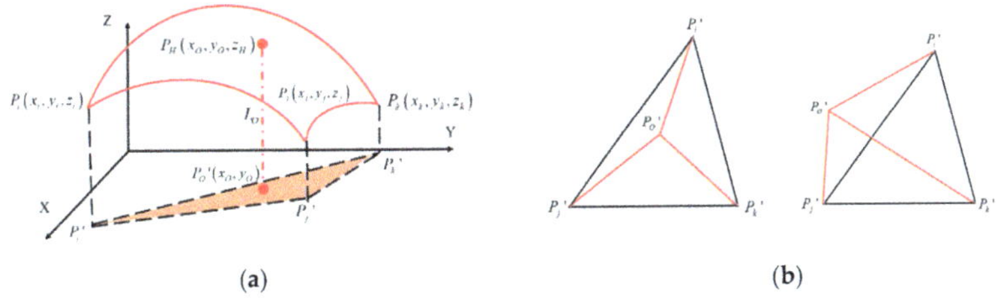

Figure 11. The vertical line intersects the curved triangle: (**a**) the projection point (Z-direction); (**b**) location of projection point.

Taking a vertical line along the Z-direction as an example, when the triangle in the Hermite model intersecting the vertical line is determined, the parameter value (u_H, v_H) of the sampled mapping point on the Hermite surface triangle can be found by Equation (16). The vertical coordinates of the sampling mapping points (e.g., Equation (16)) can be obtained by substituting (u_H, v_H) into Equation (14).

$$\begin{cases} x_O & = C_{00}^x + C_{10}^x u_H + C_{01}^x v_H + C_{20}^x {u_H}^2 + C_{11}^x u_H v_H \\ & + C_{02}^x {v_H}^2 + C_{21}^x {u_H}^2 v_H + C_{12}^x u_H {v_H}^2 + C_{30}^x {u_H}^3 + C_{03}^x {v_H}^3 \\ y_O & = C_{00}^y + C_{10}^y u_H + C_{01}^y v_H + C_{20}^y {u_H}^2 + C_{11}^y u_H v_H \\ & + C_{02}^y {v_H}^2 + C_{21}^y {u_H}^2 v_H + C_{12}^y u_H {v_H}^2 + C_{30}^y {u_H}^3 + C_{03}^y {v_H}^3 \end{cases} \tag{15}$$

$$\begin{aligned} z_H & = C_{00}^z + C_{10}^z u_H + C_{01}^z v_H + C_{20}^z {u_H}^2 + C_{11}^z u_H v_H \\ & + C_{02}^z {v_H}^2 + C_{21}^z {u_H}^2 v_H + C_{12}^z u_H {v_H}^2 + C_{30}^z {u_H}^3 + C_{03}^z {v_H}^3 \end{aligned} \tag{16}$$

It is worth noting that the procedure calculates all the sampled mapping points on the surface triangles intersecting the vertical line and takes the point closest to the sampled point as the error calculation point.

In the Z-direction of the STL fitted model, the triangles with vertices $P_i(x_i, y_i, z_i)$, $P_j\left(x_j, y_j, z_j\right)$ and $P_k(x_k, y_k, z_k)$, where the sampled mapping point $P_S(x_O, y_O, z_S)$ is located, should satisfy the following conditions.

$$\begin{vmatrix} x - x_i & y - y_i & z - z_i \\ x_j - x_i & y_j - y_i & z_j - z_i \\ x_k - x_i & y_k - y_i & z_k - z_i \end{vmatrix} = 0 \tag{17}$$

The vertical coordinate of this sample mapping point can be described as:

$$z_S = \frac{-a(x_O - x_i) - b(y_O - y_i)}{c} + z_i \tag{18}$$

where

$$\begin{aligned} a &= (y_j - y_i)(z_k - z_i) - (y_k - y_i)(z_j - z_i) \\ b &= (z_j - z_i)(x_k - x_i) - (z_k - z_i)(x_j - x_i) \\ c &= (x_j - x_i)(y_k - y_i) - (x_k - x_i)(y_j - y_i) \end{aligned} \tag{19}$$

After obtaining the sampling mapping points on the fitted model, the error between the fitted model and the original model can be obtained by calculating the distance between the sampling points and the mapping points, and the unit of distance is by millimeter. In order to make a comprehensive evaluation of the error of the fitted model, the evaluation method of calculating the fitting error in terms of mean deviation and variance is proposed. In the Z-direction, the mean deviation (mm) and variance (mm^2) of the sampled points can be described as:

$$\overline{e_Z} = \frac{|H_1^Z| + |H_2^Z| + \ldots + |H_n^Z|}{n} \tag{20}$$

$${s_z}^2 = \frac{\left(|H_1^Z| - \overline{e_z}\right)^2 + \left(|H_2^Z| - \overline{e_z}\right)^2 + \ldots + \left(|H_n^Z| - \overline{e_z}\right)^2}{n-1} \tag{21}$$

where $H_{ij}^Z(x, y) = M_i(x, y) - M_j(x, y)$. M_i and M_j are the original model and the fitted model, respectively. H is the distance difference between the models at the point (x, y). n is the number of sampling point. Similarly, the fitting errors in the X and Y directions can be solved by the above method.

5. Numerical Cases

Case 1. *In this section, we take the Stanford classical rabbit model with high complexity morphology as an example and compare the fitting error of its Hermite surface model and the 3D printing generic STL model. As shown in Figure 12, the solid rabbit model in step format was imported into Hypermesh software, and the sampled data points with 702 located on the model surface were obtained by meshing.*

(a) (b) (c)

Figure 12. Sampling point acquisition for rabbit model: (**a**) solid model; (**b**) grid division; (**c**) sampling data points.

The rabbit model was fitted using the Hermite surface triangles constructed in this paper and the planar triangles of the STL, respectively. The number of triangular surface slices is controlled by changing the approximation tolerance and mesh density of the fit. And finally, five groups of Hermite and STL fitted models were obtained, corresponding to the number of triangular facets of 16,862, 18,519, 22,820, 25,323 and 27,239, respectively. Table 1 shows the sampling mean deviation and variance of the original model and the fitted model in the X, Y, and Z directions.

Table 1. Results of model fitting errors.

Number	Models	Number of Facets	Z-Direction		Y-Direction		X-Direction	
			Mean Deviation (mm)	Variance (mm^2)	Mean Deviation (mm)	Variance (mm^2)	Mean Deviation (mm)	Variance (mm^2)
1	Hermite STL	16,862	0.29 13.17	4.15 5561	0.89 3.47	16.20 59.78	0.64 2.31	25.06 102.10
2	Hermite STL	18,519	0.27 22.19	3.92 215,450	1.77 3.12	160.76 53.73	0.79 3.50	38.81 174.12
3	Hermite STL	22,820	0.50 2.06	16.88 57.23	0.92 3.16	23.84 57.51	0.43 2.10	9.66 112.78
4	Hermite STL	25,323	0.26 2.59	3.70 306.76	0.60 2.64	9.13 40.47	0.47 1.81	10.15 67.52
5	Hermite STL	27,239	0.25 2.76	3.67 259.87	0.84 2.69	21.53 45.57	0.38 2.84	8.80 461.31

Theoretically, as the number of fitted model facets increases, the accuracy of the fit should also improve, i.e., the mean deviation will be reduced. However, as shown in Table 1, there is no such rule between the obtained data. The main reason for the existence of this phenomenon is that when the fitted model is not finely drawn to the local features of the original model, the triangle that should be used to calculate the fitting error at its closest distance does not intersect with the vertical line after some of the sampling points make a vertical line along the error calculation direction. This will cause an offset in the mapping point selection, resulting in a large sampling error.

Tables 2 and 3 show the distribution of the sampling mean deviation of the data in groups 1 and 5, respectively. As shown in Table 2, the maximum fitting error of the Hermite surface model in the Z direction is around 50, and then the overall fitting error is mostly distributed below 5. In contrast, the maximum fitting error of the STL model reaches more than 800, and there is still a large distribution around 50. In the other two directions, the maximum fitting errors of the two are similar, but the Hermite surface model has a smaller value of sampling error and a relatively concentrated distribution. In contrast, the fitting errors of the STL model are relatively discrete in distribution and have larger values. Similarly, as shown in Table 3, the maximum fitting error is similar in the Y direction, but in the Z and X directions, the maximum fitting error of the STL model is much larger than that of the Hermite surface model, and the overall error distribution is more discrete in all three directions.

As shown in Tables 1–3, the overall fitting accuracy of the Hermite surface model is higher compared to the STL model, but the obtained fitting error fluctuates more due to the existence of mapping point bias, which does not truly reflect the deviation between the fitted model and the original model. In order to evaluate the fitting ability of the two models more accurately and objectively, the distance-weighted nearest neighbor algorithm [28] is used to process the data in Table 1. This method achieves the screening and removal of data with more discrete distribution by calculating the distance between objects and assigning larger weights to closer distances. Table 4 and Figure 13 show the fitting errors and the variation in mean deviation of mean deviation after removing the "noise points", respectively.

Table 2. Mean deviation in the three directions in group 1.

	Hermite Model	STL Model
Z-direction	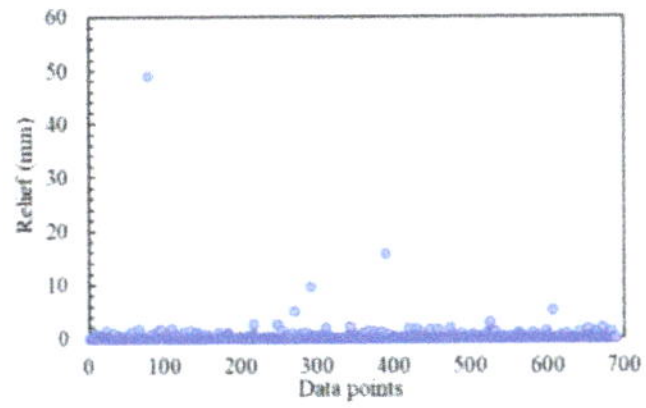	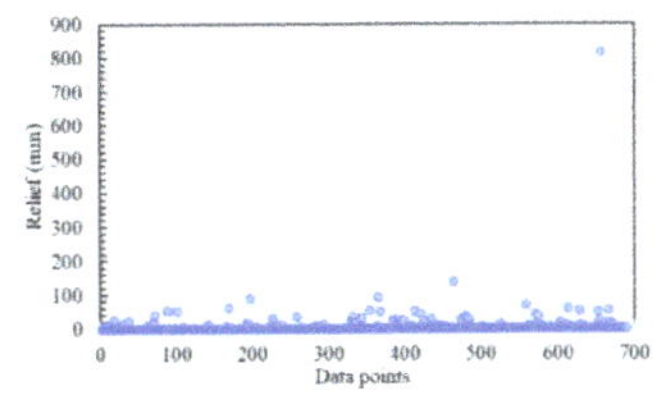
X-direction	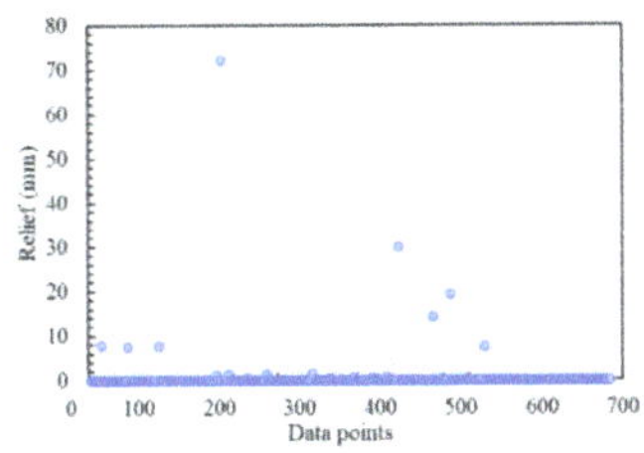	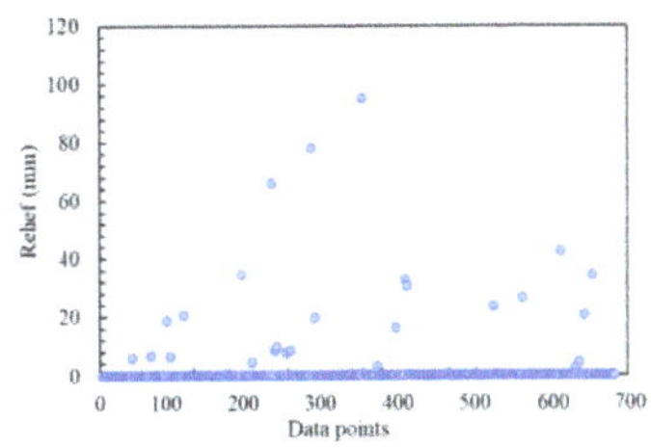
Y-direction	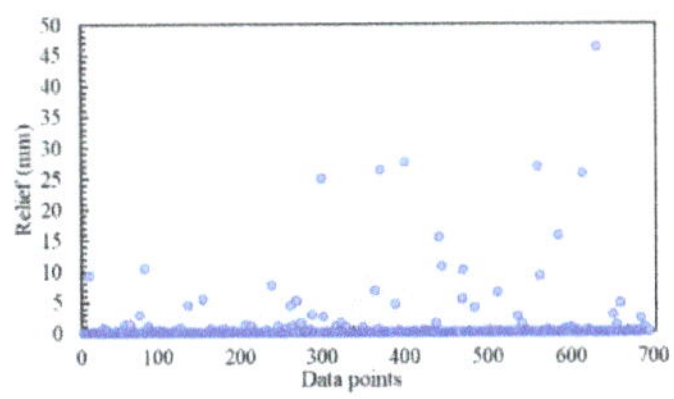	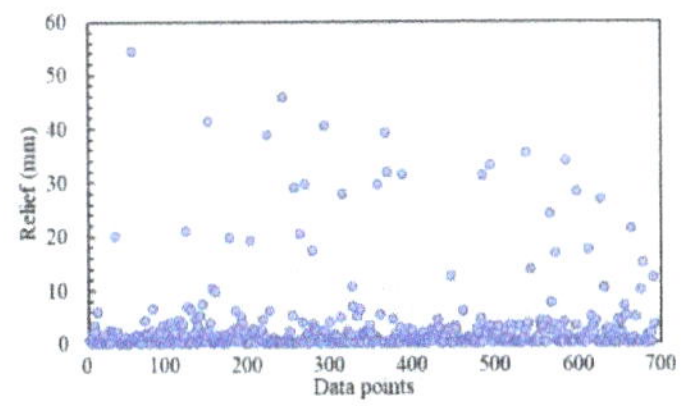

As can be seen from Table 4 and Figure 13, the mean deviation between the Hermite and STL models in the Z, X, and Y directions gradually decrease as the number of model facets increases, and gradually stabilize at the 4th and 5th groups. It shows that the accuracy of the fit of the 2 to the model is gradually improving with the increase in the number of face slices. The fitting error of the STL model fluctuates in the Y-direction, e.g., the error becomes larger in group 3. The main reason for this phenomenon is that STL uses planar triangles for fitting, which is more likely to produce an offset in the mapping points. At the same time, the overall fitting accuracy of the Hermite surface model is much better than that of the STL planar model, which can preserve the original model characteristics and accuracy as much as possible with a smaller number of face slices. This not only can effectively solve the problem of excessive data volume or data redundancy caused by the continuous subdivision of the face slices in the STL model when improving the fitting accuracy, but also can further improve the processing efficiency and manufacturing accuracy of complex models in the 3D printing process.

Table 3. Mean deviation in the three directions in group 5.

	Hermite Model	STL Model
Z-direction	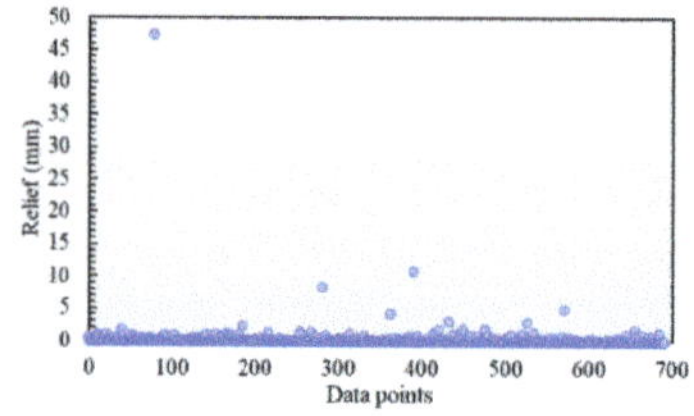	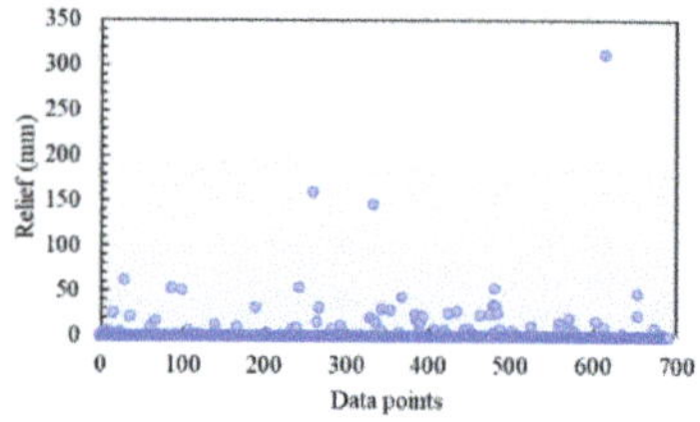
X-direction	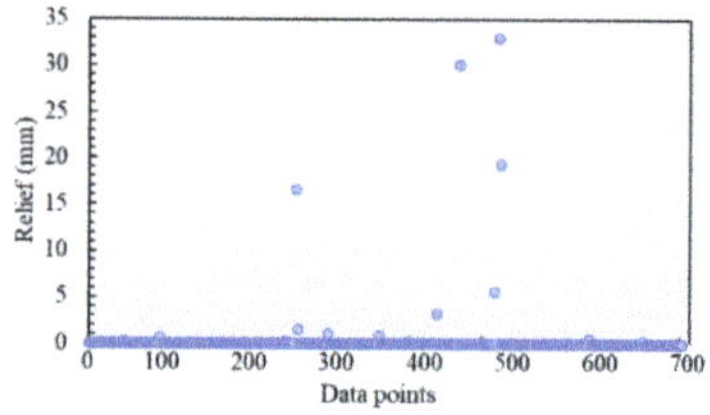	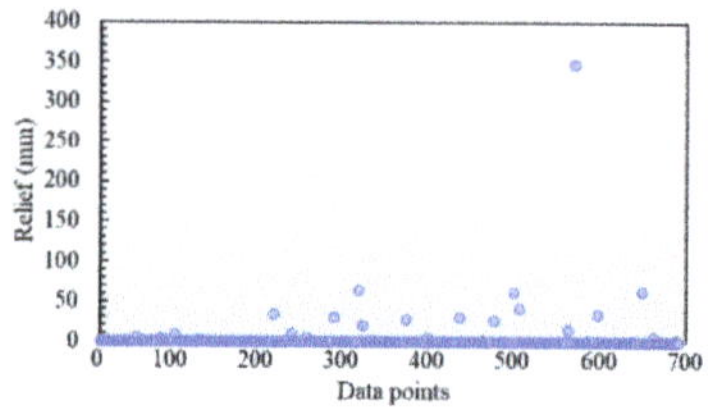
Y-direction	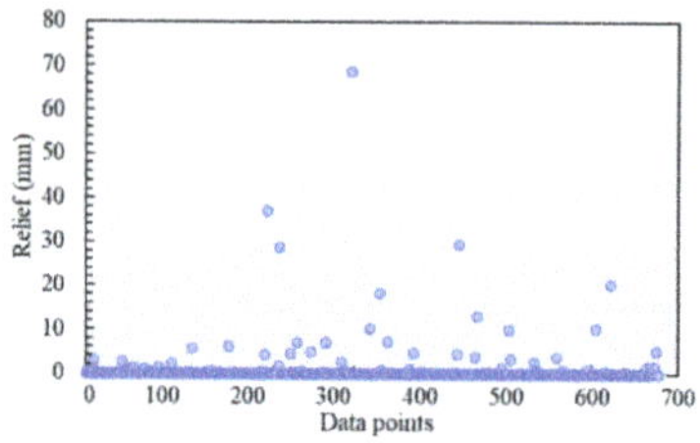	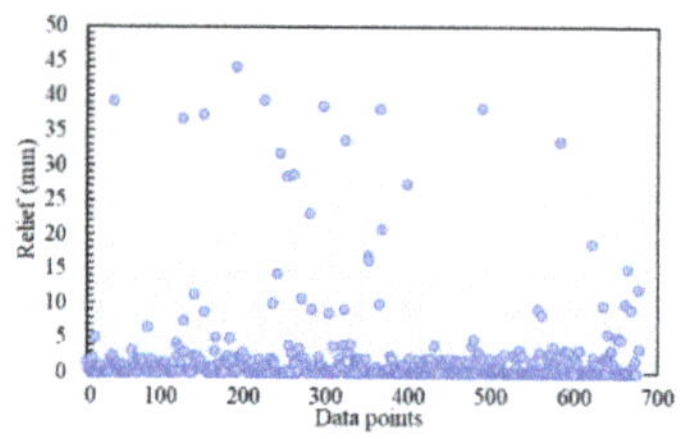

Table 4. Fitting error results after removing the "noise" from the rabbit model.

Number	Models	Number of Facets	Z-Direction Mean Deviation (mm)	Z-Direction Variance (mm^2)	Y-Direction Mean Deviation (mm)	Y-Direction Variance (mm^2)	X-Direction Mean Deviation (mm)	X-Direction Variance (mm^2)
1	Hermite STL	16,862	0.18 0.40	0.23 2.33	0.32 1.39	1.43 3.21	0.15 0.28	0.91 1.80
2	Hermite STL	18,519	0.16 0.38	0.22 2.15	0.33 1.20	1.32 2.86	0.13 0.27	0.74 1.87
3	Hermite STL	22,820	0.10 0.33	0.49 2.01	0.23 1.44	0.71 4.04	0.02 0.21	0.04 1.32
4	Hermite STL	25,323	0.09 0.29	0.31 1.74	0.24 1.32	0.65 3.81	0.02 0.17	0.17 1.09
5	Hermite STL	27,239	0.08 0.29	0.29 1.73	0.28 1.20	1.07 3.22	0.03 0.16	0.05 0.84

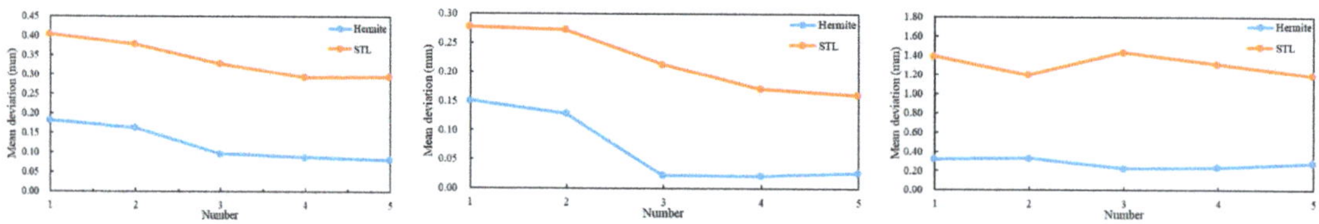

(a) (b) (c)

Figure 13. Variation in mean deviation: (**a**) Z-direction; (**b**) X-direction; (**c**) Y-direction.

Case 2. *The turbine model is used as another example to verify the effectiveness of the proposed method. A Hermite surface model and a 3D printing generic STL model are used to fit it, and the fitting error is analyzed. The solid model of the turbine and the 950 sampled data points extracted based on the grid division are shown in Figure 14.*

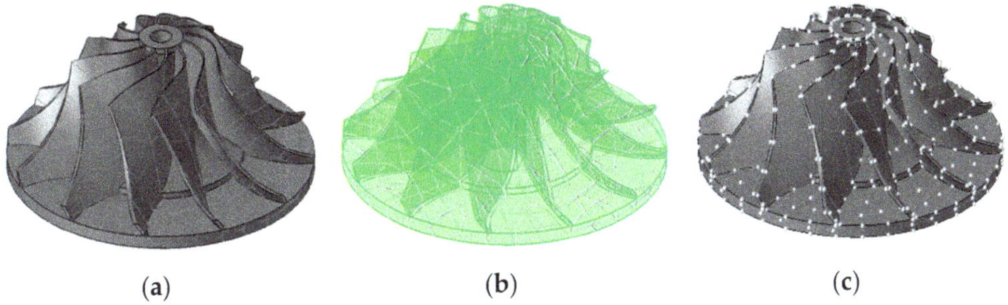

Figure 14. Sampling point acquisition for turbine model: (**a**) CAD model; (**b**) grid division; (**c**) sampling data points.

Table 5. Fitting error results after removing the "noise" from the turbine model.

Number	Models	Number of Facets	Z-Direction		X-Direction		Y-Direction	
			Mean Deviation (mm)	Variance (mm^2)	Mean Deviation (mm)	Variance (mm^2)	Mean Deviation (mm)	Variance (mm^2)
1	Hermit STL	18,272	0.25 0.35	0.08 0.13	0.18 0.30	0.07 0.19	0.24 0.33	0.07 0.16
2	Hermit STL	20,540	0.26 0.38	0.08 0.13	0.12 0.24	0.02 0.12	0.25 0.36	0.07 0.13
3	Hermit STL	22,752	0.23 0.37	0.05 0.14	0.13 0.26	0.06 0.18	0.24 0.33	0.07 0.16
4	Hermit STL	24,336	0.20 0.35	0.07 0.12	0.12 0.25	0.08 0.12	0.23 0.33	0.08 0.19
5	Hermit STL	28,592	0.21 0.34	0.07 0.12	0.11 0.25	0.10 0.12	0.20 0.32	0.04 0.08

The Hermite and STL fitting errors were compared by five groups of models with face piece numbers of 18,272, 20,540, 22,752, 24,336, and 28,592, respectively. Table 5 shows the mean deviation and variance of the fitted model in three directions after removing the "noise". The variation of the mean deviation is shown in Figure 15.

As is shown in Table 4 and Figure 13, the fitting error is gradually decreasing as the number of model facets increases in the Z and Y directions. The model fit error was minimized in group 5. The mean deviation was about 0.2 and 0.35 for Hermite and STL, respectively. In the X-direction, the fitting error of Hermite ranged from 0.2 to 0.1, and gradually tended to 0.1 as the number of model facets increased. The fitting error of the STL model showed a gradual decrease with the increase in the number of model facets, except for a slight fluctuation due to the mapping point offset on group 3. However, its lowest error among the 5 groups is still around 0.25. The overall fitting accuracy of the Hermite model for this case is still better than that of the STL model, which verifies the effectiveness of the proposed method.

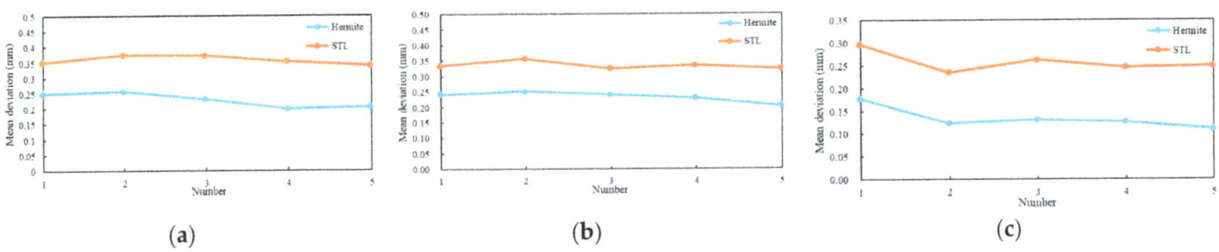

Figure 15. Variation in mean deviation: (**a**) Z-direction; (**b**) X-direction; (**c**) Y-direction.

6. Conclusions

In this paper, we propose a Hermite surface triangle model construction method considering the high-precision fitting of 3D printing models. The mapping relationship between different surface triangles and characteristic triangles is established by radial variation. Using the vertex and tangent vector information, a cubic Hermite curve model with adjustable accuracy of the model local fitting is constructed based on the general parametric cubic surface model. The model effectively reduces the parameter variables, simplifies the complexity of the calculation, and achieves the specification of the solution problem. A model fitting error calculation and evaluation method based on sampling mapping points is proposed. It transforms the continuous integration into a discrete summation problem, effectively solving the problem that is difficult to express uniformly in mathematical formulas due to the uncertainty of the original model. The effectiveness of the proposed method in improving the model fitting accuracy was verified by using rabbit and turbine models with five different sets of face slices. In the future research work, we will carry out a systematic study of $\boldsymbol{a}$. Considering the different characteristics of the model, we will study an adaptive control method of $\boldsymbol{a}$ that can satisfy high-precision fitting. In addition, based on the constructed Hermite surface model, we will study the adaptive layering technique of the surface model and surface path planning method considering the layered slicing and path planning errors. This will provide vital support to further enhance efficient and high-precision manufacturing of 3D printing.

Author Contributions: Conceptualization, R.L.; formal analysis, R.L.; investigation, J.F.; methodology, J.F.; writing—original draft, R.L. and J.F.; writing—review and editing, S.J., Y.C. and R.L. All authors have read and agreed to the published version of the manuscript.

Funding: The work described in this paper was jointly funded by the National Key Research and Development Program of China under Grant No. 2017YFB1102804.

Data Availability Statement: Not applicable.

Conflicts of Interest: The authors declare no conflict of interest.

References

1. Fu, J.; Li, H.; Song, X.; Fu, M. Multi-scale defects in powder-based additively manufactured metals and alloys. *J. Mater. Sci. Technol.* **2022**, *27*, 165–199. [CrossRef]
2. Gardan, J. Additive manufacturing technologies: State of the art and trends. *Int. J. Prof. Bus. Rev.* **2015**, *54*, 3118–3132.
3. Johnson, N.; Vulimiri, P.; To, A.; Zhang, X.; Brice, C.; Kappes, B.; Stebner, A. Invited review: Machine learning for materials developments in metals additive manufacturing. *Addit. Manuf.* **2020**, *36*, 101641. [CrossRef]
4. Singamneni, S.; Roychoudhury, A.; Diegel, O.; Huang, B. Modeling and evaluation of curved layer fused deposition. *J. Mater. Process. Technol.* **2012**, *212*, 27–35. [CrossRef]
5. Blakey-Milner, B.; Gradl, P.; Snedden, G.; Brooks, M.; Pitot, J.; Lopez, E.; Leary, M.; Berto, F.; Plessis, A. Metal additive manufacturing in aerospace: A review. *Mater. Design* **2021**, *12*, 110008. [CrossRef]
6. Gradl, P.; Teasley, T.; Protz, C.; Katsarelis, C. Process Development and Hot-fire Testing of Additively Manufactured NASA HR-1 for Liquid Rocket Engine Applications. In Proceedings of the AIAA Propulsion and Energy 2021 Forum, Reston, VA, USA, 9–11 August 2021.
7. Gardner, L.; Kyvelou, P.; Herbert, G.; Buchanan, C. Testing and initial verification of the world's first metal 3D printed bridge. *J. Constr. Steel Res.* **2020**, *172*, 106233. [CrossRef]
8. Armstrong, M.; Mehrabi, H.; Naveed, N. An overview of modern metal additive manufacturing technology. *J. Manuf. Process.* **2022**, *84*, 1001–1029. [CrossRef]
9. Jafari, D.; Vaneker, T.; Gibson, I. Wire and arc additive manufacturing: Opportunities and challenges to control the quality and accuracy of manufactured parts. *Mater. Des.* **2021**, *202*, 109471. [CrossRef]
10. Gao, W.; Zhang, Y.; Ramanujan, D.; Ramani, K.; Chen, Y.; Williams, C.; Wang, C.; Shin, Y.; Zhang, S.; Zavattieri, D. The status, challenges, and future of additive manufacturing in engineering. *Comput. Aided Des.* **2015**, *69*, 65–89. [CrossRef]
11. Zhao, D.; Guo, W. Shape and Performance Controlled Advanced Design for Additive Manufacturing: A Review of Slicing and Path Planning. *J. Manuf. Sci. E* **2020**, *142*, 010801. [CrossRef]
12. Kim, Y.; Nguyen, P.; Kim, H.; Choi, Y. Multi-morphology cellular structure design with smooth transition of geometry and homogenized mechanical properties between adjacent cells. *Mater. Des.* **2022**, *218*, 110727. [CrossRef]
13. Zhu, Z.; Anwer, N.; Mathieu, L. Shape Transformation Perspective for Geometric Deviation Modeling in Additive Manufacturing. *Procedia CIRP* **2018**, *75*, 75–80. [CrossRef]
14. Navangul, G.; Paul, R.; Anand, S. Error minimization in layered manufacturing parts by stereolithography file modification using a vertex translation algorithm. *J. Manuf. Sci. Eng.* **2013**, *135*, 031006. [CrossRef]

15. Zha, W.; Anand, S. Geometric approaches to input file modification for part quality improvement in additive manufacturing. *J. Manuf. Process.* **2015**, *20*, 465–477. [CrossRef]
16. Vlachos, A.; Peters, J.; Boyd, C.; Mitchell, J. Curved PN triangles. *Symp. Inter. Graph.* **2001**, *3*, 159–166.
17. Hamann, B.; Trotts, I.; Farin, G. On approximating contours of the piecewise trilinear interpolant using triangular rational quadratic Bezier patches. *Vis. Comput. Graph. IEEE Trans.* **1997**, *3*, 215–227. [CrossRef]
18. Márta, S.; Ildikó, S. C1-continuous Coons-type blending of triangular patches. *KoG* **2005**, *9*, 29–34.
19. Barnhill, R.; Birkhoff, G.; Gordon, W. Smooth interpolation in triangles. *J. Approx. Theory* **1973**, *8*, 114–128. [CrossRef]
20. Hagen, H. Geometric surface patches without twist constraints. *Comput. Aided Geom. Des.* **1986**, *3*, 179–184. [CrossRef]
21. *ISO/ASTM52915–2020*; Standard Specification for Additive Manufacturing File Format (AMF) Version 1.2. ASTM International: West Conshohocken, PA, USA, 2020.
22. Technical Committee AMT/8. IBS EN ISO/ASTM 52900: Additive manufacturing–general principles–terminology. *Int. Stand.* **2017**, *8*, 1–30.
23. Li, J.; Liu, C. Cubic Trigonometric Hermite Interpolation Curve: Construction, Properties, and Shape Optimization. *J. Funct. Spaces* **2022**, *2022*, 7525056 . [CrossRef]
24. DeBoor, C. High accuracy Hermite interpolation. *Comput. Aided Geom. Des.* **1987**, *4*, 269–278. [CrossRef]
25. Hollig, K.; Koch, J. Geometric Hermite interpolation. *Comput. Aided Geom. Des.* **1995**, *13*, 67–580. [CrossRef]
26. Imre, J. Cubic parametric curve of given tangent and curvature. *Comput. Aided Geom. Des.* **1998**, *30*, 1–9.
27. Stearns, C.; Kannappan, K. *Method for 2-D Affine Transformation of Images: US5475803 A*; LSI Logic Corporation: Milpitas, CA, USA, 1995.
28. Parvinnia, E.; Sabeti, M.; Jahromi, M.; Boostani, R. Classification of EEG Signals using adaptive weighted distance nearest neighbor algorithm. *J. King Saud. Univ.* **2014**, *26*, 1–6. [CrossRef]

axioms

MDPI

Article

An Asymmetric Model Position Dependent Mass: Quantum Mechanical Study

Biswanath Rath [1,*], Pravanjan Mallick [1,*], Jihad Asad [2,*], Rania Wannan [3], Rabab Jarrar [2] and Hussein Shanak [2]

[1] Department of Physics, Maharaja Sriram Chandra Bhanja Deo University, Takatpur, Baripada 757003, Odisha, India
[2] Department of Physics, Faculty of Applied Sciences, Palestine Technical University—Kadoorie, Tulkarm P305, Palestine; r.jarrar@ptuk.edu.ps (R.J.); h.shanak@ptuk.edu.ps (H.S.)
[3] Department of Applied Mathematics, Faculty of Applied Sciences, Palestine Technical University—Kadoorie, Tulkarm P305, Palestine; r.wannan@ptuk.edu.ps
* Correspondence: biswanathrath10@gmail.com (B.R.); pravanjanphy@gmail.com (P.M.); j.asad@ptuk.edu.ps (J.A.)

Abstract: We propose an asymmetric model position dependent mass and study its quantum mechanical behaviour on different potentials such as harmonic oscillator potential, double well potential, Gaussian single well potential and triangular single well model potential. It is observed from our study that the model asymmetric mass works well for weak coupling preserving the symmetric phase portrait. However, the dominance of asymmetric feature of the mass in the system clearly visible for higher values of the constant associated with the mass. Though, both position dependent mass and potential have significant role in controlling the spectral feature of the system, one may dominate over other for certain cases.

Keywords: PDM; phase portrait; quantum study; asymmetry; real spectra

PACS: 03.65. Ge

Citation: Rath, B.; Mallick, P.; Asad, J.; Wannan, R.; Jarrar, R.; Shanak, H. An Asymmetric Model Position Dependent Mass: Quantum Mechanical Study. *Axioms* **2023**, *12*, 318. https://doi.org/10.3390/axioms12040318

Academic Editor: Leonid Plotnikov

Received: 31 January 2023
Revised: 11 March 2023
Accepted: 15 March 2023
Published: 23 March 2023

1. Introduction

The study of the problems associated with position dependent mass (PDM) have continued to attract the attention of scientific community due to their relevance in various branches of physics and allied areas of science [1,2]. Further, the identification of wave function in a complex environment could be possible by solving Schrödinger equation with PDM [3]. The majority of such studies dedicated to the problems relevance to semiconductor physics and solid state physics [4–7]. The PDM involved in various problems can either be symmetric or asymmetric in nature. Further, the asymmetric forms of PDM have successfully been explained certain features related to semiconductor physics. For example; the propagation of electron through the abrupt interface of a semiconductor heterostructure [8] as well as optomechanical features of resonator [9] can be shown to explain by the PDM of type.

$$m(x) = \frac{m}{(1+\gamma x)^2} \tag{1}$$

Further, in a recent work, da Costa et al. [10] has investigated the coherent state nature using the above PDM. El-Nabulsi has studied the system involving the PDM of the type.

$$m(x) = m(1+\gamma x)^k \tag{2}$$

And reported some of its implications to semiconductors, quantum dots, crystalline solid in the presence of impurity [4,5]. In addition, the transport of electrons in a semiconductor can be tailored by considering the PDM of the form [11].

$$m(x) = me^{ax+\frac{1}{2}bx^2} \tag{3}$$

Several authors also explained different aspects of the quantum systems considering asymmetric PDM. For example; Dong et al. [12] obtained the eigenvalues and eigenfunctions of the asymmetric model singular mass oscillator with mass of the type.

$$m(x) = \frac{1}{\tau^{\alpha}(x+a)^{\alpha}} \tag{4}$$

Asad et al. [1] studied the phase portrait and stability of the harmonic like oscillator associated with asymmetric PDM of the type.

$$m(x) = \frac{m}{1+e^{-x-\lambda x^2}} \tag{5}$$

Recently, Dong et al. [3] have used an asymmetric model PDM

$$m(x) = \frac{\alpha e^{-\alpha x}}{(1-e^{-\alpha x})} \tag{6}$$

And reported the exact solution of the Schrodinger equation for few typical potentials. One can extend such asymmetric PDM for understanding the properties of solid state and semiconductor physics. Further, it is worth mentioning here that the some of the properties of the semiconductor has also been studied using symmetric PDM [13]. For example; El-Nabulsi [14] has studied the dynamics of electron with PDM of type.

$$m(x) = me^{-ax^2} \tag{7}$$

Silva et al. studied the electronic properties of electrons on a bilayer graphene catenoid bridge characterized by the PDM of the form [15].

$$m(x) = m\left(1+\frac{\lambda R^2}{(x^2+R^2)^2}\right) \tag{8}$$

Further, the vibrational inversion modes of NH_3 molecule has been explained by using the PDM of the form [16]

$$m(x) = m\left(\frac{1-\eta a^2x^2}{1-a^2x^2}\right) \tag{9}$$

In view of the importance of the PDM, several studies [17–23] also report different features of the systems in which the PDM varies either symmetric or asymmetric. The sgn(x) unction shows the asymmetry character which can suitably be used in formulating quantum mechanical problem involving double well [24,25].

In the present study, we designed a new type mass which varies asymmetrically with position in view of the importance of asymmetric PDM in explaining different features of semiconductor physics and study the spectral characteristics of the system by varying model parameter associated with the PDM as well as potentials. Our study thus suggested that both the PDM and potential have significant role in controlling the spectral feature of the system. Further, one may dominate over other for certain case.

2. Characteristic Features of New Asymmetric PDM

The asymmetric PDM used in this work is constructed as

$$m(x) = \frac{m}{1 + \lambda(\text{sgn}(x)) + \lambda^2 x^2} \tag{10a}$$

In the above, the sign function, $\text{sgn}(x)$ is defined as

$$\text{sgn}(x) = \begin{cases} 1, x \succ 0 \\ 0, x = 0 \\ -1, x \prec 0 \end{cases} \tag{10b}$$

The above PDM is very sensitive to the parameter, λ and its behaviour changes dramatically upon the change of λ value. Figures 1 and 2 show the behaviour of the PDM with distance for $\lambda = 0.01$ and 0.1 respectively. The nature of the graph changes dramatically for higher values of λ.

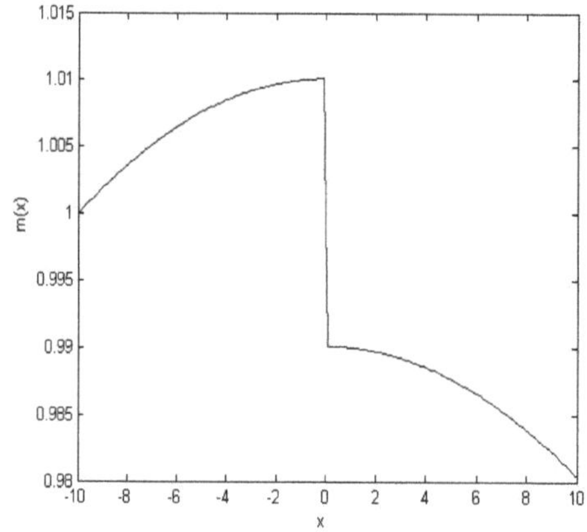

Figure 1. Variation of $m(x)$ (Equation (10a)) with position for $\lambda = 0.01$.

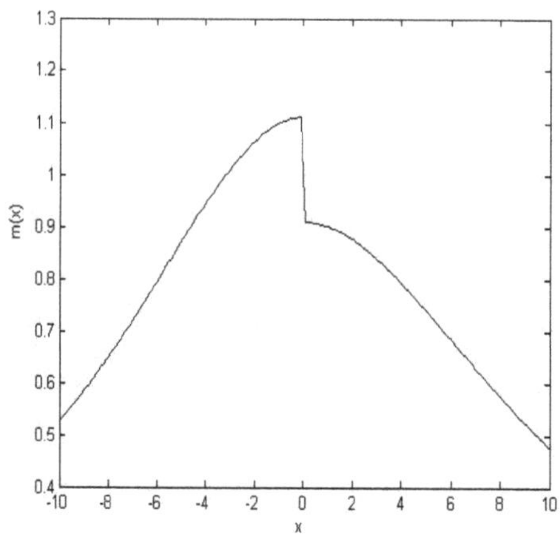

Figure 2. Variation of $m(x)$ (Equation (10a)) with position for $\lambda = 0.1$.

3. Quantum Mechanical Study on the New PDM Systems

Here, we solve the eigenvalue relation [26],

$$H|\Psi\rangle = E|\Psi\rangle \tag{11a}$$

where

$$|\Psi\rangle = \sum_n A_n|\phi_n\rangle \tag{11b}$$

In the above, $|\phi_n\rangle$ satisfies the relation

$$H_o|\phi_n\rangle = (2n+1)|\phi_n\rangle \tag{12a}$$

and

$$|\phi_n\rangle = N_n e^{-\frac{x^2}{2}} H_n(x) \tag{12b}$$

where N_n is the normalization constant such that

$$\langle\phi_n|\phi_m\rangle = \delta_{mn} \tag{13}$$

The Hamiltonian considered here is

$$H = T + m(x)V(x) \tag{14}$$

where,

$$T = \frac{1}{[m(x)]^{\frac{1}{4}}} p \frac{1}{[m(x)]^{\frac{1}{2}}} p \frac{1}{[m(x)]^{\frac{1}{4}}} \tag{15}$$

The above expression is considered due to the mass and the said kinetic energy is due to von Roos model operator [27,28].

4. Effect of Potential

Here, we consider different forms of potential in order to study their effect on the energy eigenvalues of the Hamiltonian (Equation (14)) associated with PDM (Equation (10a)). We have seen that the change of potential also affect the spectral features of the Hamiltonian. The details of these studies are discussed in the followings.

4.1. Single Well Potential

We consider the single potential as

$$V(x) = x^2 \tag{16}$$

And study the behaviour of the PDM Hamiltonian as stated above. On solving the Hamiltonian with the potential (Equation (16), Figure 3), we obtained the closed phase portrait for $\lambda = 0.01$ (Figure 4) and 0.1 (Figure 5) along with the stable real energy level. The representative real energy spectra for the studied system with $\lambda = 0.01$ is shown in Figure 6. It should be noted here that the circular symmetric nature of the phase portrait is evident for $\lambda = 0.01$ (Figure 4) and the effect of mass becomes significant at higher values of λ i.e., $\lambda = 0.1$. The effect of asymmetry associated with the PDM is clearly visible in the asymmetric nature of phase portrait (Figure 5).

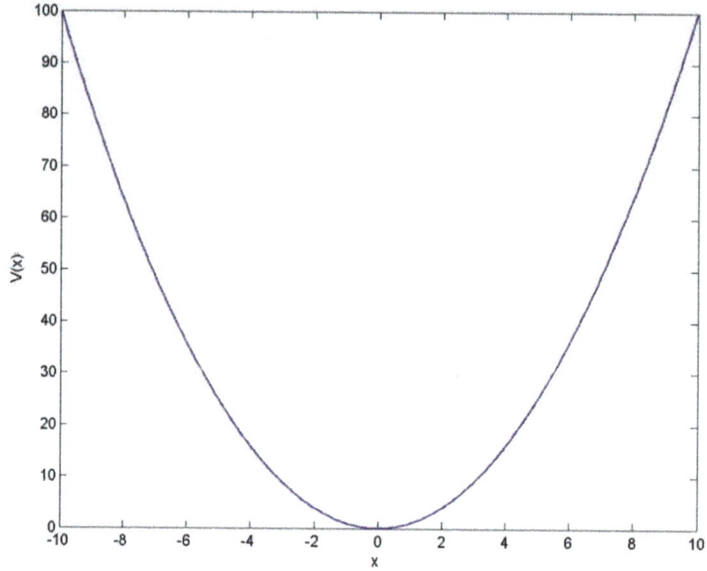

Figure 3. Variation of $V(x)$ (Equation (16)) with x.

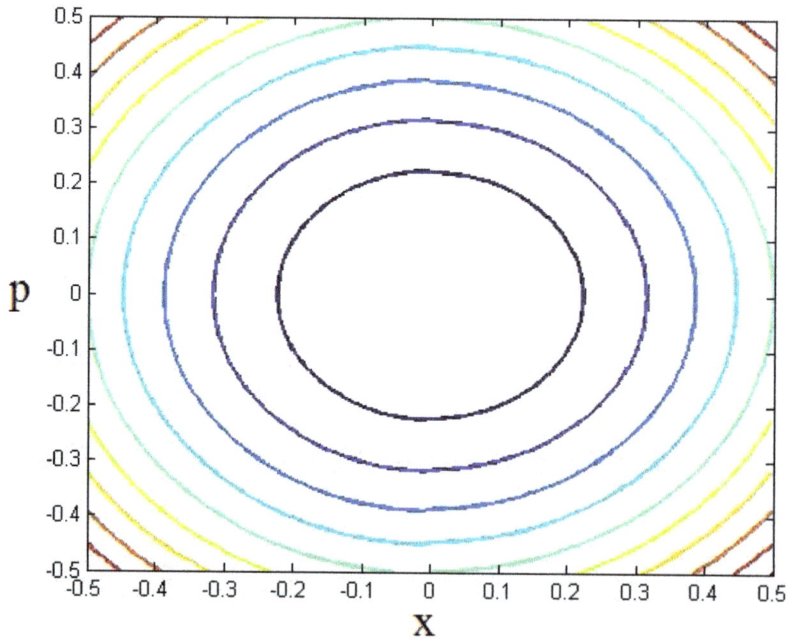

Figure 4. Phase portrait of the PDM Hamiltonian associated with single well potential for $\lambda = 0.01$.

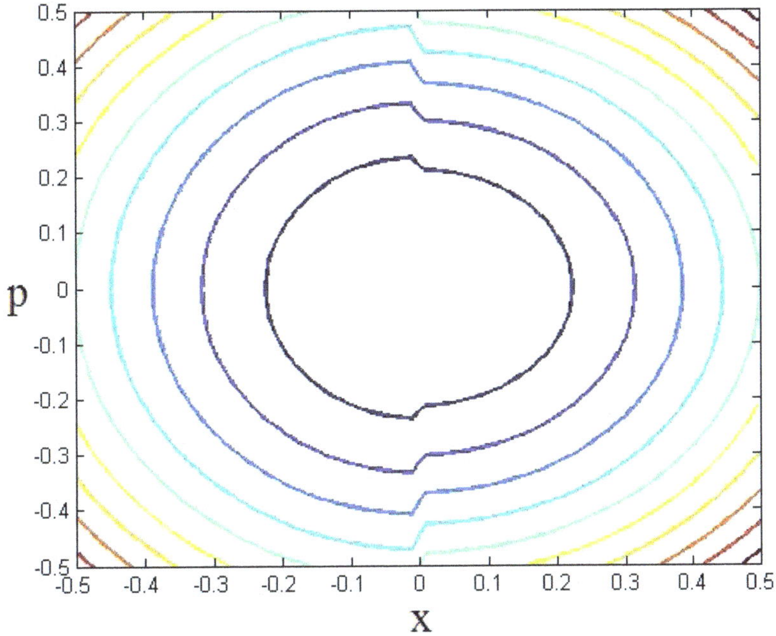

Figure 5. Phase portrait of the PDM Hamiltonian associated with single well potential for $\lambda = 0.1$.

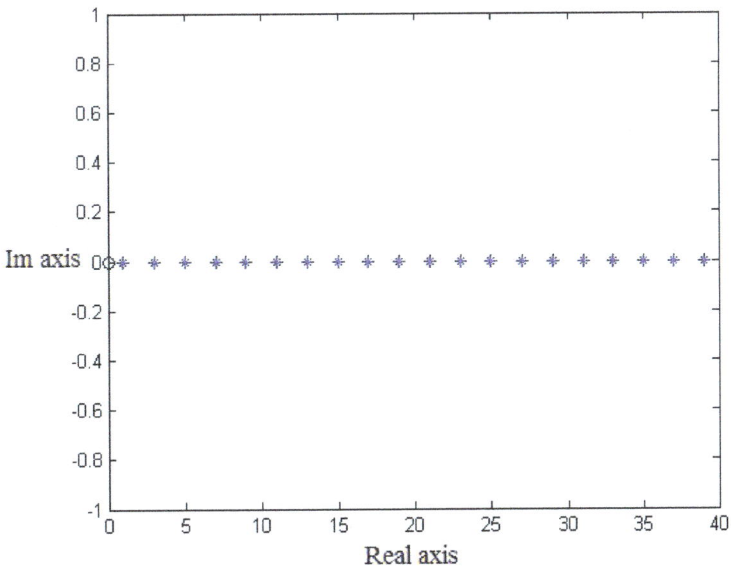

Figure 6. Energy eigenvalues of the PDM Hamiltonian associated with single potential for $\lambda = 0.01$.

4.2. Double Well Potential

We consider the single potential as [29]

$$V(x) = x^4 - 3x^2 \tag{17}$$

And study the behaviour of the PDM Hamiltonian as stated above. On solving the Hamiltonian with the potential (Equation (17), Figure 7), we obtained the closed phase portrait for $\lambda = 0.01$ (Figure 8) and 0.1 (Figure 9) along with the stable real energy level. The representative real energy spectra for the studied system with $\lambda = 0.01$ is shown in Figure 10. Like the single well case, the asymmetry associated with the PDM is also clearly visible in the asymmetric nature of phase portrait for $\lambda = 0.1$ (Figure 9).

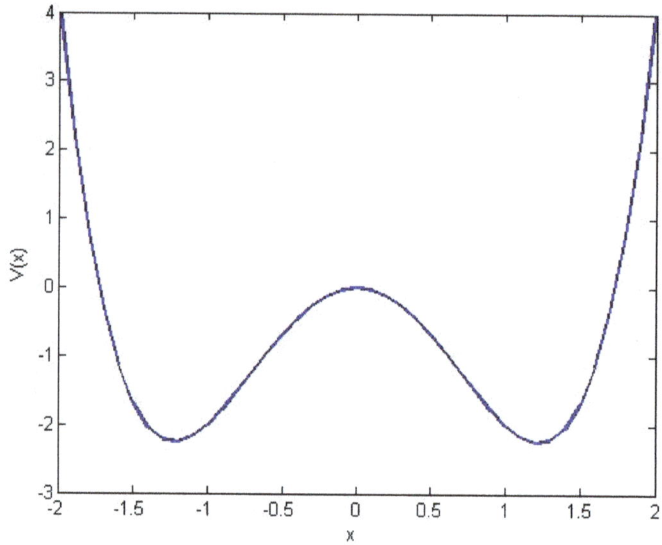

Figure 7. Variation of $V(x)$ (Equation (17)) with x.

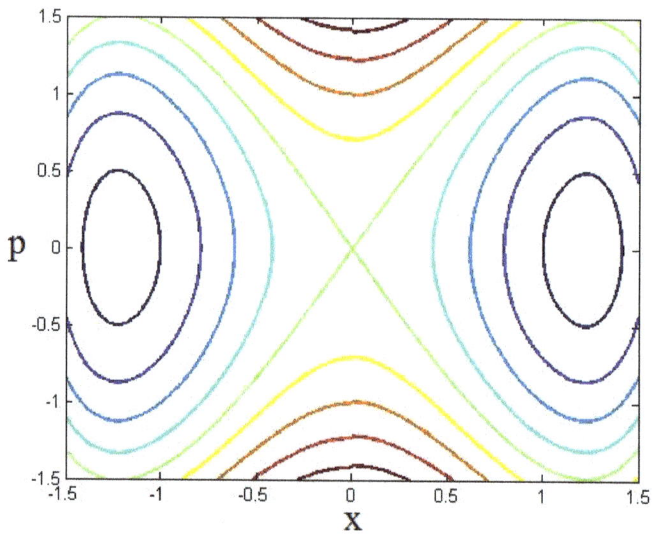

Figure 8. Phase portrait of the PDM Hamiltonian associated with double well potential for $\lambda = 0.01$.

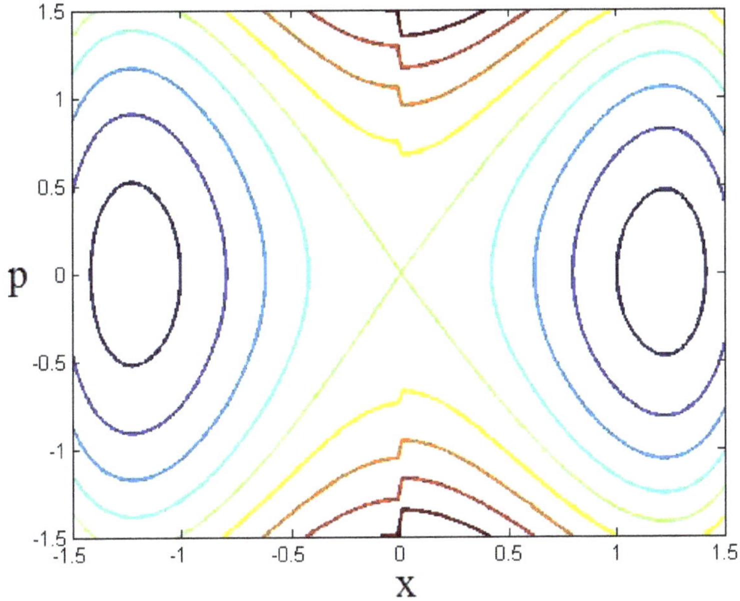

Figure 9. Phase portrait of the PDM Hamiltonian associated with double well potential for λ = 0.1.

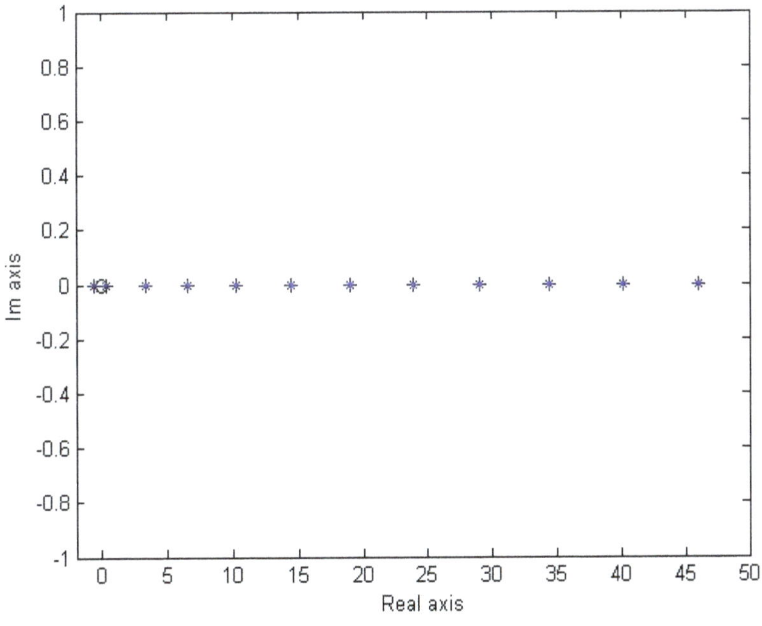

Figure 10. Energy eigenvalues of the PDM Hamiltonian associated with double well potential for $\lambda = 0.01$.

4.3. Gaussian Single Well Potential

Here, we consider the potential as [30]

$$V(x) = -100e^{-x^2} \tag{18}$$

To study the behaviour of the PDM Hamiltonian as stated above. On solving the Hamiltonian with the potential (Equation (18), Figure 11), we obtained the closed phase portrait for $\lambda = 0.01$ and 0.1 (Figure 12) with the stable real energy level (Figure 13). In this case, a typical type symmetric phase portrait is seen for $\lambda = 0.01$. However, the same showed distortion with appearance of asymmetry for $\lambda = 0.1$ like the previous cases. This result also indicates the dominance of the potential for low values of λ i.e., for $\lambda = 0.01$.

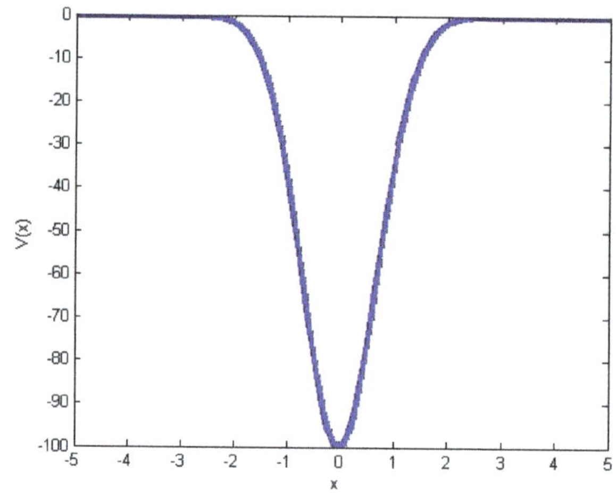

Figure 11. Variation of $V(x)$ (Equation (18)) with x.

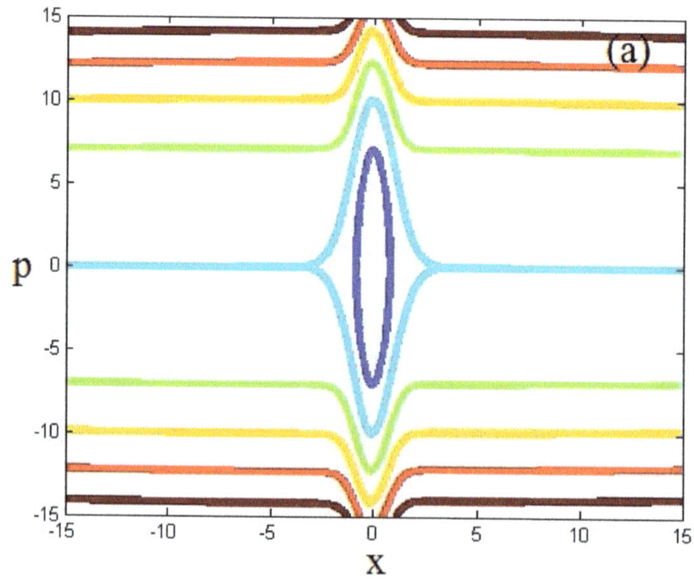

Figure 12. *Cont.*

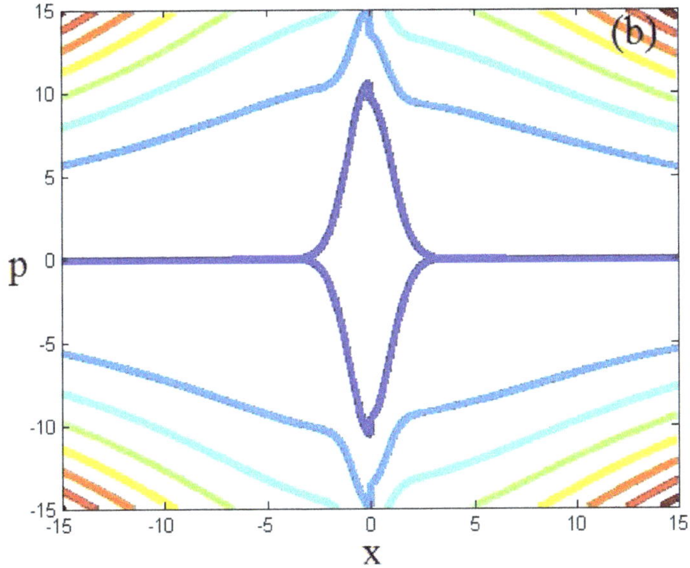

Figure 12. Phase portrait of the PDM Hamiltonian associated with Gaussian single well type potential for (**a**) $\lambda = 0.01$ and (**b**) $\lambda = 0.1$.

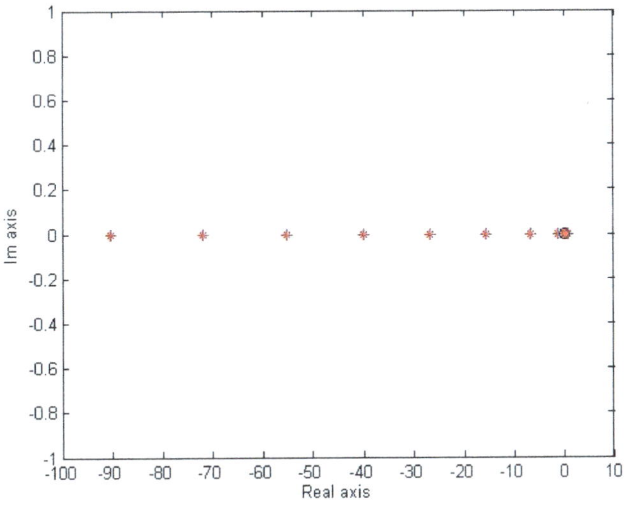

Figure 13. Energy eigenvalues of the PDM Hamiltonian associated with Gaussian single well type potential for $\lambda = 0.01$.

4.4. Rath Triangular Potential

Here, we consider the potential as [31]

$$V(x) = 100(1 - \exp(-0.02|x|)) \tag{19}$$

To study the behaviour of the PDM Hamiltonian as stated above. On solving the Hamiltonian with the potential (Equation (19), Figure 14), we obtained the closed phase portrait for $\lambda = 0.01$ and 0.1 (Figure 15) with the stable real energy level (Figure 16). It is to be noted here that the phase portrait still shows the symmetric loop in this case as well as for single well and double well potential cases for very low values of λ i.e., for $\lambda = 0.01$ due to the dominating nature of respective potentials (single well, double well and Rath triangular potentials) at low value of constant ($\lambda = 0.01$) associated with the PDM. The symmetric nature of the phase portrait starts distorting and the asymmetry becomes clearly visible for $\lambda = 0.1$ due to the dominating nature of asymmetric PDM. The present study thus suggests that both potential and PDM have the significant role in controlling the spectral feature of the Hamiltonian. However, one may dominate over other for certain case.

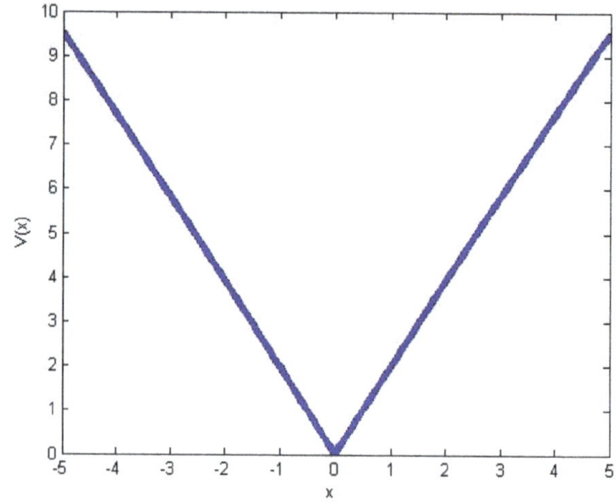

Figure 14. Variation of $V(x)$ (Equation (20)) with x.

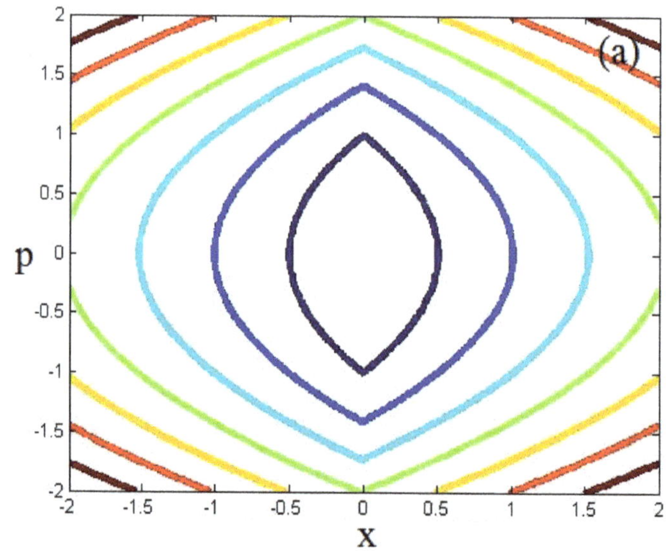

Figure 15. *Cont.*

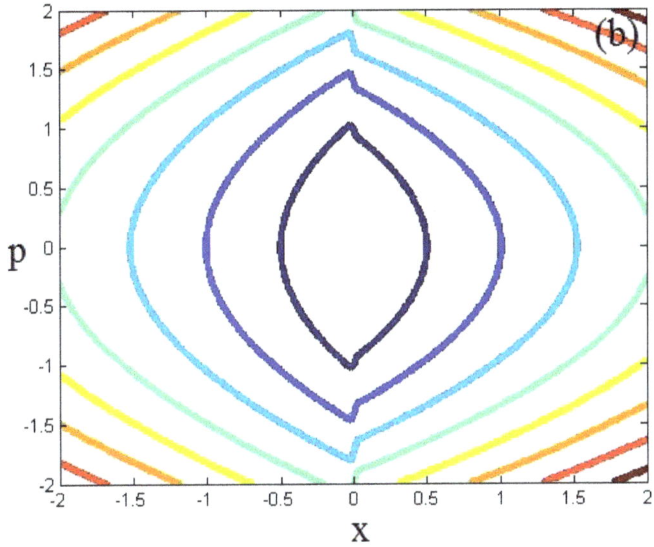

Figure 15. Phase portrait of the PDM Hamiltonian associated with Rath potential for (**a**) λ = 0.01 and (**b**) λ = 0.1.

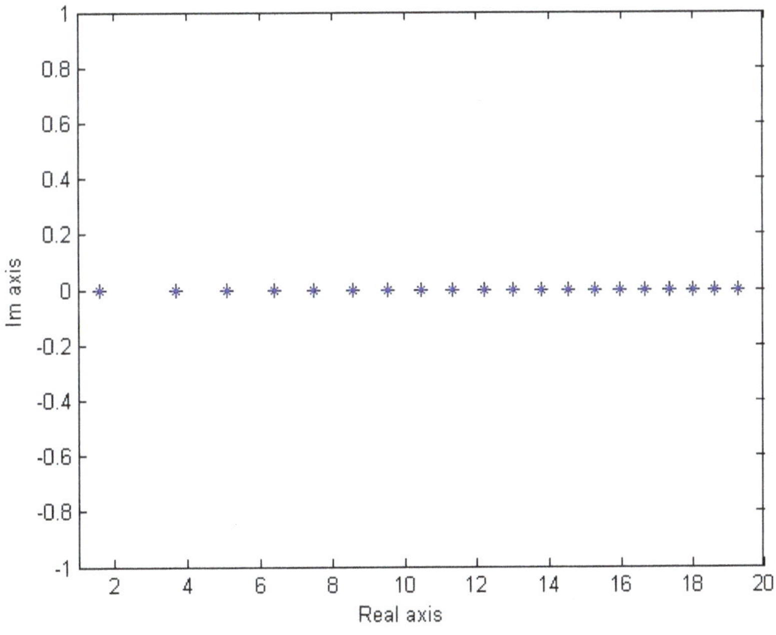

Figure 16. Energy eigenvalues of the PDM Hamiltonian associated with Rath potential for $\lambda = 0.01$.

5. Validity of Uncertainty Relation

All the PDM Hamiltonian discussed above would satisfy the uncertainty relation [32]. Here, we consider a typical harmonic oscillator type potential (single well potential (16)) to calculate the uncertainty relation as follows

$$\Delta x = \sqrt{\langle x^2 \rangle - \langle x \rangle^2} = 0.7137 \quad (20)$$

$$\Delta p = \sqrt{\langle p^2 \rangle - \langle p \rangle^2} = 0.7035 \quad (21)$$

$$\Delta x \Delta p = 0.5057 \quad (22)$$

The above analysis indicates that the uncertainty product ($\Delta x \Delta p$) is greater than that of simple harmonic oscillator.

6. Discussion and Conclusions

We model a new PDM in view of its importance in realizing different features associated with semiconductor physics. The effect of PDM on the spectral feature of the associated Hamiltonian was studied by varying potential term and constant parameter of the PDM using the matrix diagonalization method as started above. It is worth mentioning here that the method of calculation has also been tested for other systems [1,2,26]. Further, the result of da Costa et al. [33] have successfully been reproduced and reflected in our recent work [34]. The present PDM contain a sensitive asymmetric term i.e., sgn(x) function and a symmetric function ($\lambda^2 x^2$). Without the sgn(x) function, the mass function remains the same as that of Mathew's Lakshmanan [35] PDM which is quite symmetric about the origin. However, with the inclusion of sgn(x) function, we noticed that the mass function shows asymmetric character. In fact the study of sgn(x) function is crucial. The sgn(x) function has previously been used in supersymmetry [24] where the shape invariant property cannot be verified. Secondly, the sgn(x) function is also used in double well potential [25]. We therefore introduce sgn(x) function in designing a new mass and study its spectral behaviour associated with different potentials. It is worth mentioning here that the triangular model potential proposed by Rath [31] is an alternative to model scattering potential [36] for the study of spectral feature. Since the PDM used in the present study is asymmetric in nature, the phase portrait is expected to preserve the asymmetric feature. The phase portrait of different systems for different potentials show the symmetric behaviour for weak coupling limit i.e., for $\lambda = 0.01$. The symmetric nature of the phase portrait starts distorting and the asymmetry becomes clearly visible for $\lambda = 0.1$ due to the dominating nature of asymmetric PDM. However, the closed phase portraits of the studied systems reflect the unbroken nature of spectra irrespective of symmetric or asymmetric nature. This feature thus signifies the stability of the system. In order to study the spectral nature, we used matrix diagonalization method [26] and noticed that the spectral feature (energy levels) remains invariant for different size of the matrix. We feel that the interested readers will be motivated by the present investigation. Our study thus suggested that both PDM and potential have significant role in controlling the spectral feature of the system. Further, one may dominate over other for certain case. We believe that the results of all the PDM Hamiltonian discussed above would be within the preview of usual uncertainty relation.

Author Contributions: B.R.: Conceptualization, Methodology, Data curation, Writing—Reviewing and Editing, Validation. P.M.: Investigation, Writing—Original draft preparation, Writing—Reviewing and Editing, Validation. J.A.: Writing—Reviewing and Editing, Validation. R.W.: Writing—Reviewing and Editing, Validation. R.J.: Writing—Reviewing and Editing, Validation. H.S.: Writing—Reviewing and Editing, Validation. All authors have read and agreed to the published version of the manuscript.

Funding: The authors did not receive support from any organization for the submitted work.

Data Availability Statement: All data generated or analysed during this study are included in this published article.

Acknowledgments: The authors (Jihad Asad, Rania Wannan, Rabab Jarrar and Hussein Shanak) would like to thank Palestine Technical University—Kadoorie for financial support.

Conflicts of Interest: The authors have no competing interest to declare that are relevant to the content of this article. The authors declare they have no financial interests.

References

1. Asad, J.; Mallick, P.; Samei, M.E.; Rath, B.; Mohapatra, P.; Shanak, H.; Jarrar, R. Asymmetric variation of a finite mass harmonic like oscillator. *Results Phys.* **2020**, *19*, 103335. [CrossRef]
2. Rath, B.; Mallick, P.; Mohapatra, P.; Asad, J.; Shanak, H.; Jarrar, R. Position-dependent finite symmetric mass harmonic like oscillator: Classical and quantum mechanical study. *Open Phys.* **2021**, *19*, 266–276. [CrossRef]
3. Dong, S.-H.; Huang, W.-H.; Sedaghatnia, P.; Hassanabadi, H. Exact solutions of an exponential type position dependent mass problem. *Results Phys.* **2022**, *34*, 105294. [CrossRef]
4. El-Nabulsi, R.A. A generalized self-consistent approach to study position-dependent mass in semiconductors organic heterostructures and crystalline impure materials. *Phys. E* **2020**, *134*, 114295. [CrossRef]
5. El-Nabulsi, R.A. A new approach to Schrodinger equation with position-dependent mass and its implications in quantum dots and semiconductors. *J. Phys. Chem. Sol.* **2020**, *140*, 109384. [CrossRef]
6. Peter, A.J. The effect of position dependent effective mass of hydrogrnic impurities in parabolic GaAs/GaAlAs quantum dots in a strong magnetic field. *Int. J. Mod. Phys. B* **2009**, *23*, 5109. [CrossRef]
7. Sinha, A. Scattering states of a particle, with position-dependent mass, in a double heterojunction. *Eur. Phys. Lett.* **2011**, *96*, 20008. [CrossRef]
8. Costa Filho, R.N.; Almeida, M.P.; Farias, G.A.; Andrade, J.S., Jr. Displacement operator for quantum systems with position-dependent mass. *Phys. Rev. A* **2011**, *84*, 050102. [CrossRef]
9. Ullah, K.; Ullah, H. Enhanced optomechanically induced transparency and slow/fast light in a position-dependent mass optomechanics. *Eur. Phys. J. D* **2020**, *74*, 197. [CrossRef]
10. Da Costa, B.G.; Gomez, I.S.; Rath, B. Exact solution and coherent states of an asymmetric oscillator with position-dependent mass. *J. Math. Phys.* **2023**, *64*, 012102. [CrossRef]
11. Biswas, K.; Saha, J.P.; Patra, P. On the position-dependent effective mass Hamiltonian. *Eur. Phys. J. Plus* **2020**, *135*, 457. [CrossRef]
12. Cruz y Cruz, S.; Rosas-Ortiz, O. Position-dependent mass oscillators and coherent states. *J. Phys. A Math. Theor.* **2009**, *42*, 185205. [CrossRef]
13. Sari, H.; Kasapoglu, E.; Sakiroglu, S.; Sökmen, I.; Duque, C.A. Effect of position-dependent effective mass on donor impurity- and exciton-related electronic and optical properties of 2D Gaussian quantum dots. *Eur. Phys. J. Plus* **2022**, *137*, 341. [CrossRef]
14. El-Nabulsi, R.A. Dynamics of position-dependent mass particle in crystal lattices microstructures. *Phys. E* **2021**, *127*, 114525. [CrossRef]
15. Silva, J.E.G.; Furtado, J.; Ramos, A.C.A. Position-dependent mass effects on a bilayer graphene catenoid bridge. *Eur. J. Phys. B* **2021**, *94*, 127. [CrossRef]
16. Roy-Layinde, T.O.; Omoteso, K.A.; Oyero, B.A.; Laoye, J.A.; Vincent, U.E. Vibrational resonance of ammonia molecule with doubly singular position-dependent mass. *Eur. J. Phys. B* **2022**, *95*, 80. [CrossRef]
17. El-Nabulsi, R.A. Quantum dynamics in low-dimensional systems with position-dependent mass and product-like fractal geometry. *Phys. E* **2021**, *134*, 114827. [CrossRef]
18. Ghosh, A.P.; Mandal, A.; Sarkar, S.; Ghosh, M. Influence of position-dependent effective mass on the nonlinear optical properties of impurity doped quantum dots in presence of Gaussian white noise. *Opt. Commun.* **2016**, *367*, 325–334. [CrossRef]
19. Alpdogan, S.; Havare, A. Dirac Particle for the Position Dependent Mass in the Generalized Asymmetric Woods-Saxon Potential. *Adv. High Energy Phys.* **2014**, *2014*, 973847. [CrossRef]
20. Aydogdu, O.; Arda, A.; Sever, R. Effective-mass Dirac equation for Woods-Saxon potential: Scattering, bound states, and resonances. *J. Math. Phys.* **2012**, *53*, 042106. [CrossRef]
21. Rajashabala, S.; Navaneethakrishnan, K. Effects of dielectric screening and position dependent effective mass on donor binding energies and on diamagnetic susceptibility in a quantum well. *Superlattices Microstruct.* **2008**, *43*, 247–261. [CrossRef]
22. Amir, N.; Iqbal, S. Coherent states for nonlinear harmonic oscillator and some of its properties. *J. Math. Phys.* **2015**, *56*, 062108. [CrossRef]
23. Dos Santos, M.A.; Gomez, I.S.; da Costa, B.G.; Mustafa, O. Probability density correlation for PDM-Hamiltonians and superstatistical PDM-partition functions. *Eur. Phys. J. Plus* **2021**, *136*, 96. [CrossRef]
24. Chen, Y.; Yan, J.; Mihalache, D. Stable flat-top solitons and peakons in the $\mathcal{PT}$-symmetric δ-signum potentials and nonlinear media. *Chaos* **2019**, *29*, 083108. [CrossRef]
25. Marques, F.; Negrini, O.; da Silva, A.J. A new simple class of superpotentials in SUSY quantum mechanics. *J. Phys. A Math. Theor.* **2012**, *45*, 115307. [CrossRef]
26. Rath, B.; Mallick, P.; Akande, J.; Mohapatra, P.P.; Adja, D.K.K.; Koudahoun, L.H.; Kpomahou, Y.J.F.; Monsia, M.D.; Sahoo, R.R. A General type of Liénard Second Order Differential Equation: Classical and Quantum Mechanical Study. *Proc. Indian Natl. Sci. Acad.* **2017**, *83*, 935–940.

27. Von Roos, O. Position-dependent effective masses in semiconductor theory. *Phys. Rev. B* **1983**, *27*, 7547. [CrossRef]
28. Von Roos, O.; Mavromatis, H. Position-dependent effective masses in semiconductor theory. II. *Phys. Rev. B* **1985**, *31*, 2294. [CrossRef]
29. Rath, B.; Mavromatis, H.A. Energy-level calculation through modified Hill determinant approach: For general oscillator. *Indian J. Phys.* **1999**, *73B*, 641.
30. Killingbeck, J.P.; Scott, T.; Rath, B. A matrix method for power series potentials. *J. Phys. A Math. Gen.* **2000**, *33*, 6999. [CrossRef]
31. Jones, H.F. Comment on Solvable model of bound states in the continuum (BIC) in on dimension (2019, 94, 105214). *Phys. Scr.* **2021**, *96*, 087001. [CrossRef]
32. Zettili, N. *Quantum Mechanics: Concepts and Applications*, 2nd ed.; John Wiley: New York, NY, USA, 2001; p. 37.
33. Da Costa, B.G.; Da Silva, G.A.C.; Gomez, I.S. Supersymmetric quantum mechanics and coherent states for a deformed oscillator with position-dependent effective mass. *J. Math. Phys.* **2021**, *62*, 092101. [CrossRef]
34. Asad, J.; Mallick, P.; Samei, M.E.; Rath, B.; Mohapatra, P.; Shanak, H.; Jarrar, R. Reply to Comment on "Asymmetric Variation of a Finite Mass Harmonic Like Oscillator". *Results Phys.* **2022**, *32*, 105148. [CrossRef]
35. Mathews, P.M.; Lakshmanan, M. On a unique nonlinear oscillator. *Q. Appl. Math.* **1974**, *32*, 215–218. [CrossRef]
36. Ahmed, Z.; Kumar, S.; Ghosh, D.; Goswami, T. Solvable model of bound states in the continuum (BIC) in one dimension. *Phys. Scr.* **2019**, *94*, 105214. [CrossRef]

axioms

Article

Mathematical Description of the Aerodynamic Characteristics of Stationary Flows in a Vertical Conical Diffuser When Air Is Supplied through Various Tube Configurations

Leonid Plotnikov

Turbines and Engines Department, Ural Federal University named after the first President of Russia B.N. Yeltsin, Str. Mira, 19, 620002 Yekaterinburg, Russia; leonplot@mail.ru; Tel.: +7-922-291-64-50

Abstract: Conical diffusers of various configurations are used in many kinds of technical equipment and manufacturing processes. Therefore, it is a relevant objective to obtain reliable experimental and mathematical data on the aerodynamic characteristics of diffusers. This article presents experimental data on the aerodynamics of stationary flows in a vertical conical diffuser when air is supplied through tubes with various cross sections (circle, square, and triangle). Instantaneous values of air flow velocity are measured with a constant-temperature hot-wire anemometer. Data are obtained on the velocity fields and turbulence intensity along the height and the diameter of the diffuser's cylindrical part when air is supplied through tubes of various configurations. It is established that air supply through profiled tubes has a significant effect on the shape of the velocity field and turbulence intensity in a vertical conical diffuser. For example, higher values of turbulence intensity are typical of air supplied through profiled tubes (the differences reach 50%). A mathematical formulation (linear and exponential equations) of the change in the average speed and intensity of air flow turbulence along the height of the diffuser's cylindrical part for various initial conditions and supply tube configurations is presented. The obtained findings will make it possible to refine mathematical models and update algorithms for engineering the design of diffusers for various engineering processes and pieces of technical equipment.

Keywords: vertical cone diffuser; aerodynamics; stationary flow; profiled tubes; velocity field; turbulence intensity; empirical regularities

MSC: 41A30; 41A45

Citation: Plotnikov, L. Mathematical Description of the Aerodynamic Characteristics of Stationary Flows in a Vertical Conical Diffuser When Air Is Supplied through Various Tube Configurations. *Axioms* **2023**, *12*, 244. https://doi.org/10.3390/axioms12030244

Academic Editor: Yuli Chashechkin

Received: 25 January 2023
Revised: 23 February 2023
Accepted: 24 February 2023
Published: 27 February 2023

1. Introduction

In almost all branches of technology, equipment is used in which the main engineering process is associated with the movement of either liquid or gas. Examples of such equipment are heat exchangers, gas cleaning units, boilers, chemical equipment, industrial furnaces, various types of dryers, ventilation devices, and nozzle systems. Almost all these devices have a working chamber or supply channels with a conical diffuser [1]. The operation of these devices has shown that their calculated efficiency cannot be always achieved [2]. In many cases, this is due to the uneven supply of the working medium to the equipment's working area and, therefore, the flow's physical features in the diffuser and corresponding gas-dynamic losses [3,4]. All this testifies to the importance of studying the aerodynamics of process units with conical diffusers from the point of view of gas-dynamic improvement and finding ways to control flow characteristics. Moreover, it is necessary to obtain reliable experimental data on the features of gas flow dynamics and their mathematical formulation in order to improve engineering methods for calculating conical diffusers and to refine mathematical models [5,6].

Presented below is a brief review of classical (fundamental) research and updated knowledge of aerodynamic characteristics in a conical diffuser.

There are basic studies on the gas dynamics of flows in conical diffusers for various designs and at various initial levels of turbulence intensity. For example, A. Klein conducted experiments and examined experimental data on the influence of inlet conditions on conical diffuser characteristics [7]. In particular, the article describes the effects of blocking the thickness of the inlet boundary layer for various forms of inlet flow, turbulence intensity, and Reynolds numbers. K. Jeyachandran and V. Ganesan studied the effect of various gas-dynamic conditions at the diffuser inlet with physical and mathematical modelling [8]. The paper presents an analysis of data on the velocity profiles, pulse power, shape factor, performance, and efficiency of the studied diffusers. P. A. C. Okwuobi and R. S. Azad experimentally studied the flow structure and turbulence characteristics in a conical diffuser with a fully developed inlet flow [9]. As a result, the basic physical laws were established, and recommendations for the design of conical diffusers were formulated [7–9].

There are modern studies of the gas-dynamic characteristics of gas flows in various diffuser configurations aimed at clarifying calculation methods, supplementing the knowledge base about the relevant physical mechanisms, and creating original mathematical models [10–13]. A. Ferrari proposed analytical dependencies that complement the Fanno models relating to viscous adiabatic flow in a pipe with a constant cross section and an inviscid adiabatic flow in a conical diffuser [10]. These new dependences expand the set of exact solutions of gas dynamics and improve the quality of mathematical models. J. Lee and co-authors investigated the features of turbulence and coherent structures in gas flow through a conical diffuser with various opening angles (2°, 4°, and 8°) based on numerical simulation [11]. The data obtained made it possible to refine the physical features of the gas dynamics of stationary flows in diffusers. X. Wu et al. discovered an internal layer in the gas flow through an asymmetric flat diffuser based on physical and mathematical modelling [12]. Other research will improve the engineering design methods for diffusers. Using numerical simulation, F. J. De Souza and co-authors investigated the gas dynamics of an air flow with various particles in a vertical conical diffuser [13]. It was found that even a small number of particles significantly affects the flow's shape in a diffuser; the flow can stick to the wall under certain conditions.

There is a large amount of applied research on the use of diffusers in various kinds of technical equipment. For example, diffusers are actively used in the organic Rankine cycle [14–18]. In this cycle, the key device is a turbo expander equipped with an outlet diffuser to increase pressure recovery and improve turbine efficiency. A. Zou et al. optimised the diffuser design for a specific turbine based on numerical simulation [14]. B. Dong et al. evaluated the influence of gas parameters at the diffuser inlet on its gas-dynamic characteristics (in particular, the velocity coefficient) with mathematical modelling [15]. C. S. From et al. studied the influence of flow turbulence intensity at the diffuser inlet on gas dynamics and its influence on the performance of a particular turbine [17]. J. A. Keep and co-authors studied the geometry of a combined (annular–radial) diffuser in relation to the gas-dynamic characteristics of flows and turbine efficiency in the Rankine cycle [18].

Diffuser channels are actively used and studied in many other technical applications. For example, it is relevant to optimise the shape and parameters of flows in a diffuser for a mixing chamber in ejectors [19,20], to fine-tune supersonic diffusers [21], to estimate parameters in diffusers for aircraft gas turbine engines [22], to introduce a diffuser into a reactor for ammonia oxidation [23], and to model and improve the aerodynamics of diffusers in a tubular reactor [24] or an industrial furnace [25].

Thus, the performed review showed that the study of gas-dynamic effects in a conical diffuser is an urgent task at present. This is due to the wide range of applications of diffusers in various technical applications such as: the Rankine cycle, gas and steam turbines, mixing chambers, ammonia oxidation reactors, syngas production plants, industrial furnaces, and others.

Based on an analysis of the data of other authors, the following conclusions can be drawn:

- Gas-dynamic improvement and the development of ways to control the gas dynamics of flows in a conical diffuser remain urgent tasks for fundamental and applied science;
- Recent research into the gas-dynamic characteristics of flows in conical diffusers is carried out mainly through numerical simulation;
- There is a lack of reliable experimental data (and their mathematical description) on the gas-dynamic characteristics of flows in a vertical conical diffuser for mathematical model verification.

Consequently, this research's key objectives are as follows:

- To develop an experimental set-up for studying stationary flows in a vertical conical diffuser when air is supplied through tubes with various cross sections;
- To choose measuring instruments and research methods taking into account the physical features of the processes under research;
- To obtain data on the instantaneous values of the air flow velocity along the height and diameter of the diffuser's cylindrical part for various initial conditions;
- To establish the evolution of the velocity fields along the height of the diffuser's cylindrical part for various configurations of the supply tubes;
- To empirically determine the value of the drop in the average velocity along the height of the diffuser and provide a mathematical description of that process;
- To establish and mathematically describe the patterns of changes in the intensity of turbulence along the height of the conical diffuser under various initial conditions.

2. Description of the Experimental Measurement Facility

The aerodynamics of stationary flows in a vertical diffuser were studied using a laboratory set-up, the main elements of which are shown in Figure 1.

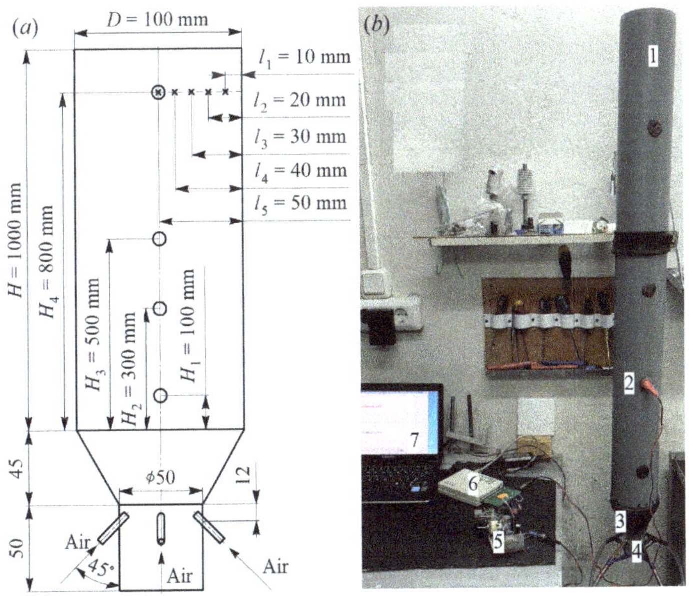

Figure 1. Size dimensions (**a**) and photograph (**b**) of the experimental set-up for studying the aerodynamic characteristics of flows in a vertical diffuser: 1—cylindrical part of the set-up; 2—hot-wire anemometer sensor; 3—diffuser; 4—base with supply tubes; 5—hot-wire anemometer for measuring instantaneous values of flow velocity; 6—analogue-to-digital converter; 7—laptop for collecting and processing experimental findings; H, $H_1 \ldots H_4$—height of the cylindrical part in the diffuser; $l_1 \ldots l_4$—linear dimension along the diameter of the cylindrical part of the diffuser.

The main purpose of the installation was to ensure the gas-dynamic characteristics of the air flow remained as similar as possible to real processes in such technical devices. The conical diffuser consisted of three main elements: (1) a base with a diameter of 50 mm, in which four supply tubes were installed at an angle of 45°; (2) a tapered section with a ratio of 1:2 (50 mm by 100 mm); and (3) a cylindrical section with a diameter of 100 mm and a height of 1000 mm. This configuration of the diffuser is used in installations for synthesis gas production. There were four control sections along the height of the diffuser at heights of 100 mm, 300 mm, 500 mm, and 800 mm. In each control section, measurements of the instantaneous values of the air flow velocity along the diameter of the cylindrical part at distances of 10 mm, 20 mm, 30 mm, 40 mm, and 50 mm (in the centre of the diffuser) were made. Measurements of instantaneous velocity values were carried out for 5–10 s for each location of the hot-wire anemometer sensor (primary data). Further processing was carried out in a specialized program (final data).

The working medium in the experiments was air with a temperature of t = 20–22 °C and a barometric pressure of p_o = 0.1013 mPa. The typical measurement in this work was taken in the control section at a height of 100 mm and the centre of the cylindrical part of the diffuser (at a distance of 50 mm). During the experiments, the average air flow velocity in the typical section varied from 4 m/s to 8 m/s (26,500 < Re < 53,500). Therefore, the flow pattern in this study was developed and turbulent.

The automated measurement system consisted of a hot-wire anemometer with sensors, an analogue-to-digital converter, and a laptop with software. Measurements of the instantaneous values of the air flow velocity w_x with a hot-wire anemometer at a constant temperature were made. A nichrome filament with a diameter of 5 μm and a length of 4 mm was used as the sensitive element of the hot-wire anemometer sensor. Five sensors with thread at distances of 10, 20, 30, 40, and 50 mm were made. The data from the hot-wire anemometer (output signal from 0 to 5 V) were received by an analogue-to-digital converter, and then they were transferred to a laptop for processing with custom-made software. The standard relative uncertainty of air flow measurement was 3.6%. The measuring system and the features of its functioning are described in more detail in articles [26,27].

The average flow velocity w at the measuring point was determined as the mathematical expectation of the function $w_x = f(\tau)$. One of the key aerodynamic characteristics of diffuser flows is the turbulence intensity TI. In this research, TI was calculated as the ratio of the root–mean–square pulsation velocity component to the average velocity of the flow under study [27].

The cross-sectional shape of pipelines has a significant effect on the structure of the flow in them [2,28,29]. Therefore, in this study, air entered the vertical diffuser through supply tubes with various cross-sectional shapes (Figure 2).

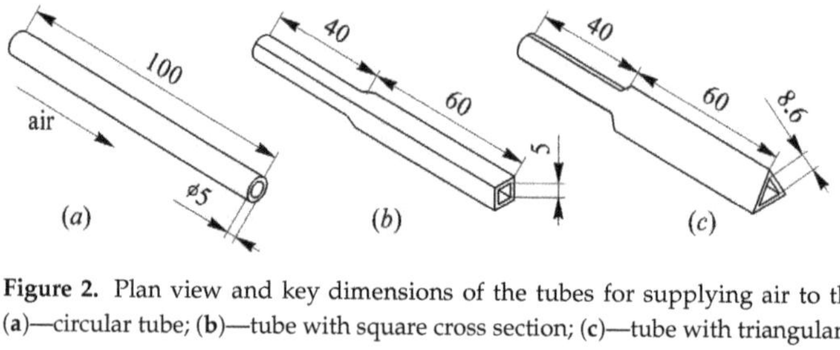

Figure 2. Plan view and key dimensions of the tubes for supplying air to the vertical diffuser: (**a**)—circular tube; (**b**)—tube with square cross section; (**c**)—tube with triangular cross section.

The size dimensions of the supply tubes' cross sections were determined based on the equality of the equivalent hydraulic diameters. Consequently, the round tubes had a diameter of d = 5 mm, the side of the square tubes was also 5 mm, and the side of the tubes with triangular cross sections was 8.6 mm.

3. Results and Analysis of Experimental Findings

Figure 3 shows the primary data as a function $w_x = f(\tau)$ for the sensor in the first section (H_1 = 100 mm) in the centre of the diffuser's cylindrical part (l_5 = 50 mm) when air is supplied through various tube configurations. The data are selected so that the average speed in the first section is approximately the same for all the data presented and is about 4.5 m/s.

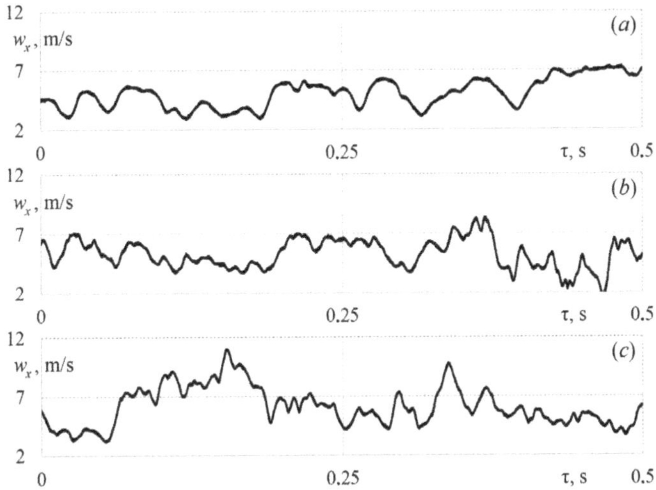

Figure 3. Dependences of local values of air flow velocity w_x on time τ in the first control section (H_1 = 100 mm) for a sensor in the centre of the installation (l_5 = 50 mm) for supply tube configurations: (**a**)—round tube (average parameters in cross section: w_1 = 4.2 m/s, TI = 0.231); (**b**)—square tube (w_1 = 4.45 m/s, TI = 0.18); (**c**)—triangular tube (w_1 = 4.48 m/s, TI = 0.285).

Figure 3 shows that the function $w_x = f(\tau)$ is white noise for all supply tube configurations. At the same time, visual observations show that the use of profiled tubes leads to the creation of small fluctuations in the $w_x = f(\tau)$ function. This can also be confirmed by the calculated data: TI for round tubes is 0.231 and 0.285 for triangular tubes (almost a 20% difference). This is due to the fact that stable and vortex structures are created in the corners of the profiled tubes, which turbulise the flow in the diffuser. Similar gas-dynamic effects were discovered in [28,29]. It is inappropriate to describe the presented dependencies mathematically, as those data are special cases.

Figure 4 also shows the primary data as the $w_x = f(\tau)$ function, but for the sensor in the third section (H_3 = 300 mm) in the centre of the diffuser's cylindrical part (l_5 = 50 mm) when air is supplied through tube configurations. Based on a comparison of the functions $w_x = f(\tau)$ in Figures 3 and 4, it is found that there is a significant decrease in small fluctuations in velocity upstream. This indicates a gradual relaxation of the flow along the height of the diffuser's cylindrical part.

Next, the data in each control section were averaged over the diffuser's height and diameter. As a result, the velocity fields in the diffuser's cylindrical part were obtained when air was supplied through tubes with various cross sections (Figures 5–7).

Figure 5 shows that there is a decrease in the average flow velocity along the height of the diffuser's cylindrical part (air supply through round tubes). A change in the velocity field's shape in the upward flow direction is also observed. Similar results in the case of air supply through profiled tubes were obtained (Figures 6 and 7). There is also a drop in velocity with the diffuser's height. When air is supplied through triangular tubes, the velocity fields are somewhat different from those in other configurations. This may be due to the influence of the vortex structures that are created at the corners of the triangular tubes.

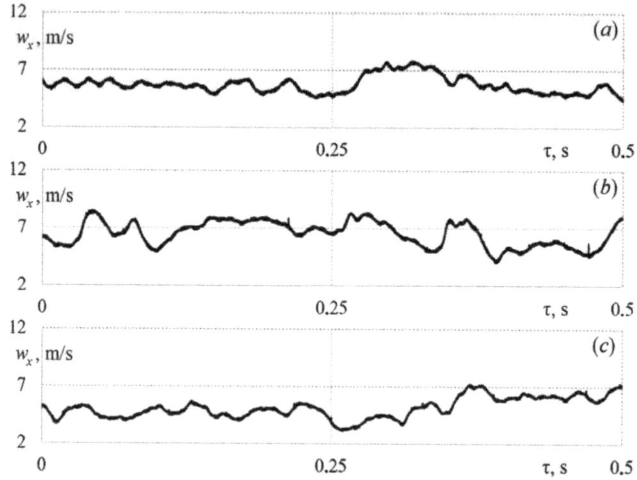

Figure 4. Dependences of local values of air flow velocity w_x on time τ in the third control section (H_3 = 500 mm) for a sensor in the centre of the set-up (l_5 = 50 mm) for various supply tube configurations: (**a**)—round tube (average parameters in the section: w_3 = 4.81 m/s, TI = 0.124); (**b**)—square tube (w_3 = 4.85 m/s, TI = 0.313); (**c**)—triangular tube (w_3 = 4.79 m/s, TI = 0.199).

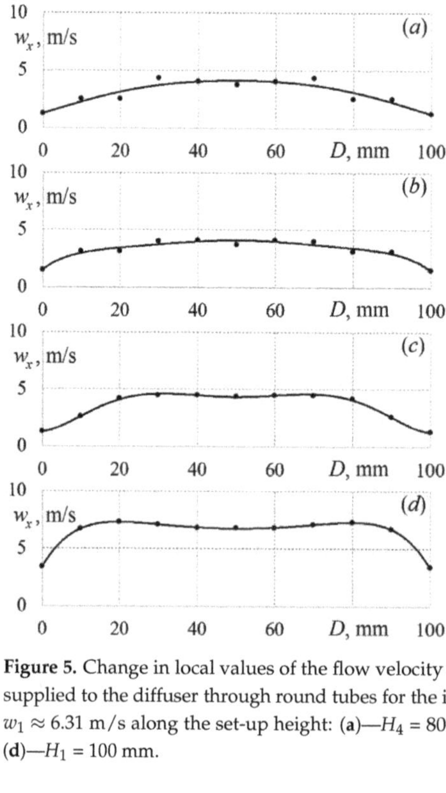

Figure 5. Change in local values of the flow velocity w_x along the cylinder diameter D when air is supplied to the diffuser through round tubes for the initial average flow velocity in the first section $w_1 \approx 6.31$ m/s along the set-up height: (**a**)—H_4 = 800 mm; (**b**)—H_3 = 500 mm; (**c**)—H_2 = 300 mm; (**d**)—H_1 = 100 mm.

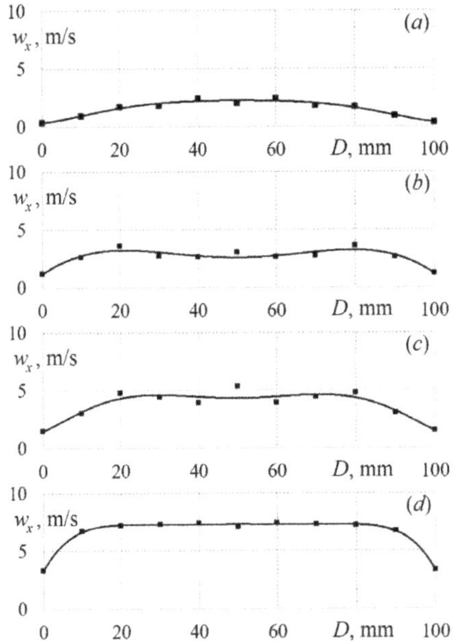

Figure 6. Change in local values of the flow velocity w_x along the cylinder diameter D when air is supplied to the diffuser through square tubes for the initial average flow velocity in the first section $w_1 \approx 6.45$ m/s along the set-up height: (**a**)—H_4 = 800 mm; (**b**)—H_3 = 500 mm; (**c**)—H_2 = 300 mm; (**d**)—H_1 = 100 mm.

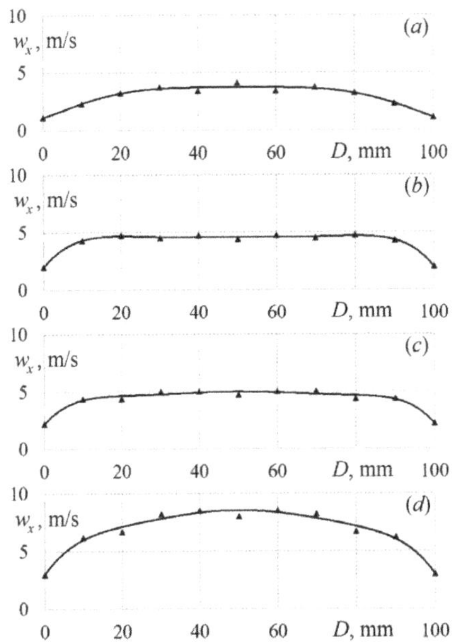

Figure 7. Change in local values of the flow velocity w_x along the cylinder diameter D when air is supplied to the diffuser through triangular tubes for the initial average flow velocity in the first section $w_1 \approx 6.65$ m/s along the set-up height: (**a**)—H_4 = 800 mm; (**b**)—H_3 = 500 mm; (**c**)—H_2 = 300 mm; (**d**)—H_1 = 100 mm.

The detected gas-dynamic effects are preserved for all initial air supply velocities under study and for all configurations of supply tubes. Accordingly, it can be assumed that there is a need for a mathematical formulation of the integral aerodynamic characteristics of flows for the studied configurations of a vertical conical diffuser. These data will make it possible to refine mathematical models and algorithms for the design of diffusers for various pieces of technical equipment.

The velocity fields are a clear illustration of the features of the gas dynamics of flows for different ways of supplying air to the diffuser. The analysis of the velocity field makes it possible to determine the gas-dynamic structure of the flow and draw a conclusion about the behaviour of the working medium in a potential technical device.

Figure 8 shows examples of changes in the turbulence intensity TI along the diameter of the diffuser's cylindrical part in the first section when air is supplied through tubes of various configurations.

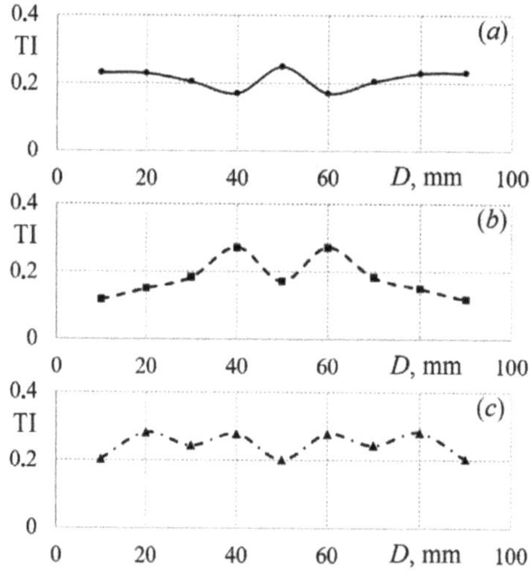

Figure 8. Change in turbulence intensity TI (in the first section H_1 = 100 mm) along the cylinder diameter D when air is supplied to the diffuser through tubes of various configurations: (**a**)—round tube ($w_1 \approx 6.65$ m/s, $TI_1 = 0.213$); (**b**)—square tube ($w_1 \approx 6.45$ m/s, $TI_1 = 0.180$); (**c**)—triangular tube ($w_1 \approx 6.65$ m/s, $TI_1 = 0.245$).

From Figure 8 and the entire array of data obtained, we can draw the following conclusions:

- No clear patterns in the change in TI along the diameter of the diffuser's cylindrical part were found;
- The influence of the tubes' cross-sectional shape on the function TI = f (D) has not been established (according to the author, the changes are random);
- The amplitude of the fluctuations of turbulence intensity values relative to the average value is ±35%;
- The mathematical formulation of the function TI = f (D) for the cases under study is inappropriate until the physical laws are established.

Data on the turbulence intensity are important for practical applications, as the TI determines the efficiency of mixing of various media, the level of pulsation of the working medium in the flow, the duration of the working medium in the diffuser, and more.

Figure 9 shows the regularities of the drop in the average velocity in the control section along the height of the diffuser's cylindrical part when air is supplied through tubes with various cross sections for various initial flow rates.

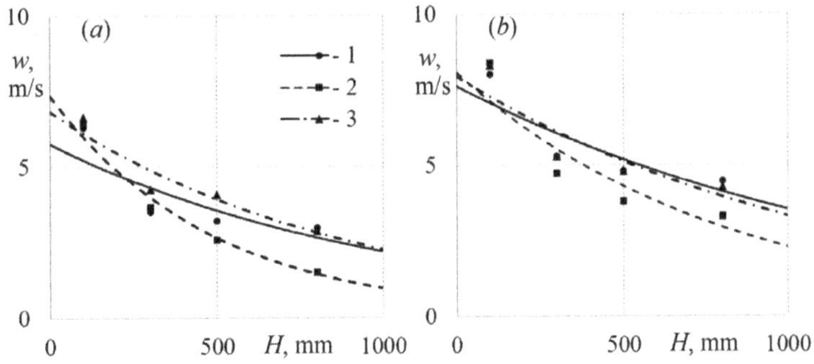

Figure 9. Dependences of the average air flow rate w along the installation height H when air is supplied through round (1), square (2), and triangular (3) tubes for the initial average flow rate in the first section: (**a**)—$w_1 \approx 6.5$ m/s; (**b**)—$w_1 \approx 8.25$ m/s.

The patterns of speed reduction in the upward flow direction are described with sufficient accuracy by the exponential dependence $w = a \cdot e^{-b \cdot H}$. At the same time, the shape of the tube cross section does not actually affect the intensity of the drop in the air flow velocity along the set-up's height.

For example, the mathematical description of the change in the average air flow velocity along the height of the diffuser's cylindrical part, when air is supplied through round tubes with an initial velocity $w_1 \approx 6.5$ m/s, is described by the following equation (approximation reliability 0.85):

$$w = 5.7e^{-0.0001 \cdot H}. \quad (1)$$

Similar relationships for air supply through square and triangular tubes are described by the following equation ($w_1 \approx 6.5$ m/s, approximation reliability 0.92):

$$w = 7.1e^{-0.00015 \cdot H}. \quad (2)$$

The mathematical description of the change in the average air flow rate along the height of the diffuser's cylindrical part, when air is supplied through round and triangular tubes in the case of an increase in the initial flow rate to $w_1 \approx 8.25$ m/s, is described by the following equation (approximation reliability 0.83):

$$w = 7.7e^{-0.00085 \cdot H}. \quad (3)$$

Similar dependencies for air supply through square tubes are as follows ($w_1 \approx 8.25$ m/s, approximation reliability 0.86):

$$w = 8.0e^{-0.001 \cdot H}. \quad (4)$$

The proposed Equations (1)–(4) for describing the regularities of the average velocity drop in the control section along the height of the diffuser's cylindrical part when air is supplied through tubes with various configurations are applicable for the indicated velocities and a set-up height of up to 1000 mm.

Figure 10 shows the dependences of the change in the turbulence intensity TI on the value of the average air flow velocity w in the control sections at distances of 100 mm and 800 mm. There is a decrease in TI with an increase in the initial air flow velocity, and this decrease in TI is well described by linear equations. At the same time, the intensity of turbulence has higher values when air is supplied through profiled tubes compared to round tubes. This is also explained by the turbulence of the flow due to the vortex structures typical of profiled channels.

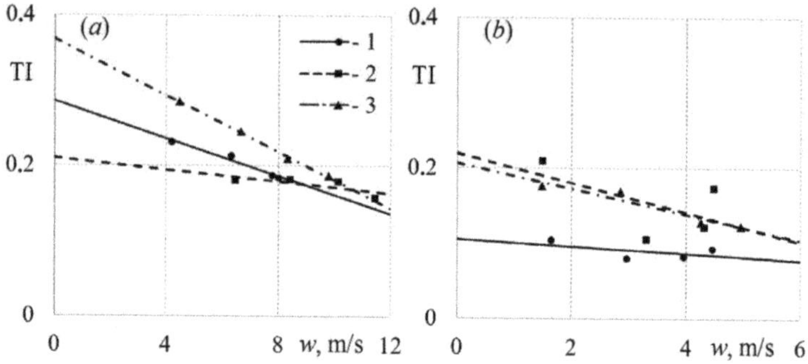

Figure 10. Dependences of the turbulence intensity TI on the average air flow velocity w when air is supplied through round (1), square (2), and triangular (3) tubes for various control sections: (**a**)—control section at the height of H_1 = 100 mm; (**b**)—H_4 = 800 mm.

The mathematical description of the function TI = f (w) for the first section (H_1 = 100 mm) when air is supplied through round (5), square (6), and triangular (7) tubes has the following form (approximation reliability not less than 0.89):

$$\mathrm{TI} = -0.125 \cdot w + 0.286. \tag{5}$$

$$\mathrm{TI} = -0.0038 \cdot w + 0.21. \tag{6}$$

$$\mathrm{TI} = -0.0188 \cdot w + 0.369. \tag{7}$$

The similar mathematical descriptions of the function TI = f (w) for the fourth section (H_4 = 800 mm) when air is supplied through round (8), square (9), and triangular (10) tubes have the forms (approximation reliability not less than 0.92):

$$\mathrm{TI} = -0.0047 \cdot w + 0.105. \tag{8}$$

$$\mathrm{TI} = -0.0196 \cdot w + 0.219. \tag{9}$$

$$\mathrm{TI} = -0.017 \cdot w + 0.207. \tag{10}$$

The proposed equations for describing the patterns of change in the intensity of turbulence from the average air flow velocity along the height of the diffuser's cylindrical part are applicable for an initial velocity range from 1 to 12 m/s and a set-up height of up to 1000 mm.

The experimental data and proposed description equations can be used to verify mathematical models for modelling aerodynamic processes in a vertical conical diffuser.

Figure 11 shows the experimental dependences of the intensity of turbulence on the height of the diffuser's cylindrical part for various initial flow velocities when air is supplied through tubes with various cross sections. The distribution of turbulence intensity along the height of the diffuser is important for the design of mixing chambers, industrial furnaces, and synthesis gas plants. Engineers need data on the level of pulsations in the installation to evaluate the efficiency and stability of operation.

Based on the analysis of the TI = f (H) functions, the following conclusions can be drawn:

- The tubes' cross-sectional shape has a significant effect on turbulence intensity along the height of the vertical diffuser's cylindrical part;
- The dependency TI = f (H) has a pronounced maximum in the region of H = 300 mm, which is typical of all initial average flow rates and all configurations of the supply tubes;

- The turbulence intensity has significantly higher values (up to 50%) when air is supplied through profiled tubes compared to round tubes, which is especially typical of a low initial w.

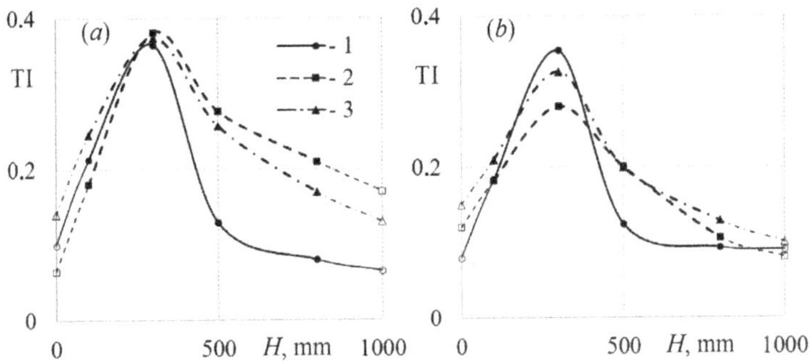

Figure 11. Dependences of the intensity of turbulence TI on the height of the diffuser H when air is supplied through round (1), square (2), and triangular (3) tubes for the initial average flow velocity in the first section: (**a**)—$w_1 \approx 6.5$ m/s; (**b**)—$w_1 \approx 8.25$ m/s.

For the convenience of the mathematical description of the experimental data, it is advisable to divide the TI = f (H) function into two sections: (1) the first section for a height from 0 to 300 mm (section of the growth of turbulence intensity) and (2) the second section for H = 300–1000 mm (TI decline section). For the first section (H = 0–300 mm), the TI = f (H) function for all tube configurations can be described by a linear equation with an approximation confidence of 0.94 (Figure 11a):

$$\mathrm{TI} = 0.00085 \cdot H + 0.1. \quad (11)$$

Consequently, for the second section (H = 300–1000 mm), the TI = f (H) function can be described by exponential equations for round (12), square (13), and triangular (14) tubes with an approximation confidence of 0.98 (Figure 11a):

$$\mathrm{TI} = 0.254\mathrm{e}^{-0.001 \cdot H}. \quad (12)$$

$$\mathrm{TI} = 0.452\mathrm{e}^{-0.0003 \cdot H}. \quad (13)$$

$$\mathrm{TI} = 0.507\mathrm{e}^{-0.001 \cdot H}. \quad (14)$$

The proposed equations for describing the patterns of changes in the intensity of turbulence along the height of the diffuser's cylindrical part when air is supplied through tubes with various configurations are applicable for an initial velocity range from 1 to 12 m/s and a set-up height of up to 1000 mm.

It should be noted that all the above formulas are valid only for the studied configuration of the experimental set-up and the initial conditions of air flow. It is necessary to additionally clarify the regularities of changes in the physical characteristics of the flow in the case of a change in geometry or gas-dynamic parameters.

Thus, the aerodynamic characteristics of flows in a vertical conical diffuser can be approximated by relatively simple mathematical expressions that are convenient to use in engineering design.

4. Conclusions

Based on the conducted experimental studies, the following main conclusions can be drawn:

1. Experimental data on the instantaneous values of the stationary flow velocity along the height and diameter of the diffuser's cylindrical part for various initial conditions and when air is supplied through tubes with different configurations are obtained.
2. The velocity fields along the height of the diffuser's cylindrical part for various initial conditions when air is supplied through tubes with cross sections in the form of a circle, a square, and a triangle are determined. The evolution of the velocity fields in the upward flow direction is shown.
3. The flow's turbulence intensity TI along the height and diameter of the diffuser is calculated for various initial conditions and when air is supplied through tubes with various configurations.
4. A mathematical description (exponential equations) of the change in the average flow velocity along the height of the diffuser's cylindrical part for various initial conditions and configurations of the supply tubes is presented.
5. The regularities of changes in the intensity of turbulence along the height of the diffuser for various initial conditions and tube configurations are established.
6. The obtained data on the aerodynamic characteristics of flows in a conical diffuser can be useful for refining and verifying mathematical models and improving engineering calculations.
7. Further research could be conducted to obtain dimensionless equations for describing the aerodynamic characteristics of flows in vertical diffusers of various designs, as well as to refine mathematical models for modelling aerodynamics.

Funding: The research funding from the Ministry of Science and Higher Education of the Russian Federation (Ural Federal University Program of Development within the Priority-2030 Program) is gratefully acknowledged.

Data Availability Statement: Not applicable.

Conflicts of Interest: The author declares no conflict of interest.

Nomenclature

w_x	local air flow velocity, m/s
w	average flow velocity, m/s
τ	time, s
p_o	barometric pressure, kPa
t	temperature, °C
(d)	diameter, mm
l	linear dimension, mm
H	height, mm
Re	Reynolds number
TI	turbulence intensity

References

1. Japikse, D.; Baines, N.C. Diffuser Design Technology. Concepts ETI: Norwich, UK, 1998; 524p.
2. Idelchik, I.E. *Aerohydrodynamics of Technological Apparatuses (Inlet, Outlet and Distribution of the Flow over the Cross Section of the Devices)*; Mashinostroenie: Moscow, Russia, 1983; 351p. (In Russian)
3. Emmons, H.W. *Fundamentals of Gas Dynamics*; Princeton University Press: Princeton, NJ, USA, 2015; 783p.
4. Hirsch, C. *Numerical Computation of Internal and External Flows: The Fundamentals of Computational Fluid Dynamics*; Elesvier: Amsterdam, The Netherlands, 2007; 696p.
5. Yavuz, M.; Sene, N. Approximate solutions of the model describing fluid flow using generalized ρ-laplace transform method and heat balance integral method. *Axioms* **2020**, *9*, 123. [CrossRef]
6. Xiao, Y.; Yue, F.; Wang, X.; Zhang, X. Reliability-Based Design Optimization of Structures Considering Uncertainties of Earthquakes Based on Efficient Gaussian Process Regression Metamodeling. *Axioms* **2022**, *11*, 81. [CrossRef]
7. Klein, A. Effects of inlet conditions on conical-diffuser performance. *J. Fluids Eng. Trans. ASME* **1981**, *103*, 250–257. [CrossRef]
8. Jeyachandran, K.; Ganesan, V. Numerical modelling of turbulent flow through conical diffusers with uniform and wake velocity profiles at the inlet. *Math. Comput. Model.* **1988**, *10*, 87–97. [CrossRef]

9. Okwuobi, P.A.C.; Azad, R.S. Turbulence in a conical diffuser with fully developed flow at entry. *J. Fluid Mech.* **1973**, *57*, 603–622. [CrossRef]
10. Ferrari, A. Exact solutions for quasi-one-dimensional compressible viscous flows in conical nozzles. *J. Fluid Mech.* **2021**, *915*, 915A1-1. [CrossRef]
11. Lee, J.; Jang, S.J.; Sung, H.J. Direct numerical simulations of turbulent flow in a conical diffuser. *J. Turbul.* **2012**, *13*, 1–29. [CrossRef]
12. Wu, X.; Schlüter, J.; Moin, P.; Pitsch, H.; Iaccarino, G.; Ham, F. Computational study on the internal layer in a diffuser. *J. Fluid Mech.* **2006**, *550*, 391–412. [CrossRef]
13. De Souza, F.J.; Silva, A.L.; Utzig, J. Four-way coupled simulations of the gas-particle flow in a diffuser. *Powder Technol.* **2014**, *253*, 496–508. [CrossRef]
14. Zou, A.; Chassaing, J.-C.; Li, W.; Gu, Y.; Sauret, E. Quantified dense gas conical diffuser performance under uncertainties by flow characteristic analysis. *Appl. Therm. Eng.* **2019**, *161*, 114158. [CrossRef]
15. Dong, B.; Xu, G.; Li, T.; Quan, Y.; Zhai, L.; Wen, J. Numerical prediction of velocity coefficient for a radial-inflow turbine stator using R123 as working fluid. *Appl. Therm. Eng.* **2018**, *130*, 1256–1265. [CrossRef]
16. Kim, D.-Y.; Kim, Y.-T. Preliminary design and performance analysis of a radial inflow turbine for organic Rankine cycles. *Appl. Therm. Eng.* **2017**, *120*, 549–559. [CrossRef]
17. From, C.S.; Sauret, E.; Armfield, S.W.; Saha, S.C.; Gu, Y.T. Turbulent dense gas flow characteristics in swirling conical diffuser. *Comput. Fluids* **2017**, *149*, 100–118. [CrossRef]
18. Keep, J.A.; Head, A.J.; Jahn, I.H. Design of an efficient space constrained diffuser for supercritical CO_2 turbines. *J. Phys. Conf. Ser.* **2017**, *821*, 012026. [CrossRef]
19. Wang, W.-B.; Zheng, L.-X.; Lu, W.; Zhang, W.-J.; Xie, C.-X. Study on Structural Design Optimization of Ejector Expansion Chamber. *J. Eng. Thermophys.* **2021**, *42*, 309–313.
20. Sierra-Pallares, J.; García del Valle, J.; Paniagua, J.M.; García, J.; Méndez-Bueno, C.; Castro, F. Shape optimization of a long-tapered R134a ejector mixing chamber. *Energy* **2018**, *165*, 422–438. [CrossRef]
21. Maicke, B.A.; Bondarev, G. Quasi-one-dimensional modeling of pressure effects in supersonic nozzles. *Aerosp. Sci. Technol.* **2017**, *70*, 161–169. [CrossRef]
22. Vinod, L.; Mahendra, M.A.; Mahantayya, K.H. The cfd analysis of subsonic flow around struts of airfoil and cylindrical shape attached to a conical diffuser at exhaust of a gas turbine engine. *Int. J. Mech. Prod. Eng. Res. Dev.* **2019**, *9*, 943–954.
23. Moszowski, B.; Wajman, T.; Sobczak, K.; Inger, M.; Wilk, M. The analysis of distribution of the reaction mixture in ammonia oxidation reactor. *Pol. J. Chem. Technol.* **2019**, *21*, 9–12. [CrossRef]
24. Huang, Y.; Coggon, M.M.; Zhao, R.; Lignell, H.; Bauer, M.U.; Flagan, R.C.; Seinfeld, J.H. The Caltech Photooxidation Flow Tube reactor: Design, fluid dynamics and characterization. *Atmos. Meas. Tech.* **2017**, *10*, 839–867. [CrossRef]
25. Keshavarz, E.; Toghraie, D.; Haratian, M. Modeling industrial scale reaction furnace using computational fluid dynamics: A case study in Ilam gas treating plant. *Appl. Therm. Eng.* **2017**, *123*, 277–289. [CrossRef]
26. Plotnikov, L.; Plotnikov, I.; Osipov, L.; Slednev, V.; Shurupov, V. An Indirect Method for Determining the Local Heat Transfer Coefficient of Gas Flows in Pipelines. *Sensors* **2022**, *22*, 6395. [CrossRef] [PubMed]
27. Plotnikov, L.V. Thermal-mechanical characteristics of stationary and pulsating gas flows in a gas-dynamic system (in relation to the exhaust system of an engine). *Therm. Sci.* **2022**, *26*, 365–376. [CrossRef]
28. Plotnikov, L.V. Unsteady gas dynamics and local heat transfer of pulsating flows in profiled channels mainly to the intake system of a reciprocating engine. *Int. J. Heat Mass Transf.* **2022**, *195*, 123144. [CrossRef]
29. Brodov, Y.M.; Zhilkin, B.P.; Plotnikov, L.V. Influence of Intake/exhaust Channel Lateral Profiling on Thermomechanics of Pulsating Flows. *Tech. Phys.* **2018**, *63*, 319–324. [CrossRef]

axioms

MDPI

Article

LBM-MHD Data-Driven Approach to Predict Rayleigh–Bénard Convective Heat Transfer by Levenberg–Marquardt Algorithm

Taasnim Ahmed Himika [1], Md Farhad Hasan [2,3], Md. Mamun Molla [4,5,*] and Md Amirul Islam Khan [6]

1 School of Science, Computing and Engineering Technologies, Swinburne University of Technology, Hawthorn, VIC 3122, Australia
2 Victoria State Government, Melbourne, VIC 3083, Australia
3 School of Computing, Engineering and Mathematical Sciences, La Trobe University, Melbourne, VIC 3086, Australia
4 Department of Mathematics & Physics, North South University, Dhaka 1229, Bangladesh
5 Center for Applied Scientific Computing (CASC), North South University, Dhaka 1229, Bangladesh
6 School of Civil Engineering, University of Leeds, Leeds LS2 9JT, UK
* Correspondence: mamun.molla@northsouth.edu; Tel: +880-255668200 (ext.1519); Fax: +880-255668202

Abstract: This study aims to consider lattice Boltzmann method (LBM)–magnetohydrodynamics (MHD) data to develop equations to predict the average rate of heat transfer quantitatively. The present approach considers a 2D rectangular cavity with adiabatic side walls, and the bottom wall is heated while the top wall is kept cold. Rayleigh–Bénard (RB) convection was considered a heat-transfer phenomenon within the cavity. The Hartmann (Ha) number, by varying the inclination angle (θ), was considered in developing the equations by considering the input parameters, namely, the Rayleigh (Ra) numbers, Darcy (Da) numbers, and porosity (ϵ) of the cavity in different segments. Each segment considers a data-driven approach to calibrate the Levenberg–Marquardt (LM) algorithm, which is highly linked with the artificial neural network (ANN) machine learning method. Separate validations have been conducted in corresponding sections to showcase the accuracy of the equations. Overall, coefficients of determination (R^2) were found to be within 0.85 to 0.99. The significant findings of this study present mathematical equations to predict the average Nusselt number ($\overline{Nu}$). The equations can be used to quantitatively predict the heat transfer without directly simulating LBM. In other words, the equations can be considered validations methods for any LBM-MHD model, which considers RB convection within the range of the parameters in each equation.

Keywords: lattice Boltzmann; Rayleigh–Bénard convection; magnetohydrodynamics; Levenberg–Marquardt algorithm; data-driven analysis; Nusselt number; Hartmann number; porosity; rectangular cavity

MSC: 00A72; 62-07; 76A02; 76M27; 80A20

Citation: Himika, T.A.; Hasan, M.F.; Molla, M.M.; Khan, M.A.I. LBM-MHD Data-Driven Approach to Predict Rayleigh–Bénard Convective Heat Transfer by Levenberg–Marquardt Algorithm. *Axioms* **2023**, *12*, 199. https://doi.org/10.3390/axioms12020199

Academic Editor: Leonid Plotnikov

Received: 1 January 2023
Revised: 4 February 2023
Accepted: 8 February 2023
Published: 13 February 2023

1. Introduction

The lattice Boltzmann method (LBM) is an efficient approach to investigate fluid flow through numerical simulations across different geometries at microscopic, mesoscopic, and macroscopic scales [1–8]. LBM is based on statistical mechanics and has immense potential to establish a data-driven analysis for scientific progress [9–11]. Therefore, high-dimensional nonlinear LBM data could be taken into account to calibrate any statistical model through high-performance computing (HPC). With the increasing demand for HPC, researchers have shifted their focus to fluid flow simulations by considering LBM across complicated grids [5,12–15]. LBM has been found to be efficient in flow simulations and heat transfer applications in hydrology, magnetohydrodynamics, and aerodynamics, to name a few [16–20]. Magnetohydrodynamics (MHDs) represents electrically conducting fluids in liquid metals and plasma flows. The applications of MHD have been reported in a wide range of applications, such as thermal engineering, geophysics, nuclear and hydroelectric power plants, astrophysics, and so on [18,21–24]. Therefore, the analysis of heat transfer in convective flow could be established by utilizing the LBM-MHD scheme.

The study of numerical heat transfer through different media is one of the popular fields of study among researchers [25–28]. Rayleigh–Bénard (RB) convection is one form of a phenomenon that takes place in a fluid layer assigned to a vertical temperature gradient and heated from the base [18,29–31]. The difference between buoyancy and gravity leads to fluid instabilities and convective electrical currents. This type of instability has been the subject of extensive research to identify a procedure to stabilize the system. One major reason could be the lack of understanding of the LBM data and correlations of the output with the input variables prior to the numerical modeling [32]. Therefore, it is necessary to determine the correlations to predict any upcoming phenomena linked to fluid instability. Several authors reported successful outcomes in stabilizing a system by applying an external magnetic field due to the induced electrical currents within the fluid [18,33,34]. However, most of the works were based on analysis through certain numerical parametric variations. Therefore, the prospect of establishing an LBM data-driven approach to determine correlations with the heat transfer prediction remains unnoticed.

Machine learning (ML) and deep learning (DL) are two important sectors of artificial intelligence (AI) with the ability for accurate data analysis and prediction model development [35–39]. With computational resources, ML and DL can build a multivariate model by taking high-dimensional non-linear data and developing correlations and numerical prediction models within different sectors. The model needs to be trained with a dataset to calibrate the model, and validations are performed through internal and independent datasets. However, the prediction model needs to be optimized through efficient training methods [40]. An inadequately optimized model will perform below the standard and yield noise within the model, leading to low correlation to predict the outcome. The Levenberg–Marquardt (LM) algorithm is one of the training methods for ML models, particularly for artificial neural networks (ANNs). LM develops the correlations by considering the input variables to provide a nonlinear least squares minimization (NLSM) solution. Therefore, it indicates that any numerically simulated data, including those from LBM, could be fed into the LM algorithm to understand the correlations among the variables through a quasi-ML approach along with numerical validations with the literature.

There is a shortage of literature on LBM data analysis through any efficient algorithm. However, some recent studies have reported the utilization of neural networks to optimize LBM data under the influence of MHD in natural convection. For example, Alqaed et al. [41] studied natural convection and entropy generation by applying a magnetic field with ANN and presented an equation based on the correlation development. However, the ML modeling equation lacks information on whether it can be used to predict total entropy across all geometries. In addition, the equation was not validated against any published literature to showcase the accuracy and robustness. Shah et al. [42] followed similar steps by adding radiation heat transfer. However, the validation methodology and the equations to predict the entropy remained ambiguous. On the other hand, the study presented by He et al. [43] was a much-improved one, as the correlation developments of LBM data were efficiently described through ANN and internal validations. Yet, the independent validations were still missed, and therefore, the accuracy of the correlations could not be expanded beyond the internal database.

This study aims to address the shortcomings within the literature by analyzing LBM data to establish correlations by considering the numerical variables, such as Rayleigh (Ra) number, Darcy (Da) number, Hartmann (Ha) number, inclination angle (θ), and porosity (ϵ), to predict the average rate of heat transfer ($\overline{Nu}$) by the LM algorithm. The obtained equations are presented in each section, including the statistical accuracy indicators, followed by validations within the literature in each step under various circumstances. The correlation coefficients (R^2) are found to be between 0.85 and 0.99, which provides more confidence in the accuracy of the numerical model.

2. Geometry of the Porous Cavity

The schematic diagram of the porous cavity along with associated coordinates is illustrated in Figure 1. The rectangular cavity in the 2D configuration includes the effect of a magnetic field to investigate the RB convection by considering incompressible and laminar fluid flow. The LBM data were extracted within these specifications. The cavity was assumed to be filled with electrically conducting fluid. H denotes the vertical height, and the horizontal length is denoted by L, where $L = 2H$. Two vertical side walls were considered adiabatic, i.e., no heat transfer will occur. The top and bottom walls are cold and heated, represented by T_c and T_h, respectively, where $T_h > T_c$. The LBM data were extracted through different parametric variations, such as the Rayleigh (Ra) number within the higher buoyancy range (10^4 and 10^5). Three different Darcy (Da) numbers were considered, namely, $10^{-1}, 10^{-2}, 10^{-3}$, and Hartmann ($Ha$) numbers were considered to be between 0 and 100 to investigate the impact of the magnetic field. The impact of the magnetic field was further studied along with different inclination angles (θ) ranging from 0 to 90. The porosity (ϵ) was between 0.4 and 0.9. The gravitational acceleration (g_y) was acting downward. The uniform magnetic field was considered to be B in Figure 1. The study assumes the Joule heating and viscous dissipation to be negligible to focus entirely on the impact of the magnetic field [18]. However, through the Boussinesq approximation, this particular assumption is validated, which ignores the density gradient, except from the appearance where the former is multiplied by g_y.

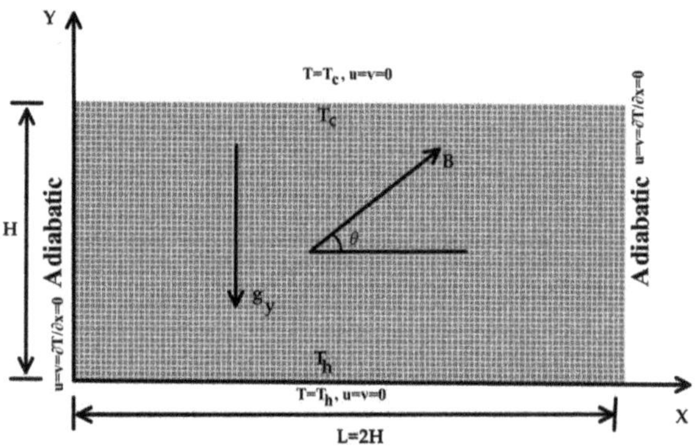

Figure 1. Considered geometry of the rectangular porous cavity along with the magnetic field.

3. Mathematical Formulations in Computation

3.1. Macroscopic Variables for Natural Convection in RB-MHD Flow in Porous Media

The formulation of LB equations using macroscopic governing equations is required to study MHD natural convection through porous media. These formulae include the energy equation, the Navier–Stokes equation with the Brinkman–Forchheimer model, and the continuity equation. However, these equations need to be converted to their non-dimensional form before being used to simulate MHD natural convection.

The dimensional equations for continuity, u-momentum, v-momentum, and energy are as follows:

$$\frac{\partial \bar{u}}{\partial \bar{x}} + \frac{\partial \bar{v}}{\partial \bar{y}} = 0 \tag{1}$$

$$\begin{aligned} &\frac{\partial \bar{u}}{\partial \bar{t}} + \frac{1}{\epsilon}\left(\bar{u}\frac{\partial \bar{u}}{\partial \bar{x}} + \bar{v}\frac{\partial \bar{u}}{\partial y}\right) \\ &= \frac{1}{\rho}\left[-\epsilon\frac{\partial \bar{p}}{\partial \bar{x}} + \bar{\mu}\left(\frac{\partial^2 \bar{u}}{\partial \bar{x}^2} + \frac{\partial^2 \bar{u}}{\partial \bar{y}^2}\right) + \sigma\epsilon B^2(\bar{v}\sin\phi\cos\phi - \bar{u}\sin^2\phi)\right] - \frac{\epsilon v}{K}\bar{u} \end{aligned} \tag{2}$$

$$
\begin{aligned}
& -\frac{1.75}{\sqrt{150\epsilon K}}|\bar{\mathbf{u}}|\bar{u} \\
& \frac{\partial \bar{v}}{\partial \bar{t}}+\frac{1}{\epsilon}\left(\bar{u}\frac{\partial \bar{v}}{\partial \bar{x}}+\bar{v}\frac{\partial \bar{v}}{\partial \bar{y}}\right) \\
= & \frac{1}{\rho}\left[-\epsilon\frac{\partial \bar{p}}{\partial \bar{y}}+\bar{\mu}\left(\frac{\partial^2 \bar{v}}{\partial \bar{x}^2}+\frac{\partial^2 \bar{v}}{\partial \bar{y}^2}\right)+g_y\epsilon\rho\beta(\bar{T}-T_c)+\epsilon\sigma B^2(\bar{u}\sin\phi\cos\phi-\bar{v}\cos^2\phi)\right]-\frac{\epsilon\nu}{K}\bar{v} \\
& -\frac{1.75}{\sqrt{150\epsilon K}}|\bar{\mathbf{u}}|\bar{v}
\end{aligned}
\tag{3}
$$

$$
\frac{\partial \bar{T}}{\partial \bar{t}}+\bar{u}\frac{\partial \bar{T}}{\partial \bar{x}}+\bar{v}\frac{\partial \bar{T}}{\partial \bar{y}}=\alpha\left(\frac{\partial^2 \bar{T}}{\partial \bar{x}^2}+\frac{\partial^2 \bar{T}}{\partial \bar{y}^2}\right)
\tag{4}
$$

Meanwhile, the dimensionless governing equations could be written as the following:

$$
\frac{\partial u}{\partial x}+\frac{\partial v}{\partial y}=0
\tag{5}
$$

$$
\begin{aligned}
& \frac{\partial u}{\partial t}+\frac{1}{\epsilon}\left(u\frac{\partial u}{\partial x}+v\frac{\partial u}{\partial y}\right) \\
= & -\epsilon\frac{\partial p}{\partial x}+\frac{\Pr}{\sqrt{Ra}}\left(\frac{\partial^2 u}{\partial x^2}+\frac{\partial^2 v}{\partial y^2}\right)+\epsilon\frac{\Pr}{\sqrt{Ra}}\mathrm{Ha}^2(u\sin\phi\cos\phi-v\sin^2\phi)-\epsilon\frac{\Pr}{\sqrt{Ra}Da}u
\end{aligned}
\tag{6}
$$

$$
\begin{aligned}
& -\frac{1.75}{\sqrt{100\epsilon Da}}|\mathbf{u}|u \\
& \frac{\partial v}{\partial t}+\frac{1}{\epsilon}\left(u\frac{\partial v}{\partial x}+v\frac{\partial v}{\partial y}\right) \\
= & -\epsilon\frac{\partial p}{\partial y}+\frac{\Pr}{\sqrt{Ra}}\left(\frac{\partial^2 v}{\partial x^2}+\frac{\partial^2 v}{\partial y^2}\right)+\epsilon\theta\Pr+\epsilon\frac{\Pr}{\sqrt{Ra}}Ha^2(u\sin\phi\cos\phi-v\cos^2\phi)-\epsilon\frac{\Pr}{\sqrt{Ra}Da}v \\
& -\frac{1.75}{\sqrt{100\epsilon Da}}|\mathbf{u}|v
\end{aligned}
\tag{7}
$$

$$
\frac{\partial T}{\partial t}+\bar{u}\frac{\partial T}{\partial x}+v\frac{\partial T}{\partial y}=\frac{1}{\sqrt{Ra}}\left(\frac{\partial^2 T}{\partial x^2}+\frac{\partial^2 T}{\partial y^2}\right)
\tag{8}
$$

Here,
ρ is the fluid density,
α is the thermal diffusivity,
ϵ is the porosity,
T_c is the cold temperature,
T_h is the hot temperature,
σ is the electrical conductivity,
μ is the dynamic viscosity,
H is the height of the cavity,
B is the magnetic field strength,
ϕ is the angle of an applied magnetic field,
β is the thermal expansion coefficient,
g_y is the gravity acting downward along the y-axis,
Da is the Darcy number,
Ha is the Hartmann number,
$\Delta T = T_h - T_c$ is the temperature gradient between the top (hot) and bottom (cold) walls

$(T_h > T_c)$,
$|\mathbf{u}| = \sqrt{u^2 + v^2}$

The relations which are implied to convert the dimensional equations to non-dimensional form are

$$\begin{aligned} x &= \frac{\bar{x}}{H} \\ y &= \frac{\bar{y}}{H} \\ u &= \frac{\bar{u}}{(\frac{\alpha}{H})\sqrt{Ra}} \\ \nu &= \frac{\mu}{\rho} \\ v &= \frac{\bar{v}}{(\frac{\alpha}{H})\sqrt{Ra}} \\ \theta &= \frac{\bar{T}-T_c}{T_h-T_c} \\ t &= \frac{\bar{t}}{\left(\frac{H^2}{\alpha}\right)}\sqrt{Ra} \\ p &= \frac{\bar{p}}{p(\frac{\alpha}{H})^2 Ra} \\ Ra &= \frac{g_y \beta \Delta T H^3}{v\alpha} \\ Pr &= \frac{v}{\alpha} \\ Da &= \frac{K}{H^2} \\ Ha &= BH\sqrt{\frac{\sigma}{\mu}} \end{aligned} \tag{9}$$

3.2. LBEs for Heat Transfer and Fluid Flow

The lattice Boltzmann method is also referred to as thermal LBM or ($TLBM$) because it simulates the fluid flow mechanics by solving both the Boltzmann and the energy equations. $TLBM$ calculates two distribution functions-f_i for fluid field, and g_i for temperature field. These distribution functions could be defined by considering the probability of particles in position x at time t moving toward each lattice direction i with speed c_i during time Δt. It enables the formulation of macroscopic fluid parameters, such as pressure, temperature, and velocity. In addition, the fluid domain is discretized into homogeneous lattice nodes. The inclusion of the BGK approximation into the LB equation results in the following equations with an external force [18]:

For the flow field:

$$f_i(\bar{x} + \bar{e}_i\Delta t, t + \Delta t) = f_i(\bar{x}, t) - \frac{f_i(\bar{x}, t) - f_i^{eq}(\bar{x}, t)}{\tau_v} + \Delta t \bar{F}_i \tag{10}$$

For the temperature field:

$$g_i(\bar{x} + \bar{e}_i\Delta t, t + \Delta t) = g_i(\bar{x}, t) - \frac{g_i(\bar{x}, t) - g_i^{eq}(\bar{x}, t)}{\tau_\alpha} \tag{11}$$

Here, $\tau_v = 3\nu + 0.5$, and $\tau_\alpha = 3\alpha + 0.5$ are the single-relaxation times (SRTs) that define the approaching rate to the equilibrium state. Meanwhile, kinematic viscosity ν and thermal diffusivity α are presented as the following:

$$\nu = \left(\tau_v - \frac{1}{2}\right) c_s^2 \Delta t \quad (12)$$

$$\alpha = \left(\tau_c - \frac{1}{2}\right) c_s^2 \Delta t \quad (13)$$

where c_s is the speed of sound, $c_s = c/\sqrt{3}$, and c is the spacing among the lattice.

The external force term $\bar{F}_i$ consists of three forces: ${F^M}_i$ (for MHD), ${F^P}_i$ (for porous media, which is the Brinkman–Forcheimer model), and finally, ${F^b}_i$ (buoyancy term):

$$F_i = {F^M}_i + {F^P}_i + {F^b}_i \quad (14)$$

On the other hand, magnetic force ${F^M}_i$ acts in x and y directions, and is expressed as the following [44]:

$${F^M}_i = F_x + F_y \quad (15)$$

$$F_x = \frac{Ha^2 \mu}{L^2} (v \sin\phi \cos\phi - u \sin^2\phi) \quad (16)$$

$$F_y = \frac{Ha^2 \mu}{L^2} (u \sin\phi \cos\phi - v \sin^2\phi) \quad (17)$$

The buoyancy force term can be expressed as

$${F^b}_i = \rho g_y \beta (T_h - T_c) \quad (18)$$

The applied magnetic field does affect the force term. The present study considers the external magnetic field is applied in different directions. The direction is horizontal, vertical, or in other angles (for example, θ = 0, 45, 90). The external MHD forces acting in x and y directions are presented as the following:

$$F_x = 3\omega_k \rho \epsilon A[(v \sin\theta \cos\theta) - (u \sin^2\theta)] \quad (19)$$

$$F_y = 3\omega_k \rho \epsilon (g_y \beta (T - T_{ref})) + A[(u \sin\theta \cos\theta) - (v \cos^2\theta)] \quad (20)$$

The magnetic buoyancy force in terms of weighting factor is written as

$${F^b}_i = 3\omega_k \rho \epsilon g_y \beta (T - T_m) \quad (21)$$

Here, $T_m = (T_h + T_c)/2$.

The body force for porous media, ${F^P}_i$, is expressed through Ergun's equation [45]:

$$\bar{F}_i = -\frac{\epsilon v_k}{K} \bar{u} - \frac{\epsilon F_\epsilon}{\sqrt{K}} |\bar{u}| \bar{u} + \epsilon \bar{G} \quad (22)$$

where, F_ϵ represents the geometric function ($F_\epsilon = \frac{1.75}{\sqrt{150}}$), K is the permeability ($K = Da \cdot H^2$) with H symbolizing the domain height, v_k represents the kinematic viscosity, and $\bar{G}$ represents the external body force term.

An alternative equation of force term, ${F^P}_i$, for porous media was proposed by Mohamad [46] to obtain the accurate solution of hydrodynamics, which is the following:

$${F^P}_i = -w_k \left[9 \left(\frac{\epsilon v}{K} \right) (u e_x + v e_y) + \frac{F_\epsilon \epsilon}{\sqrt{K}} (|\bar{u}| u e_x + |\bar{v}| v e_y) \right] \quad (23)$$

The present study considers $D2Q9$, i.e., two-dimensional nine-velocities, model [47]. Therefore, the equilibrium distribution functions (f_i^{eq}) for the $D2Q9$ model of porous media is written as the following:

$$f_i^{eq} = \omega_k \rho \left[1 + \frac{\bar{e}_i.\bar{u}}{c_s^2} + \frac{(\bar{e}_i.\bar{u})^2}{2\epsilon c_s^4} - \frac{|\bar{u}|^2}{2\epsilon c_s^2} \right] \tag{24}$$

Here, ϵ is denoted as the porosity. The discrete velocities $\bar{e}_i$ for the $D2Q9$ model have different parametric values and are expressed as mentioned by [48]:

$$\bar{e}_i = \begin{cases} (0,0) & \text{for i } = 0 \\ cos[(i-1)\pi/4], sin[(i-1)\pi/4] & \text{for i} = 1\text{– }4 \\ \sqrt{2}(cos[(i-1)\pi/4 + \pi/4], sin[(i-5)\pi/2 + \pi/4]) & \text{for i} = 5\text{–}8 \end{cases} \tag{25}$$

The values of the weighting factor ω_k are the following:

$$\omega_k = \begin{cases} 4/9 & \text{for i} = 0 \\ 1/9 & \text{for i} = 1\text{–}4 \\ 1/36 & \text{for i} = 5\text{–}8 \end{cases} \tag{26}$$

The thermal equilibrium energy function can be expressed as the following [48]:

$$g_i^{eq} = \omega_k T \left[1 + \frac{1}{c_s^2} \bar{e}_i.\bar{u} \right] \tag{27}$$

3.3. *Boundary Conditions*

Boundary conditions were defined for the four walls of the rectangular cavity for the purpose of simulation. Boundary conditions are generally described as distribution functions (DFs) in LBM. It is required to determine the DFs at the boundary nodes according to the macroscopic conditions. The procedure is attributed with ensuring the stability and accuracy of the mathematical model [49].

3.3.1. Boundary Conditions for Fluid Flow

The no-slip (also known as bounce-back) boundary condition was applied on the walls of the rectangular cavity. As an aftermath of the particles' collision, the outgoing DF goes in the reverse direction of the incoming DF at a particular position within the boundary. The following expressions represent the boundary conditions of this study:

At right wall: $f_{3,m} = f_{1,m}$, $f_{7,m} = f_{5,m}$, and $f_{6,m} = f_{8,m}$

At left wall: $f_1 = f_3$, $f_5 = f_7$ and $f_8 = f_6$

At top wall: $f_{4,n} = f_{2,n}$, $f_{8,n} = f_{6,n}$ and $f_{7,n} = f_{5,n}$

At bottom wall: $f_2 = f_4$, $f_5 = f_7$ and $f_6 = f_8$

where m and n represent the domain's lattice for length and height, respectively.

3.3.2. Thermal Boundary Conditions

As described earlier, the top (T_c) and the bottom walls (T_h) have constant temperatures, but they have different values. The other walls are adiabatic and, therefore, are not participating in the mass transfer.

At isothermal cold top wall: $g_{4,n} = -g_{2,n}$

At isothermal hot bottom wall: $g_2 = T_{wall}\ (w_2+w_4)-g_4$

Here, T_w is used for the 2nd-order Zou-He boundary conditions

At adiabatic west wall: $g_{i,0} = g_{i,1}$, for i = 1–8

At adiabatic east wall: $g_{i,m} = g_{i,m-1}$, for i = 1–8

where, m and $m-1$ are the boundary lattice and the lattice inside the enclosure near the boundary, respectively.

3.4. *Rate of Heat Transfer*

In the numerical investigation of the convective heat transfer problem, the Nusselt number (Nu) is an important parameter. The Nu number describes the heat transfer rate due to temperature gradient. The local Nu number at hot walls and the average Nu number (Nu_{avg}) calculated for the entire domain are formulated as the following [47,50]:

$$Nu(x) = -\frac{H}{\Delta T}\frac{\partial \overline{T}}{\partial \overline{y}} \tag{28}$$

$$Nu_{avg} = \frac{1}{H}\int_0^H Nu(x).dx \tag{29}$$

where L denotes the length of the cavity.

Clever and Busse [51] defined a modified formulation for the average Nusselt number, $\bar{Nu}$ in their work, and it is written as

$$\overline{Nu} = 1 + \frac{< \bar{v} \cdot \bar{T} >}{\sqrt{Ra\Delta T\alpha / H}} \tag{30}$$

where H represents the distance between the bottom and top walls, ΔT is difference between temperature of top and bottom walls, $< \cdot >$ shows the average over whole flow domain, and v denotes the velocity component of the vertical direction.

For the RB convection, He at al. [52] formulated an equation for the average Nusselt number $\overline{Nu}$ in terms of critical Rayleigh Ra_c and Rayleigh number Ra:

$$\bar{Nu}_{EM} = 1.56 \times (Ra/\mathrm{Ra}_c)^{0.296} \tag{31}$$

where $Ra_c = 1707.06$.

3.5. LM Algorithm

The LBM-MHD-RB data-driven work in this study is analyzed by the LM algorithm. It is a hybrid method that considers both the Gauss–Newton and steepest descent approaches for the convergence criteria to reach an optimal solution. There is an inherent trade-off between Gauss–Newton and the steepest descent based on the requirements of the problem. For instance, if Gauss–Newton alone cannot solve a problem, the LM algorithm links the steepest descent approach for traversing the design space and determining the optimal solution. This technique is most effective in solving non-linear equations. The correlations to predict the output parameter by considering the influential parameters of LBM are typically non-linear, and therefore, the LM algorithm was a suitable option for the surface analyses.

LM develops the trust region for different computational applications. In the LM method, the difference in the weights (w_i) is obtained by determining the following [53]:

$$\alpha'\Delta = -1/2(\nabla)\lambda \tag{32}$$

where α' is the matrix of the optimization field, and λ is the mean-squared network error. The term λ is achieved by the following equation [53]:

$$\lambda = 1/N\sum_{k=1}^{N}[\vec{q}(x_k) - \vec{d}_k]^2 \tag{33}$$

where N is the number of examples; $\vec{q}(x_k)$ is the output of the network aligning with the example x_k; and $\vec{d}_k$ is the expected outcome.

Finally, the matrix α' elements are obtained by the following [53]:

$$\alpha'_{ij} = (1 + \zeta\delta_{ij})\sum_{r=1}^{z}\sum_{k=1}^{N}\left[\frac{\partial y_r(x_k)}{\partial w_i}\frac{\partial y_r(x_k)}{\partial w_j}\right] \tag{34}$$

where z is the number of the desired output from the network.

At the commencement of the algorithm, α' and $\nabla(\lambda)$ are a major part of the evaluation, followed by the obtained solution on w_i. The LBM-obtained data were initially analyzed in the R programming environment using library packages *"dplyr"* [54], followed by *"pracma"* [55] to optimise the matrix α. The analyzed, clean dataset was fed into OriginPro to perform the non-linear surface analyses. In each step of iteration, the λ value was calculated through the model, and the iteration was not terminated unless an optimal solution was reached. The convergence criteria were set if a coefficient of determination (R^2) was reached above 0.8 for any specific function. A separate function was chosen for the iteration if the required R^2 was not obtained despite adjusting the iteration cycle ζ. It should be mentioned here that the present study does not aim to perform an AI approach, such as neural networks, due to the limited data for the analysis, and since the solution was reached by the LM algorithm by solving the non-linear least squares curve fitting, the process was terminated once a standard R^2 was achieved, subject to further validations. Therefore, once the solution at w_i was reached, the equation was obtained to validate the accuracy. The obtained equation was initially checked through the interpolated dataset, which was at least 3000, depending on the percentage of outliers. The fundamental aim was to always have R^2 more than 0.95, and hence the outlier detection test was conducted on the obtained dataset. The initial process is known as the LM iteration/learning cycle, with each step destined to reduce the error from the previous one. The ζ parameter is an adjustment at each cycle. The process keeps on running unless a good correlation coefficient is reached. Once the iteration was terminated, the equation was obtained and immediately tested for the accuracy.

3.6. Code Convergence Criteria

In the LBM simulations, all the computations were terminated when the velocity field, as well as temperature, reached the following convergence criteria:

$$\frac{\sum|\psi^{n+1} - \psi^{n}|}{\sum|\psi^{n+1}|} < 10^{-15} \tag{35}$$

where ψ is either the velocity u or the temperature T, n represents the iteration index, and finally, the summation was applied over the whole domain of interest.

Meanwhile, the LM algorithm followed an iterative method unless an accurate correlation coefficient was obtained as described in Figure 2.

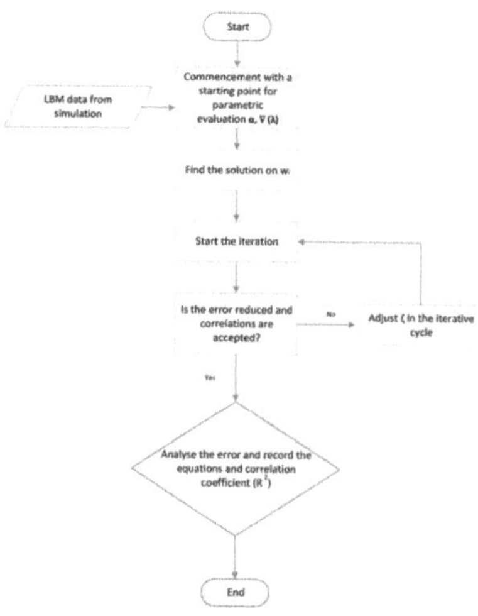

Figure 2. Flowchart of LM algorithm using LBM data for the correlation development.

4. Materials and Methods

LBM simulations for MHD-RB convection were performed in Fortran 90 [56] by using Microsoft Visual Studio Code™. Boundary conditions, collision operators, streaming functionality, and convergence criteria were all included as subroutines. The base code considered all those subroutines for the iterations. The iterations continued until the convergence criteria were obtained. The computations were performed on a Windows 10 computer in an 11th Gen Intel(R) Core™ i9 2.60 GHz processor with 64 GB RAM. The streamlines and isotherms were visualized by using Tecplot 360 2022 R1 version (https://www.tecplot.com).

As mentioned earlier, the LM algorithm was performed through the R programming language [57] by RStudio™ open source software using library packages *pracma* [55], followed by data organising, equations validation, and correlation development in OriginPro [58]. The library package *pracma* determines a large number of functions from the numerical analysis for any math function. Prior to that, the popular *dplyr* [54] package was used for data manipulation and visualization. It should be mentioned here that the same computer was utilized for LM algorithm development, which was also used for the LBM simulations. However, both RStudio and OriginPro were operated with NVIDIA RTX A3000 GPU power for fast implementation of the model optimization and correlations development.

5. Results

The primary purpose of the present study is to develop the correlation among the important variables to quantitatively predict $\overline{Nu}$ number, which is representative of the average heat transfer rate. Therefore, the results will discuss the numerical correlations based on LBM-MHD data interpretation through the LM algorithm. Each segment demonstrates the obtained outcome from the non-linear surface analysis, followed by validations with literature to showcase the accuracy of the obtained equations. However, two separate comprehensive analyses are first conducted to pinpoint some of the significant changes in the streamlines and isotherms.

5.1. Effect of Numerical Parameters on Streamlines

The impact of Ra and Ha numbers, as well as ϵ on streamlines, is illustrated in Figure 3 under a constant $\theta = 0$. The combined analysis will depict each variable's influence on the streamlines' pattern.

Figure 3a is assigned with $Ra = 10^5$, $Da = 10^-2$, $\epsilon = 0.4$, and $Ha = 0$, leaving entirely no impact of the external magnetic field. As per Figure 3a, three separate circular rolls distributed within a symmetry within the cavity were observed. However, the circular rolls in the left and right locations of the cavity exhibited almost similar characteristics with the maximum contour values at the center. However, the circular roll in the middle demonstrated the opposite and negative contour values. This behavior could be attributed to side-heated adiabatic walls and the top and bottom walls being active in the heat transfer process. Therefore, the circular roll in the middle directly depicted the effect of convective characteristics instead of conductive ones [18]. However, as the Ha number increased from 0 to 50, the shape and contour values changed significantly, as seen in Figure 3b. The Bénard cell reduced from 3 to 1 and started to stretch from the central region. The maximum contour value also reduced from 8 to 6, which is almost a 25% reduction due to the 50% augmentation in the Ha number. Therefore, it was expected that increasing Ha number would keep on reducing the heat conduction. The hypothesis was confirmed from Figure 3c, where the maximum contour value at the center plummeted to 1. By increasing the Ha number from 50 to 100, the maximum contour value reduced by approximately 83.33%, indicating the negative influence of the external magnetic field and the existence of restriction within the cavity to reduce the heat transfer mobility.

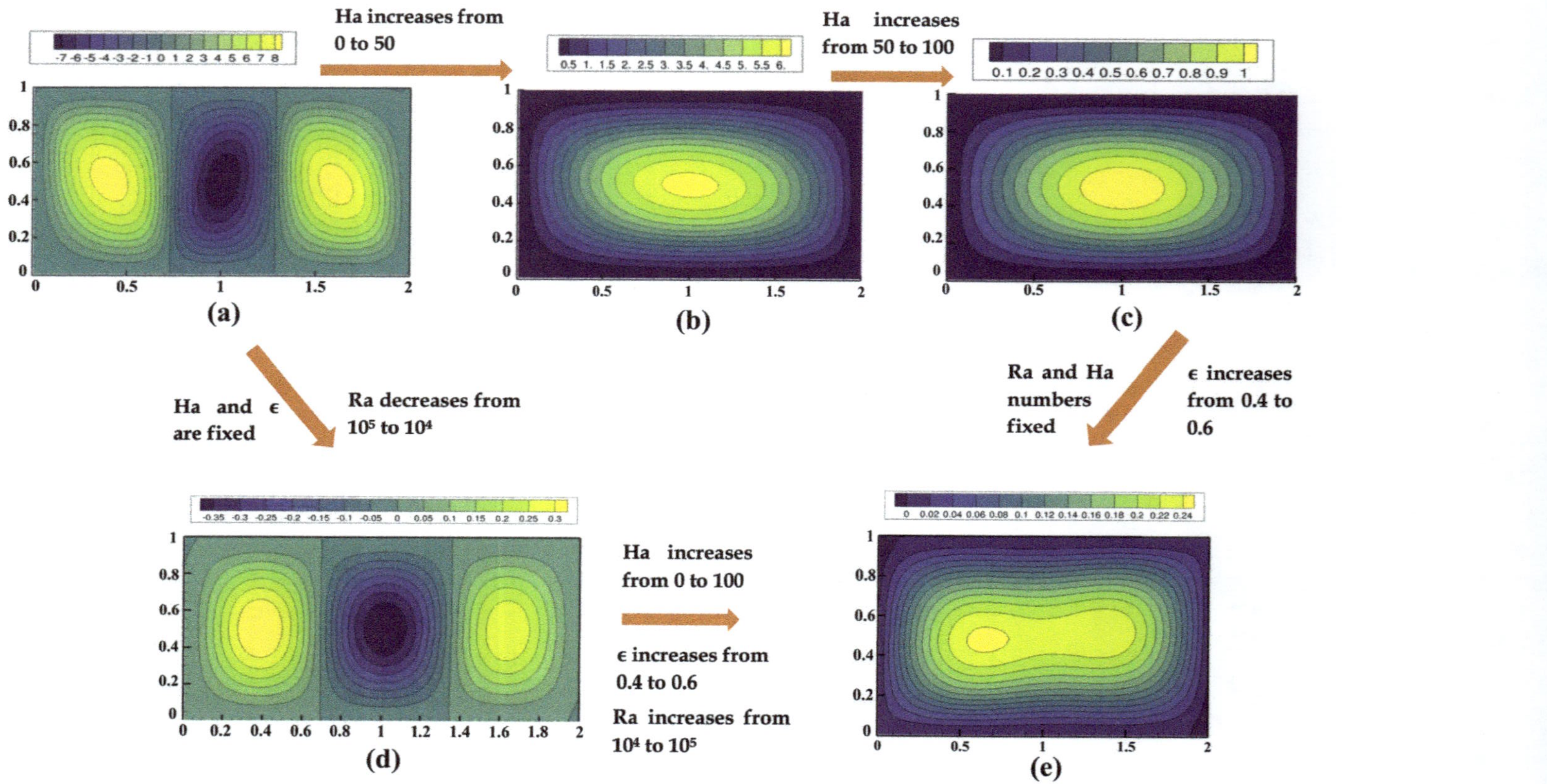

Figure 3. Illustrations of streamlines as Ra, Ha numbers, and ϵ varied under $\theta = 0$, and $Da = 10^{-2}$: (**a**) $Ra = 10^5$, $\epsilon = 0.4$, and $Ha = 0$, (**b**) $Ra = 10^5$, $\epsilon = 0.4$, and $Ha = 50$, (**c**) $Ra = 10^5$, $\epsilon = 0.4$, and $Ha = 100$, (**d**) $Ra = 10^4$, $\epsilon = 0.4$, and $Ha = 0$, (**e**) $Ra = 10^5$, $\epsilon = 0.6$, and $Ha = 100$.

In the second part of this analysis, the influence of Ra numbers and ϵ was observed. By comparing Figure 3a and Figure 3d, the impact of the Ra numbers could be observed by keeping Ha and ϵ constant. As Ra decreased from 10^5 to 10^4, the maximum contour value decreased from 8 to 0.3, which is a rapid 96.25% reduction. With this observation, the impact of buoyancy in the RB convection could be understood. In the next part, all the variables, namely Ra and Ha numbers, and ϵ were increased concurrently as shown in Figure 3e. According to Figure 3e, three Bénard cells reduced to one but demonstrated strong attraction toward the thermal walls by showing similarity with the thermal dipole. The maximum contour value was 20% reduced from 0.3 to 0.24, and no negative value was recorded. While the increased Ha number was directly responsible for negative contour values, the increased ϵ and Ra numbers enhanced the heat transfer phenomena, leading to 0 as the lowest contour value within the cavity. The impact of ϵ was also tested by keeping Ha and Ra numbers constant. By comparing Figure 3c,e, the influence of ϵ could be analyzed, where the maximum contour value decreased from 1 to 0.24. However, in both Figure 3c,e, the Ha number is 100, which repelled the heat transfer application. Therefore, the contour value reduced significantly by about 140%.

5.2. Isothermal Changes

The final section of the results focuses on the changes in the isotherms, similar to the previous section. Figure 4 represents such changes in five different frames.

The impact of Ha can be observed from Figure 4a–c by increasing from 0 to 100 in three separate frames. The distribution of isotherms is kept within 0 to 1. As per Figure 4a, the isothermal lines demonstrate an oscillating pattern due to the heat transfer within the cavity without the influence of Ha number. The pattern within 1 to 1.5 of the horizontal axis is the opposite of what was observed within 0 to 1 of the same axis. This behavior could be linked with the conduction and convective rolls observed in Figure 3a, where the middle convective rolls represent the negative contour values. Therefore, the pattern of the isotherm from 1 to 1.5 on the horizontal axis is the opposite. As the Ha number increases from 0 to 50, the isothermal lines exhibit uniformity within the cavity, as the oscillation disappears and all the lines start to become quasi-linear as seen in Figure 4b. The presence of the Ha number leads to the presence of a magnetic field which develops the Lorentz force within the cavity. Therefore, the instability within the thermal walls is reduced. Further decreasing oscillation could be observed from Figure 4c, where the isothermal lines edge closer to the linearity. While a wavy pattern could be seen at the lowest contour, the isothermal lines are quite linear at the maximum contour values, which are closer to the horizontal axis.

Meanwhile, the effect of plummeting Ra numbers could be observed from Figure 4d, where decreasing Ra from 10^5 to 10^4 significantly impacts the isothermal patterns. It could be observed that due to the decreased buoyancy, the isothermal lines show minor oscillation with a minimal peak in each line. The isothermal line close to the horizontal axis show a linear pattern due to the lack of buoyancy strength within the cavity. However, as Ha is increased from 0 to 100, Ra is increased from 10^4 to 10^5, and finally, ϵ is also increased from 0.4 to 0.6. The isothermal lines are almost linear throughout the cavity due to the strong influence of the Ha number in particular as seen in Figure 4e. In fact, keeping $Ha = 100$ constant and increasing ϵ from 0.4 to 0.6 does not significantly impact the isothermal lines either, due to the existence of the Lorentz force. By comparing Figure 4c,e, the impact of the Ha number in the RB convection could be well understood.

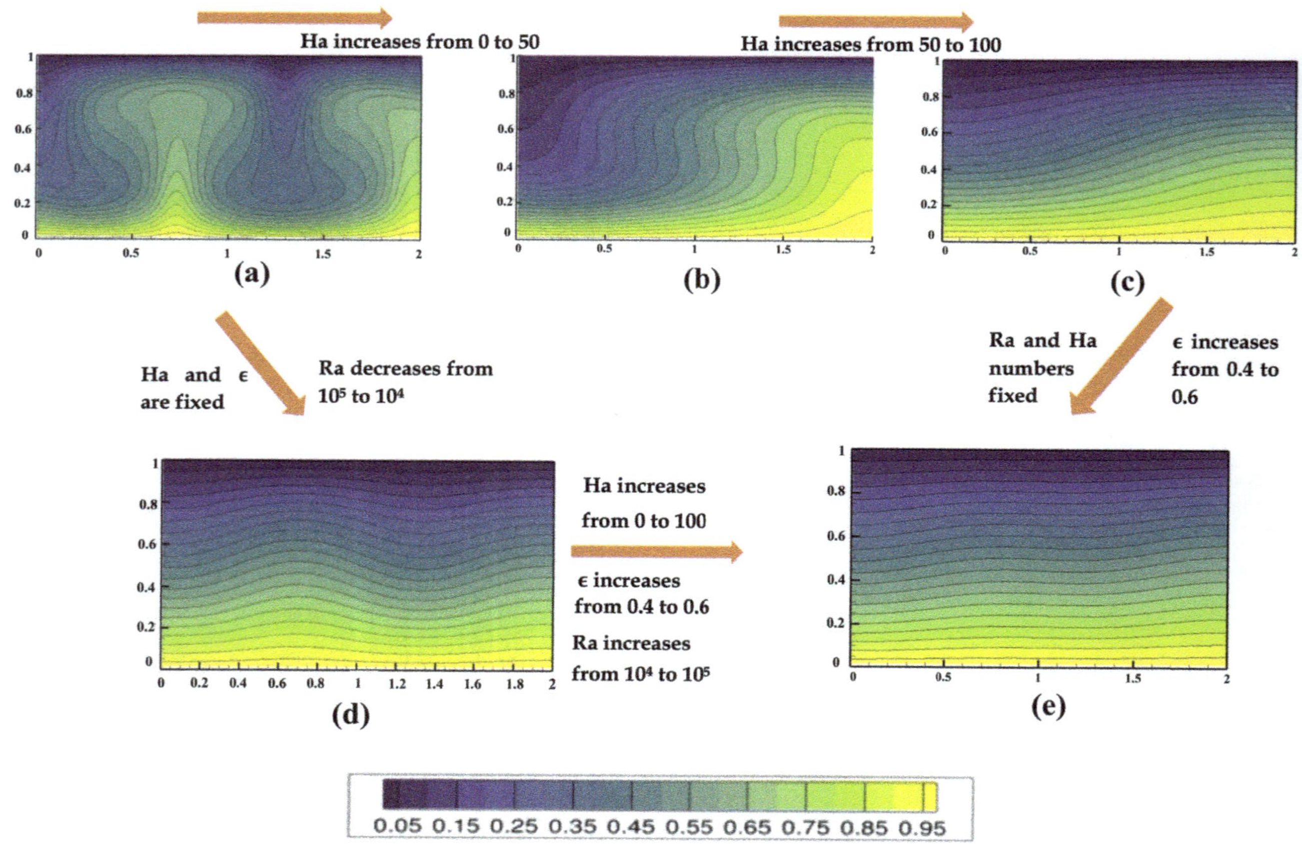

Figure 4. Representation of isotherms as Ra, Ha numbers, and ϵ changed under constant $\theta = 0$, and $Da = 10^{-2}$: (**a**) $Ra = 10^5$, $\epsilon = 0.4$, and $Ha = 0$, (**b**) $Ra = 10^5$, $\epsilon = 0.4$, and $Ha = 50$, (**c**) $Ra = 10^5$, $\epsilon = 0.4$, and $Ha = 100$, (**d**) $Ra = 10^4$, $\epsilon = 0.4$, and $Ha = 0$, (**e**) $Ra = 10^5$, $\epsilon = 0.6$, and $Ha = 100$.

5.3. Predicting $\overline{Nu}$ from Ha Number and θ

In this part of this study, individual equations to predict $\overline{Nu}$ under the influence of an external magnetic field in an inclined rectangular cavity are developed for $Ra = 10^4$ and $Ra = 10^5$. The key element of this analysis is the consideration of the electrically conducting fluid in RB convection. Different simulations were conducted at Ha number $\in [0, 100]$ based on the LBM model at $\theta \in [0, 90]$. The LM analysis was performed to build the correlation, followed by the validation with well-cited literature from the past and the recent. In general, good accuracy was established. The correlation is only valid for $Ra = 10^4, 10^5$, which are the most preferred ones for laminar flows as per the data from the literature.

5.3.1. Development of Correlation and Surface Analysis

Non-linear parabolic and power correlations were found to be the best-fitting ones among different functions for $Ra = 10^4$ and $Ra = 10^5$, respectively, under the LM algorithm. The correlation coefficients (R^2) were found to be 0.89 and 0.966 for $Ra = 10^4$ and $Ra = 10^5$, respectively. Figure 5 depicts the 3D contours for better visualization. It could be observed that the high density of the points was more aligned with low θ, as most of the important transition in the heat transfer takes place under low Ha and θ. This behavior could be attributed to the effect of the magnetic field, which is directly controlled by the Ha number. At an increasing Ha number, the rate of heat transfer declines due to the existence of both an electric field and magnetic field, leading to the presence of a Lorentz force. Consequently, increasing the Ha number lowers the values of $\overline{Nu}$. However, as part of the validation, a wide range of Ha numbers and θ was considered to demonstrate the accuracy of the model and its ability to predict the heat transfer value outside the calibration zone.

As presented in Figure 5, $\overline{Nu}$ between 0.1140 and 5.280 was obtained from the surface analysis. The equation, however, is expected to be valid to predict $\overline{Nu}$ beyond the obtained range in the analysis due to the consideration of a broader range of Ha and θ. In order to obtain the equation from the best-fitting simulated contour from the LM algorithm, the LBM-simulated data were subject to several surface analyses for the purpose of interpolation within the user-defined range, and the following equation provided the best R^2:

$$\overline{Nu} = f + aexp(-Ha/b)exp(-\theta/c) \tag{36}$$

where f, a, b, and c are fitting parameters with assigned values specifically under the aforementioned condition. Table 1 contains the values of the fitting parameters obtained through the LM algorithm with the best R^2 value.

Table 1. Fitting parameters to predict $\overline{Nu}$ from Ha number and inclination angle θ at $Ra = 10^4$.

Empirical Parameters	Fitting Values
f	0.87404
a	1.73303
b	30.0595
c	127.50
Statistical Accuracy Indicators	**Values**
R^2	0.966
p	< 0.05

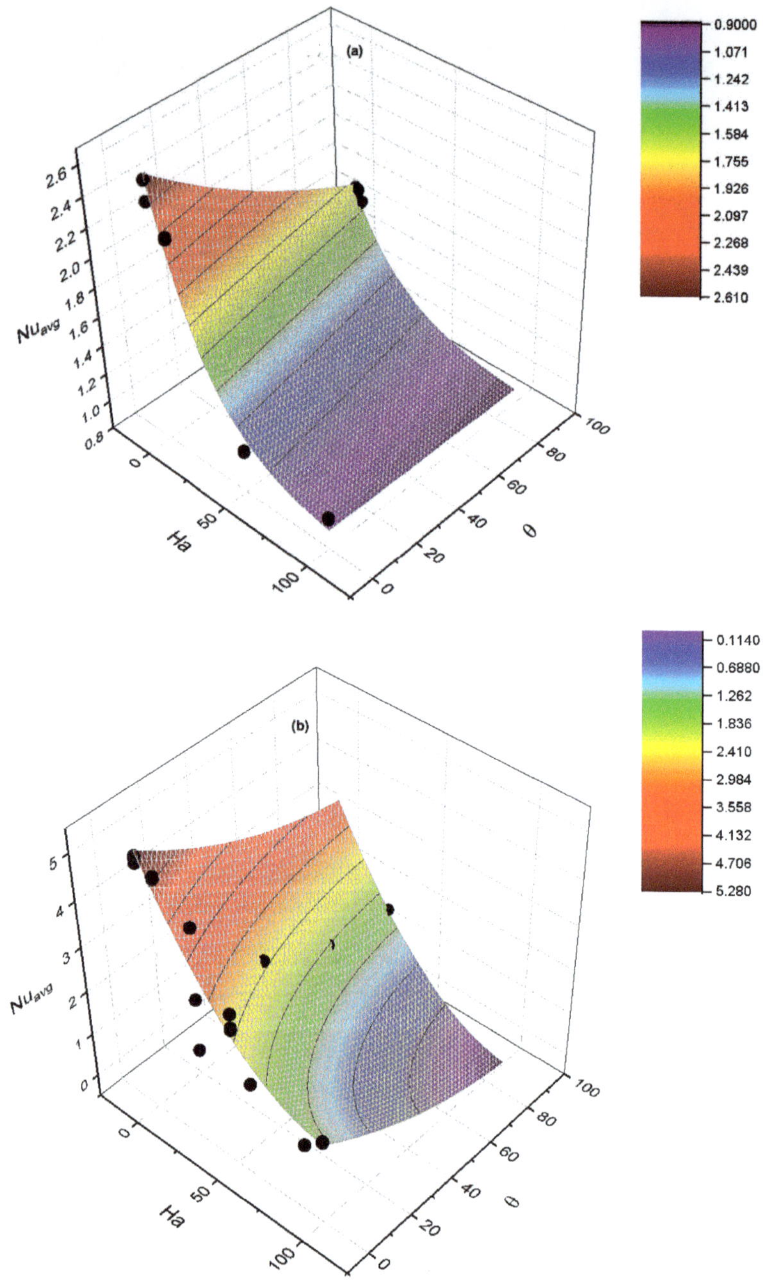

Figure 5. Development of correlation through fitting surfaces for (**a**) $Ra = 10^4$, and (**b**) $Ra = 10^5$ under the influence of external magnetic field at different inclination angles for electrically conducting fluid.

As Ra increases from 10^4 to 10^5, the impact of buoyancy inside the enclosure augments significantly. Therefore, Equation (36) is not an appropriate option to predict $\overline{Nu}$ as a function of Ha numbers and θ. In fact, different functions were considered to determine the best-fitting surface to obtain the equation to predict $\overline{Nu}$ at $Ra = 10^5$. The following equation provided the best coefficient of determination to predict $\overline{Nu}$:

$$\overline{Nu} = f + aHa + b\theta + c(Ha)^2 + d(\theta^2) \quad (37)$$

Table 2. Fitting parameters to predict $\overline{Nu}$ from Ha number and inclination angle θ at $Ra = 10^5$.

Empirical Parameters	Fitting Values
f	5.26782
a	−0.07321
b	−0.03499
c	3.43333
d	0.000162
Statistical Accuracy Indicators	**Values**
R^2	0.897
p	<0.05

It should be mentioned here that the p-value outlines the significance of the statistical study implemented in this section. The lowest p-value indicates that the null hypothesis was rejected, and the correlation is statistically significant. Overall, $p < 0.05$ was considered to be a good indicator to validate the model's accuracy.

5.3.2. Cross-Validation with Literature

The cross-validation was conducted with the literature with a similar objective. However, none of the literature provided any clear mathematical correlation among the parameters. The data from the literature were not considered to calibrate the LM model. Hence, the cross-validation serves as an independent validation to show unbiased agreement with the LBM and LM data within the considered range of input parameters.

The validation plots presented in Figure 6 demonstrate good agreement between $\overline{Nu}$ predicted from LBM and LM simulations. The majority of the points were obtained to be within the ±5% error lines. To build the model, the validation dataset contained a similar geometry considered in this study. The separate validation plots represent the agreement for two different Ra numbers ($Ra = 10^4, 10^5$) considered in developing Equations (36) and (37). The empirical parameters presented in Tables 1 and 2 were considered to obtain the $\overline{Nu}$ as presented in Figure 6a,b, respectively. A separate Table 3 is presented to indicate the accuracy individually with the literature data considered for the validation. As mentioned in the caption of Table 3, some outliers were ignored in the individual R^2 calculation since it was already considered for the overall R^2 determination. It should be mentioned that filtering the outlier point is a common practice in statistical analysis, and hence the influential negative point can be ignored.

Table 3. Obtained R^2 in each validation with literature individually. *Detected outliers were removed for the correlation development.*

Ra	Rudraiah et al. [59]	Kefayati [60]	Sheikholeslami et al. [61]	Sajjadi et al. [62]	Ahmed et al. [16]
10^4	-	-	0.9789	0.963	0.9789
10^5	0.9862	0.9878	0.9662	0.967	0.9858

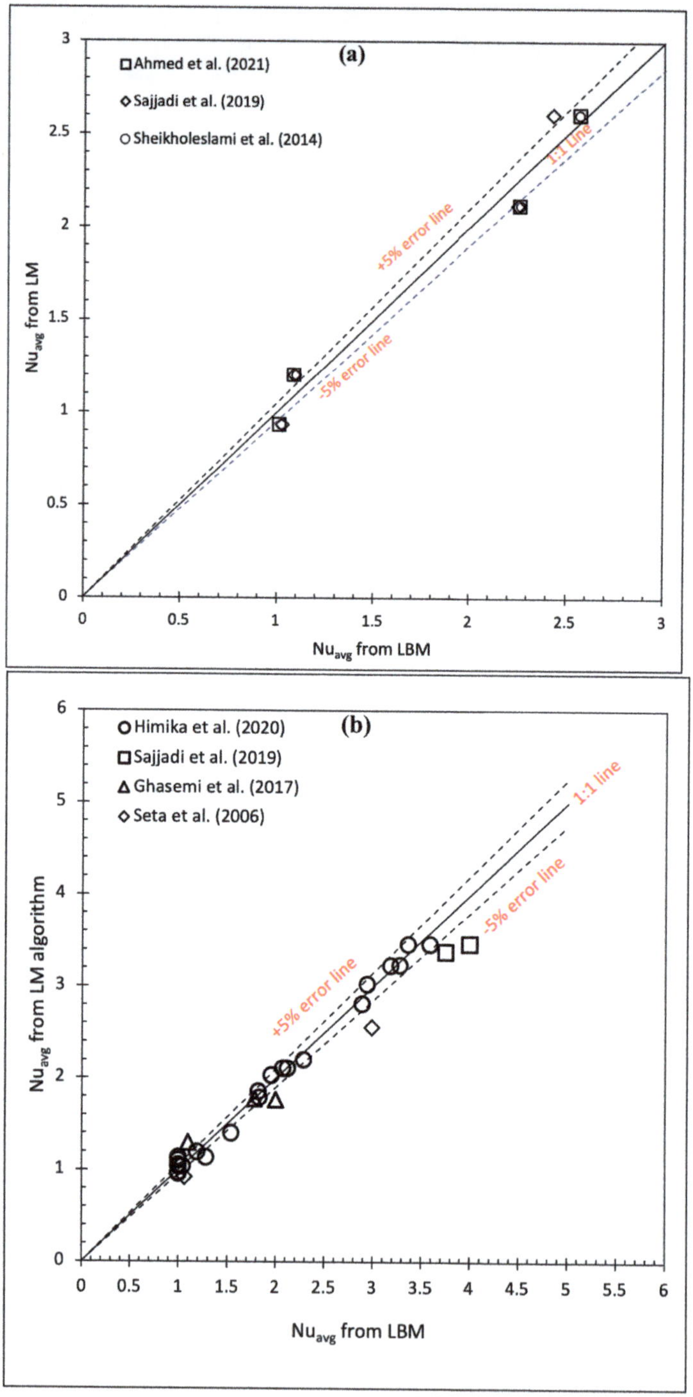

Figure 6. Cross validations with LBM data from the literature (**a**) $Ra = 10^4$ [16,61,62], and (**b**) $Ra = 10^5$ [18,62–64] under the influence of external magnetic field at different inclination angles for electrically conducting fluid.

5.4. Correlations among $\overline{Nu}$, *Ha* Numbers, and *Da* Numbers Under Constant Porosity

The major focus of this section is to predict $\overline{Nu}$ as a function of the *Da* number ($0.1 \leq Da \leq 0.0001$) and the *Ha* number ($Ha \leq 30$), at constant variables, such as porosity ($\epsilon = 0.4$) and inclination angle ($\theta = 45$). The data were obtained through LBM RB simulation, and a correlation was developed through the LM algorithm. Two different *Ra* numbers $Ra = 10^4, 10^5$ were considered in this section of the study.

5.4.1. 3D Fitting Curves and Statistical Parameters

Repeated LM algorithm simulations were performed to obtain the best-fitting results. As per Figure 7, the fitting curves are presented for $Ra = 10^4$ (Figure 7a) and $Ra = 10^5$ (Figure 7b), where the distribution of the LBM-obtained data is shown. At higher *Ha* numbers, it was anticipated to have the lower $\overline{Nu}$; therefore, $Ha \leq 30$ was considered for the model calibration. However, the independent validation was conducted with the published literature, where the *Ha* number was expanded up to 50, and still, significant agreement was established. On the other hand, a wide range of *Da* numbers was considered in this part of the simulation. Therefore, the model could still develop a correlation at a higher *Da* number.

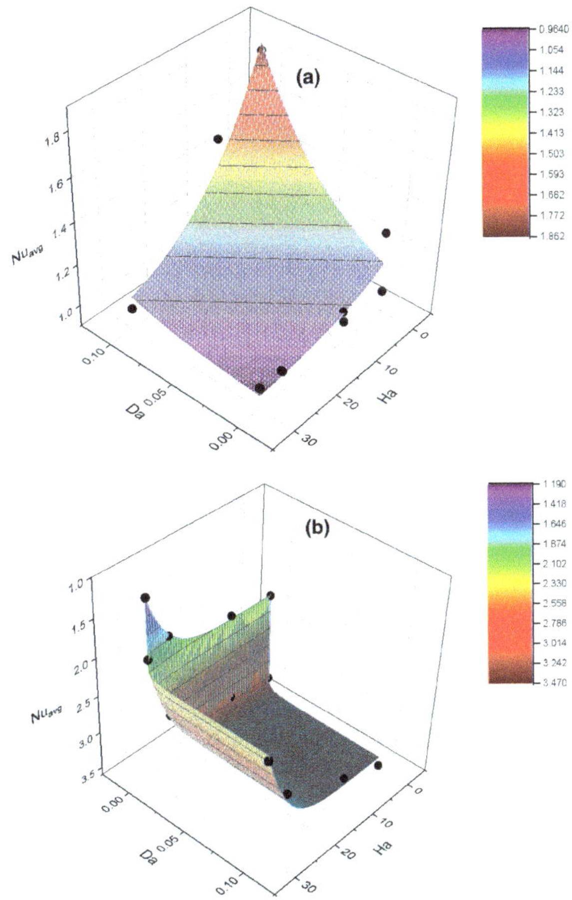

Figure 7. Independent validation plots by comparing with LBM data: (**a**) $Ra = 10^4$, and (**b**) $Ra = 10^5$, while $\epsilon = 0.4$, $\theta = 45$ were constant.

The obtained equation to predict $\overline{Nu}$ is the following (for $Ra = 10^4$):

$$\overline{Nu} = f + aexp(-Da/b)exp(-Ha/c) \tag{38}$$

The equation to predict $\overline{Nu}$ at $Ra = 10^5$ yielded the best R^2 under the power function, which was found as the following:

$$\overline{Nu} = f + a^{bDa} + c^{dHa} + e^{bdDaHa} \tag{39}$$

where f, a, b, c, d, and e are empirical parameters to adjust the fitting surface and build the correlation. Tables 4 and 5 show the quantitative values of those parameters and statistical information of the LM model, which were the foundations of obtaining Equations (38) and (39).

Table 4. Quantitative values of empirical parameters to predict $\overline{Nu}$ from Ha and Da numbers at $\epsilon = 0.4$, $\theta = 45$, and $Ra = 10^4$.

Empirical Parameters	Fitting Values
f	0.93997
a	0.19833
b	−0.06512
c	14.6836
Statistical Accuracy Indicators	**Values**
R^2	0.90
p	< 0.05

According to Tables 4 and 5, the numerical values obtained from the LM algorithm are presented. The correlations were obtained to be $R^2 = 0.9$ and $R^2 = 0.99$, for $Ra = 10^4$ and $Ra = 10^5$, respectively.

Table 5. Quantitative values of empirical parameters to predict $\overline{Nu}$ from Ha and Da numbers at $\epsilon = 0.4$, $\theta = 45$, and $Ra = 10^5$.

Empirical Parameters	Fitting Values
f	3.47307
a	−0.00128
b	−0.75661
c	-3.4519×10^{-9}
d	5.79634
e	8.81812×10^{-13}
Statistical Accuracy Indicators	**Values**
R^2	0.99
p	< 0.05

5.4.2. Independent Validation

The independent validation was conducted with the literature to showcase the ability of the correlation with data. The purpose of such an approach is to validate the present approach with well-cited data from the past and the recent, collectively.

Figure 8 presents the validation results obtained through the present approach for $Ra = 10^4$ (Figure 8a) and $Ra = 10^5$ (Figure 8b), respectively. It was found that the majority of the points were near the 1:1 line, and the agreement was within the ±5% error lines. The agreement plot also demonstrated the range of the $\overline{Nu}$ being as high as 4, which is mostly observed at a lower or no Ha number. Therefore, the present approach was able to predict LBM results within multifarious ranges with a low percentage of error. The R^2 of each comparison in the validation is presented in Table 6, where $0.9113 \leq R^2 \leq 0.9928$ was obtained, which provides more confidence in the accuracy of the present approach. Due to the limited data in the literature, the number of points presented in Figure 8a is less than those of Figure 8b.

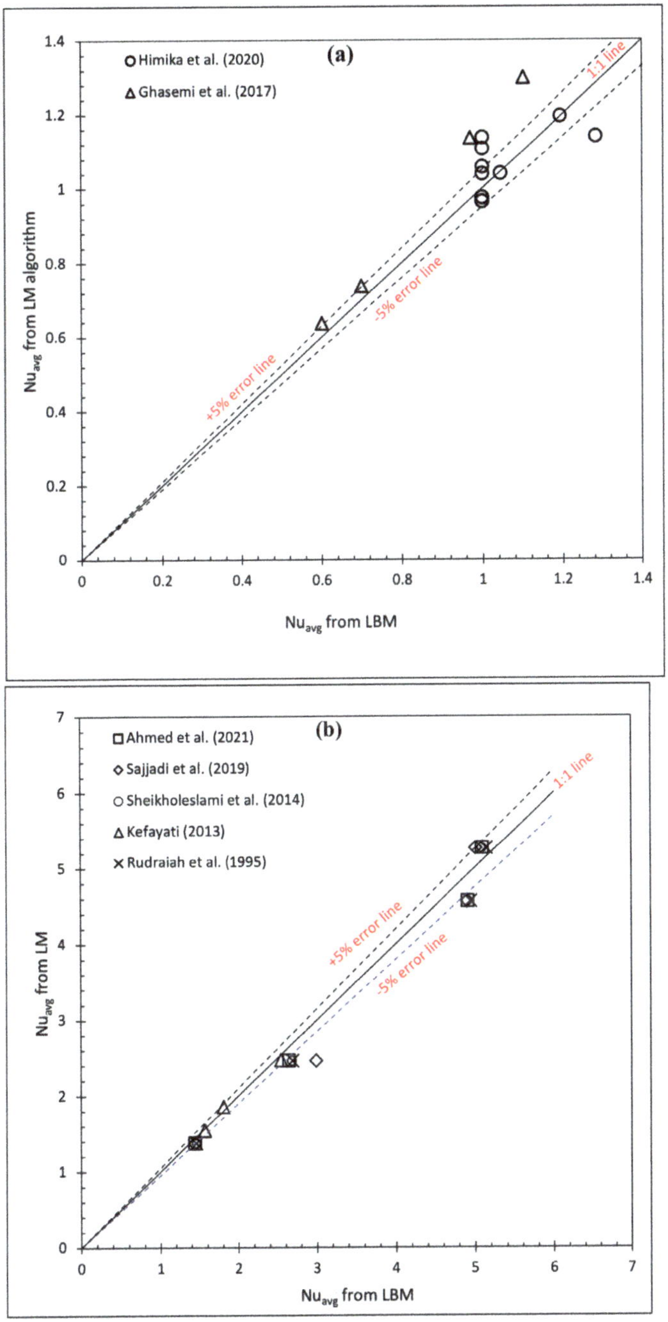

Figure 8. Surface fitting curves obtained from LBM data by LM algorithm for (**a**) $Ra = 10^4$ [18,64], and (**b**) $Ra = 10^5$ [16,44,59,61,62], while $\epsilon = 0.4$, $\theta = 45$ were constant.

Table 6. Obtained R^2 in each validation segment.

Ra	Seta et al. [63]	Ghasemi et al. [64]	Sajjadi et al. [62]	Himika et al. [18]
10^4	-	0.9946	0.9113	-
10^5	0.9994	0.9362	0.967	0.9928

5.5. Equation to Predict $\overline{Nu}$ under Variable Porosity

In this segment of statistical analysis, porosity (ϵ) is considered as a variable under the fixed θ, and Ra numbers. The primary focus is to establish a correlation under variable porosity (ϵ), which is quite sensitive to other variables concurrently. Therefore, $Ra = 10^5$ and $Da = 10^{-1}$ are considered for the sensitivity analysis.

5.5.1. 3D Fitting over a Planar Surface

It was anticipated that under constant Da and Ra numbers, and θ, the increasing rate of $\overline{Nu}$ will be linear as a function of ϵ and Ha, considering the fact that Ha remains constant in each step while ϵ varies. For example, it was pinpointed earlier that the increasing Ha number significantly plummets the heat transfer rate. However, if Ha remains unchanged, yet ϵ increases, the $\overline{Nu}$ will increase linearly due to the improved convection inside the cavity since the Da number is also unchanged.

Figure 9 serves as a testimony of the aforementioned statement, where a planar correlation was obtained through the LM algorithm. The $R^2 = 0.91$ depicts the accuracy of the correlation, which can be improved further with more relevant data within the plane. The distribution of the points within the surface implies that a suitable range was taken into account to predict $\overline{Nu}$ over the multifarious $0.4 \leq \epsilon \leq 1.0$ and $0 \leq Ha \leq 50$.

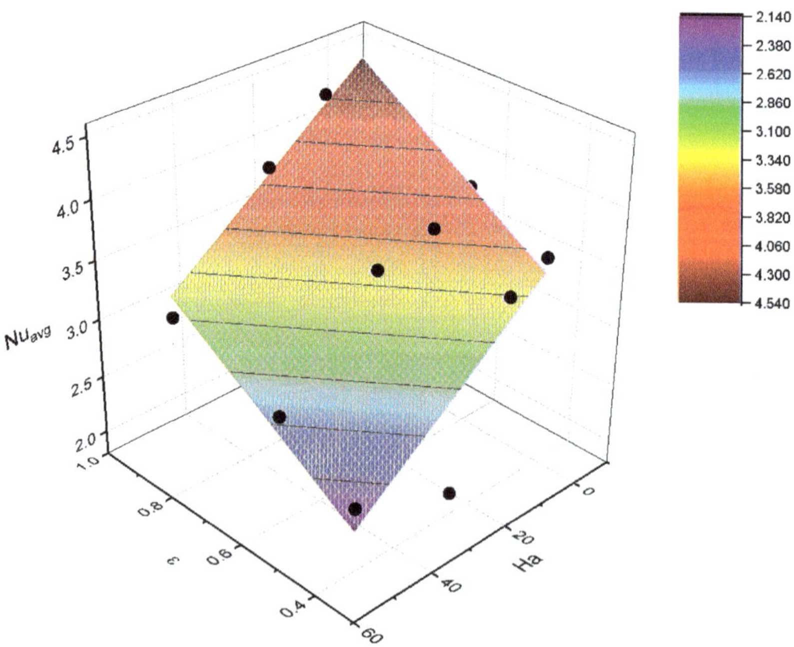

Figure 9. Fitting parametric development over a planar surface to predict $\overline{Nu}$ under constant $Ra = 10^5, \theta = 60$, and $Da = 10^{-1}$.

The equation to predict $\overline{Nu}$ under this circumstance was obtained to be the following:

$$\overline{Nu} = f + a\epsilon + bHa \tag{40}$$

The values of the empirical parameters from Equation (40) are presented in Table 7. The simplified version of the equation was also found to be effective at elevated *Ha* numbers; however, due to the insufficient data in the literature, the validation was not conducted beyond $Ha = 50$.

Table 7. Parametric values to predict $\overline{Nu}$ from Equation (40).

Empirical Parameters	Fitting Values
f	2.6105
a	2.13773
b	−0.02639
Statistical Accuracy Indicators	**Values**
R^2	0.91
p	<0.05

5.5.2. Validation Result

The immediate validation was conducted to assess the accuracy of Equation (40). The independent validation data were obtained from the literature reported by Himika et al. [18] and Sajjadi et al. [62]. The ±5% error lines were also included in a similar manner for better visualization of the agreement.

Figure 10 illustrates the agreement between the literature (LBM data) and the present LM method. In general, most of the points were found within the error lines. The statistical accuracy indicators presented in Table 7 suggest that the agreement was still acceptable. The range of $\overline{Nu}$ was found to be close to 5 (LBM-obtained result), which was, in fact, obtained at $Ha = 0$ and the highest porosity considered in this research pipeline, i.e., $\epsilon = 0.9$.

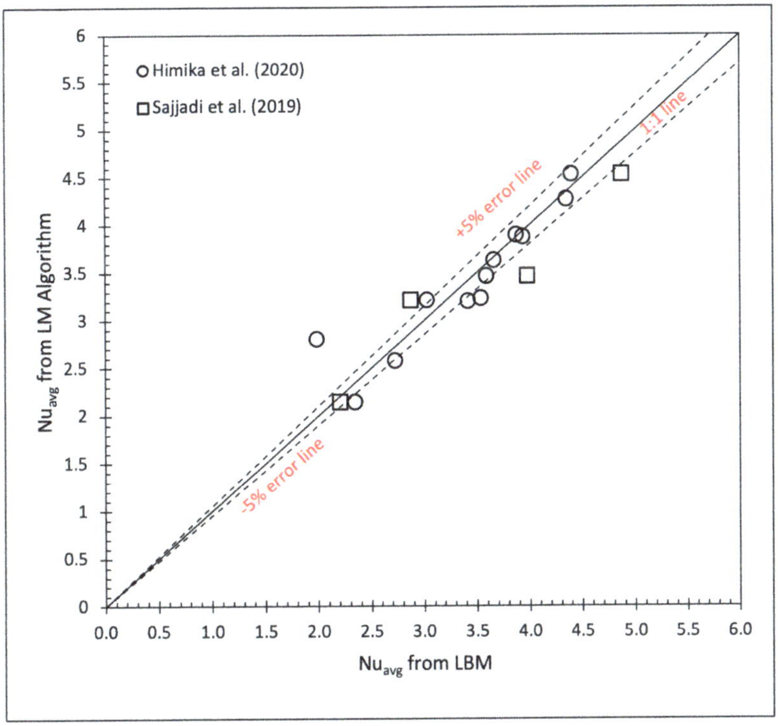

Figure 10. Fitting parametric development over a planar surface to predict $\overline{Nu}$ under constant $Ra = 10^5$, $\theta = 60$, and $Da = 10^{-1}$ [18,62] .

Table 8 shows the R^2 values found from the agreement presented in Figure 10. The values of R^2 were found between 0.85 and 0.93 in comparison with the literature data mentioned earlier. All the relevant points were considered for validation. The specific point located beyond the +5% error line was still considered, and hence the R^2 was found to be 0.85 when comparing with Himika et al. [18]. However, if those specific data were left out, the R^2 increased significantly to 0.92.

Table 8. Coefficient correlations (R^2) obtained in comparison with individual literature as presented in Figure 10.

Sajjadi et al. [62]	Himika et al. [18]
0.93	0.85

6. Discussion

6.1. Significance of the Study

The LBM is a powerful and efficient alternative to solving fluid dynamics problems. LBM simulations can provide an accurate outcome within a shorter time scale compared to other relevant techniques, such as finite difference, finite element, and finite volume. All of the methods are accurate and have pros/cons. However, the explanation of the correlation among the input variables to quantitatively predict heat transfer was hardly reported. The present study utilizes input variables, such as Ra number, Da number, ϵ, Ha number, and θ to predict MHD-RB convection within a porous enclosure. The input parameters in the relevant study were mostly chosen to showcase the impact of buoyancy or porosity for the purpose of visualization. Nevertheless, there was always a gap in having equations numerically predict $\overline{Nu}$, which can be further validated with well-cited literature.

LBM data are highly non-linear. The current study considered the LM algorithm as one of the ML training methods to build non-linear surface analyses to develop three-dimensional correlation among the variables to predict $\overline{Nu}$. In each segment of the correlation development, validations were conducted, and statistical parameters were mentioned as part of the accuracy indicators. Furthermore, the empirical parametric values are provided in a tabular form after the statistical analyses, which will allow the researchers to reproduce the work based on the requirements. The reproducibility of the work by the LM algorithm could be established in different ways. One of them could be to perform direct LBM simulations by varying the input parameters to build the dataset, followed by correlation development to predict $\overline{Nu}$. In that case, the process could be time-consuming. The dataset could be split into 80% for the model development, and the rest 20% for the validation. However, validations with relevant well-cited literature need to be performed to understand the efficacy of the approach. It is recommended to consider a sample size of 100 for this approach. Alternatively, the interpolation of a limited dataset could be another option. However, this part also requires the initial model development and creating a dataset through simulations. However, the approach should consider the highest and lowest possible range of each parameter within the geometry to interpolate. The LM algorithm combined with the R library package *"dplyr"* assists the data manipulation without any involving any complicated steps. It is possible to establish a dataset with 3000 points through interpolation without the need to run LBM simulations varying each parameter. The present study considered both aforementioned approaches.

The current study aims to fulfill such requirements by interpolating datasets for the purpose of validations. The validations by comparing with identical geometries and physical properties with relevant literature provide another form of evidence on a high level of accuracy of the present approach. Some of the similar approaches in fluid dynamics study are worth mentioning. Recently, Islam et al. [6] presented correlations to predict $\overline{Nu}$ by considering input variables, such as Darcy (Da) numbers, Rayleigh (Ra) numbers, and porosities to predict $\overline{Nu}$ by the LM algorithm. At the same time, the equation and associated empirical parameters were utilized to validate the GPU-optimized LBM model.

However, the work of Islam et al. [6] was fairly restricted to heat-transfer phenomena of the nanofluid without the influence of an external magnetic field or representative non-dimensional parameter, such as the *Ha* number. Meanwhile, the present approach did not consider any machine learning or deep learning approach for the data-driven approach, but the inclusion of artificial intelligence in fluid dynamics study has become widely popular recently and should represent the state-of-the-art approach. It can be stated that no approach can be considered a direct validation yardstick, as numerical methods are not 100% correct, but the concept of correlation development as an additional form of validation has been reported. For example, the ANN modeling of nanofluid under magnetic field influence by Shah et al. [42], Alqaed et al. [41], and He et al. [43] attest to the purpose of the data-driven approach being included in LBM simulations of fluid flow. Nevertheless, this is the first work which has provided validations in each segment to showcase the potential of the data-driven approach in fluid dynamics by correlations development through the LM algorithm without the need for implementing machine learning methods or high computational resources.

Experimental fluid dynamics is time-consuming, delicate, and expensive [65,66]. If the boundary conditions, simulated data, and final representative contours are not fully understood numerically, the experimental approaches could cost a fortune. In an industrial setup, it will always be beneficial if numerical modeling contains correlations that are repeatable and reproducible. Based on the equations, the researchers will be able to test and tune the parameters within the domain to obtain the best-performing model to meet the scientific requirements. Then the final model development can be established through the original fluid flow simulations with the best-fitting input parameters to save time and increase productivity as well as profitability. The equations presented in this study had coefficients of determination (R^2) between 0.85 and 0.99, which are within the standard acceptable range.

6.2. Factors Affecting the Accuracy of the Equations

The present study is based on LBM data. To build a proper correlation, a wide range of datasets is required. While the present study established the correlations with a large number of the dataset, those did not cover the whole domain of the input parameters due to a lack of data for the validation. For example, the *Ha* number could be as large as 200 or even more, which could reduce the $\overline{Nu} < 0.001$. The LM algorithm will not be a viable option for this approach. A machine learning or deep learning approach should be an appropriate method for such an option. Figure 5 correlations represented $0 \leq Ha \leq 100$, and while model calibration was feasible, there were no relevant data found in the literature for comparison. Furthermore, the present study considered laminar flow only. The *Ra* number could be more than 10^7, and it can cover a wide range of turbulence through the increased buoyancy. Nevertheless, a lack of efficient data in the literature within the considered geometry restricted the present study to explore further options. Since machine learning or deep learning was not considered in this study, further extension of the input parameters was not explored. Some of the references had only 3–4 relevant points (for example, Figure 6a), and therefore, the limited data could have affected the accuracy of the respective equation. While the surface analyses were conducted based on interpolation of the dataset with at least 3000 data, more validation data would have improved the R^2 values of independent validations. For instance, Figure 8b contained 31 points for the independent validation, and $R^2 = 0.99$ was obtained by comparing with three of the references [18,62,63].

In addition, the LM algorithm is susceptible to noise and may downgrade the efficiency of the neural network. However, those impacts are most visible for complex geometries. Since the present study considers a 2D porous cavity, the error percentage was acceptable. The re-tuning of the parameter would have been more efficient by considering any step associated with supervised machine learning model development. However, machine learning models require more data to train and test and cannot provide any equation for the direct implementation in the study. The data considered in this study were sufficient for the LM algorithm and correlations development. The good agreement with the independent validations attest to such a statement.

6.3. Future Recommendations

The key findings of this study offer a major hint that a machine learning model can be developed to train LBM data into a well-suited model. Therefore, building correlations to predict an output parameter by considering multiple input parameters to build multifarious LBM models is highly recommended and is under strong consideration by the authors. Furthermore, the present study aimed to provide equations under different input parameters to predict $\overline{Nu}$. In most of the fluid dynamics research, entropy generation is also determined to demonstrate the fluid irreversibility. Both heat transfer and fluid friction are responsible for fluid irreversibility, particularly for nanofluid study. The correlation development to predict total entropy will also lead to another highly correlated parameter, the Bejan number (*Be*), which is the ratio of the heat transfer irreversibility to total irreversibility due to heat transfer and fluid friction. Therefore, determining the total entropy generation is one of the major indices in heat transfer application and should require involving more input parameters. Considering the multifarious input parameters involved in affecting entropy generation, machine learning is a suitable method and will be further explored.

Finally, the present study was validated with the simulated results from CPU-based computing. Considering the multivariate model development discussed above, CPU-based computing will be tedious in terms of the machine learning approach. In addition, a 3D implementation will be required to replace the 2D model, which could also consider more complicated lattice models, such as *D3Q15* and *D3Q19*, replacing the *D2Q9* of the present study. The possible implications of GPU-based LBM simulations will significantly reduce the computational time and increase the efficiency of the model. Considering a parallel computing platform, such as the Compute Unified Device Architecture (CUDA), could be a better method for implementing the MHD-LBM hybrid machine learning model. Some of the relevant works minus the machine learning have been published by the authors [14,67]. Therefore, developing a machine learning model through GPU computing for LBM cross-validations, possibly in a 3D geometry, will be a milestone within LBM research.

7. Conclusions

This study developed an LBM-MHD data-driven method to numerically predict the average Nu number ($\overline{Nu}$) as a non-dimensional representative value of the average rate of heat transfer by the LM algorithm. The mathematical correlations to predict $\overline{Nu}$ by considering Ha numbers, Ra numbers, inclination angles (θ), Da numbers, and ϵ were explored, followed by validations with the literature. The coefficients of determinations were found within $0.85 \leq R^2 \leq 0.99$, and this provides compelling evidence for the accuracy of the equations. The streamlines and isotherms were also presented to visually demonstrate the impact of the above-mentioned numerical parameters on the heat transfer phenomena. The equations presented in this study could be taken into account to validate any existing LBM-MHD model which considers RB convection within a 2D rectangular porous cavity. More options could be explored by directly developing a machine learning model to add extra features within the LBM model to establish benchmark solutions, which are under strong consideration for future study.

Author Contributions: Conceptualization, T.A.H., M.F.H., M.M.M. and M.A.I.K.; methodology, T.A.H., M.F.H. and M.M.M.; software, T.A.H., M.F.H. and M.M.M.; validation, T.A.H. and M.F.H.; formal analysis, T.A.H., M.F.H., M.M.M. and M.A.I.K.; investigation, T.A.H. and M.F.H.; resources, T.A.H., M.F.H. and M.M.M.; data curation, T.A.H. and M.F.H.; writing—original draft preparation, T.A.H. and M.F.H.; writing—review and editing, M.M.M. and M.A.I.K.; visualization, T.A.H. and M.F.H.; supervision, M.M.M. and M.A.I.K.; project administration, M.M.M. and M.A.I.K.; funding acquisition, M.M.M. and M.A.I.K. All authors have read and agreed to the published version of the manuscript.

Funding: (i) North South University, Bangladesh, Grant No. CTRG-22-SEPS-09 (ii) Ministry of Science and Technology, Bangladesh Government, Grant No. EAS/SRG-222427.

Data Availability Statement: Data available on request.

Conflicts of Interest: The authors declare no conflict of interest. The funders had no role in the design of the study; in the collection, analyses, or interpretation of data; in the writing of the manuscript; or in the decision to publish the results.

Abbreviations

The following abbreviations are used in this manuscript, shown in the order that they appear in the texts:

LBM	Lattice Boltzmann method
MHD	Magnetohydrodynamics
RB	Rayleigh–Bénard
LM	Levenberg–Marquardt
2D	Two-dimensional
Ha	Hartmann
Ra	Rayleigh
Darcy	Da
ANN	Artificial Neural Network
Nu	Nusselt
HPC	High-performance computing
AI	Artificial intelligence
ML	Machine learning
NLSM	Nonlinear least squares minimization
TLBM	Thermal LBM
BGK	Bhatnagar–Gross–Krook
SRT	Single-relaxation times
DF	Distribution functions
Nomenclature	
English symbols	
a, b, c, d, e, f	Fitting parameters for LM-obtained equations
B	Magnitude of magnetic field
C_i	Lattice speed
C_s	Speed of sound
$\overline{d_k}$	Expected outcome
$\overline{e_i}$	Discrete velocities
f_i	Distribution function for flow fields
f_i^{eq}	Equilibrium distribution function
F_i	Force terms
F_i^M	Force term for MHD
F_i^P	Force term for porous media
F_i^b	Buoyancy term
g_i	Distribution function for temperature fields
g_y	Gravitational force acting in y-direction
g_i^{eq}	Thermal equilibrium function
H	Height of the cavity
K	Permeability
m	Lattice on the boundary
N	Sample number
n	Iteration index

Nu	Nusselt number
$\overline{Nu}$	Average Nusselt number
t	Time
Δt	Time interval
T	Temperature
ΔT	Temperature difference
T_c	Cold temperature
T_h	Hot temperature
v	Velocity component
w_i	Solution of the interpolation
x_k	Random example for the output network
z	Number of anticipated outcome from the network
Greek symbols	
α	Thermal diffusivity
α'	Optimization field matrix
β	Thermal expansion coefficient
ϵ	Porosity
λ	Mean-squared network error
μ	Dynamic viscosity
ν	Kinematic viscosity
ω	Weighting factor
ϕ	Angle of inclination
ψ	Either velocity or temperature in the convergence
ρ	Fluid density
σ	Electrical conductivity
θ	Dimensionless angle of inclination
ζ	Adjustment parameter in each iteration cycle

References

1. Benhamou, J.; Jami, M. Three-dimensional numerical study of heat transfer enhancement by sound waves using mesoscopic and macroscopic approaches. *Heat Transf.* **2022**, *51*, 3892–3919. [CrossRef]
2. Huang, R.; Wu, H.; Adams, N.A. Mesoscopic lattice Boltzmann modeling of the liquid-vapor phase transition. *Phys. Rev. Lett.* **2021**, *126*, 244501. [CrossRef] [PubMed]
3. Buzzicotti, M.; Tauzin, G. Inertial range statistics of the entropic lattice Boltzmann method in three-dimensional turbulence. *Phys. Rev. E* **2021**, *104*, 015302. [CrossRef]
4. Kefayati, G. An immersed boundary-lattice Boltzmann method for thermal and thermo-solutal problems of Newtonian and non-Newtonian fluids. *Phys. Fluids* **2020**, *32*, 073103. [CrossRef]
5. Zhang, Y.; Dong, B.; An, X.; Wang, Y.; Zhou, X.; Li, W. Phase-field-based lattice Boltzmann model for ternary fluid flows considering the wettability effect. *Appl. Math. Model.* **2022**, *103*, 195–220. [CrossRef]
6. Islam, M.; Hasan, M.F.; Bhowmick, S.; Kamrujjaman, M.; Molla, M.M. GPU-optimized LBM-MRT Simulation of Free Convection and Entropy Generation of Non-Newtonian Power-law Nanofluids in a Porous Enclosure at REV scale. *Int. J. Amb. Energy* **2022**, 1–33. [CrossRef]
7. Jalali, A.; Amiri Delouei, A.; Khorashadizadeh, M.; Golmohamadi, A.; Karimnejad, S. Mesoscopic simulation of forced convective heat transfer of Carreau-Yasuda fluid flow over an inclined square: Temperature-dependent viscosity. *J. Appl. Comput. Mech.* **2020**, *6*, 307–319.
8. Afra, B.; Delouei, A.A.; Mostafavi, M.; Tarokh, A. Fluid-structure interaction for the flexible filament's propulsion hanging in the free stream. *J. Mol. Liq.* **2021**, *323*, 114941. [CrossRef]
9. Zarei, A.; Karimipour, A.; Isfahani, A.H.M.; Tian, Z. Improve the performance of lattice Boltzmann method for a porous nanoscale transient flow by provide a new modified relaxation time equation. *Phys. A Stat. Mech. Appl.* **2019**, *535*, 122453. [CrossRef]
10. Succi, S. *The Lattice Boltzmann Equation: For Complex States of Flowing Matter*; Oxford University Press: Oxford, UK, 2018.
11. Wang, M.; Feng, Y.; Pande, G.; Zhao, T. A coupled 3-dimensional bonded discrete element and lattice Boltzmann method for fluid-solid coupling in cohesive geomaterials. *Int. J. Num. Analyt. Meth. Geomech.* **2018**, *42*, 1405–1424. [CrossRef]
12. Hosseini, S.A.; Huang, F.; Thévenin, D. Lattice Boltzmann model for simulation of flow in intracranial aneurysms considering non-Newtonian effects. *Phys. Fluids* **2022**, *34*, 073105. [CrossRef]
13. An, S.; Zhan, Y.; Yao, J.; Yu, H.W.; Niasar, V. A greyscale volumetric lattice Boltzmann method for upscaling pore-scale two-phase flow. *Adv. Water Res.* **2020**, *144*, 103711. [CrossRef]
14. Rahman, A.; Nag, P.; Molla, M.M. Non-Newtonian effects on MHD thermosolutal free convection and entropy production of nanofluids in a rectangular enclosure using the GPU-based mesoscopic simulation. *Waves Random Complex Media* **2022**, 1–33. [CrossRef]

15. Shao, X.; Santasmasas, M.C.; Xue, X.; Niu, J.; Davidson, L.; Revell, A.J.; Yao, H.D. Near-wall modeling of forests for atmosphere boundary layers using lattice Boltzmann method on GPU. *Eng. Appl. Comput. Fluid Mech.* **2022**, *16*, 2142–2155. [CrossRef]
16. Ahmed, T.; Hassan, S.; Hasan, M.F.; Molla, M.M.; Taher, M.A.; Saha, S.C. Lattice Boltzmann Simulation of Magnetic Field Effect on Electrically Conducting Fluid at Inclined Angles in Rayleigh-Bénard Convection. *Energy Eng.* **2021**, *118*, 15–36. [CrossRef]
17. Nemati, M.; Sani, H.M.; Jahangiri, R.; Chamkha, A.J. MHD natural convection in a cavity with different geometries filled with a nanofluid in the presence of heat generation/absorption using lattice Boltzmann method. *J. Therm. Anal. Cal.* **2022**, *147*, 9067–9081. [CrossRef]
18. Himika, T.A.; Hassan, S.; Hasan, M.; Molla, M. Lattice Boltzmann Simulation of MHD Rayleigh–Bénard Convection in Porous Media. *Arab. J. Sci. Eng.* **2020**, *45*, 9527–9547. [CrossRef]
19. Hartono, A.D.; Sasaki, K.; Sugai, Y.; Nguele, R. Computational Performance of Disparate Lattice Boltzmann Scenarios under Unsteady Thermal Convection Flow and Heat Transfer Simulation. *Computation* **2021**, *9*, 65. [CrossRef]
20. Trunk, R.; Weckerle, T.; Hafen, N.; Thäter, G.; Nirschl, H.; Krause, M.J. Revisiting the homogenized lattice Boltzmann method with applications on particulate flows. *Computation* **2021**, *9*, 11. [CrossRef]
21. Weichman, P.B.; Marston, J.B. Statistical Equilibrium Principles in 2D Fluid Flow: From Geophysical Fluids to the Solar Tachocline. *Entropy* **2022**, *24*, 1389. [CrossRef]
22. Manna, N.K.; Biswas, N. Magnetic force vectors as a new visualization tool for magnetohydrodynamic convection. *Int. J. Therm. Sci.* **2021**, *167*, 107004. [CrossRef]
23. Arshad, M.; Hussain, A.; Elfasakhany, A.; Gouadria, S.; Awrejcewicz, J.; Pawłowski, W.; Elkotb, M.A.; Alharbi, F.M. Magnetohydrodynamic flow above exponentially stretchable surface with chemical reaction. *Symmetry* **2022**, *14*, 1688. [CrossRef]
24. Lebedev, S.; Frank, A.; Ryutov, D. Exploring astrophysics-relevant magnetohydrodynamics with pulsed-power laboratory facilities. *Rev. Modern Phys.* **2019**, *91*, 025002. [CrossRef]
25. Habibishandiz, M.; Saghir, M. A critical review of heat transfer enhancement methods in the presence of porous media, nanofluids, and microorganisms. *Therm. Sci. Eng. Prog.* **2022**, *30*, 101267. [CrossRef]
26. Ali, H.M. Heat transfer augmentation of porous media (metallic foam) and phase change material based heat sink with variable heat generations: An experimental evaluation. *Sustain. Energy Technol. Assess.* **2022**, *52*, 102218. [CrossRef]
27. Yaseen, M.; Rawat, S.K.; Shafiq, A.; Kumar, M.; Nonlaopon, K. Analysis of heat transfer of mono and hybrid nanofluid flow between two parallel plates in a Darcy porous medium with thermal radiation and heat generation/absorption. *Symmetry* **2022**, *14*, 1943. [CrossRef]
28. Tayebi, T. Analysis of the local non-equilibria on the heat transfer and entropy generation during thermal natural convection in a non-Darcy porous medium. *Int. Comm. Heat Mass Transf.* **2022**, *135*, 106133. [CrossRef]
29. Chaabane, R.; Kolsi, L.; Jemni, A.; Alshammari, N.K.; D'Orazio, A. Numerical study of the Rayleigh–Bénard convection in two-dimensional cavities heated by elliptical heat sources using the lattice Boltzmann method. *Phys. Fluids* **2021**, *33*, 123605. [CrossRef]
30. Li, T.F.; Luo, K.; Yi, H.L. Effect of unipolar charge injection direction on the onset of Rayleigh-Bénard convection: A lattice Boltzmann study. *Int. Comm. Heat Mass Transf.* **2020**, *112*, 104496. [CrossRef]
31. Yang, X.H.; Liu, J. Probing the Rayleigh–Benard convection phase change mechanism of low-melting-point metal via lattice Boltzmann method. *Num. Heat Transf. Part A Appl.* **2018**, *73*, 34–54. [CrossRef]
32. Tahmooressi, H.; Kasaeian, A.; Yavarinasab, A.; Tarokh, A.; Ghazi, M.; Hoorfar, M. Numerical simulation of nanoparticles size/aspect ratio effect on thermal conductivity of nanofluids using lattice Boltzmann method. *Int. Comm. Heat Mass Transf.* **2021**, *120*, 105033. [CrossRef]
33. Zainal, N.A.; Nazar, R.; Naganthran, K.; Pop, I. Unsteady MHD stagnation point flow induced by exponentially permeable stretching/shrinking sheet of hybrid nanofluid. *Eng. Sci. Technol. Int. J.* **2021**, *24*, 1201–1210. [CrossRef]
34. Tucs, A.; Bojarevics, V.; Pericleous, K. Magnetohydrodynamic stability of large scale liquid metal batteries. *J. Fluid Mech.* **2018**, *852*, 453–483. [CrossRef]
35. Fukami, K.; Fukagata, K.; Taira, K. Assessment of supervised machine learning methods for fluid flows. *Theor. Comput. Fluid Dyn.* **2020**, *34*, 497–519. [CrossRef]
36. Brunton, S.L.; Noack, B.R.; Koumoutsakos, P. Machine learning for fluid mechanics. *Ann. Rev. Fluid Mech.* **2020**, *52*, 477–508. [CrossRef]
37. Wan, Z.Y.; Sapsis, T.P. Machine learning the kinematics of spherical particles in fluid flows. *J. Fluid Mech.* **2018**, *857*. [CrossRef]
38. Farook, T.H.; Ahmed, S.; Jamayet, N.B.; Rashid, F.; Barman, A.; Sidhu, P.; Patil, P.; Lisan, A.M.; Eusufzai, S.Z.; Dudley, J.; et al. Computer-aided design and 3-dimensional artificial/convolutional neural network for digital partial dental crown synthesis and validation. *Sci. Rep.* **2023**, *13*, 1561. [CrossRef]
39. García Cabello, J. Mathematical Neural Networks. *Axioms* **2022**, *11*, 80. [CrossRef]
40. Xie, J.; Li, S. Training Neural Networks by Time-Fractional Gradient Descent. *Axioms* **2022**, *11*, 507. [CrossRef]
41. Alqaed, S.; Mustafa, J.; Sharifpur, M. Numerical investigation and optimization of natural convection and entropy generation of alumina/H2O nanofluid in a rectangular cavity in the presence of a magnetic field with artificial neural networks. *Eng. Anal. Bound. Elem.* **2022**, *140*, 507–518. [CrossRef]

42. Shah, M.A.; Pan, K.; Ibrahim, M.; Saeed, T. Use of neural network and machine learning in optimizing heat transfer and entropy generated in a cavity filled with nanofluid under the influence of magnetic field: A numerical study. *Eng. Anal. Bound. Elem.* **2022**, *139*, 113–131. [CrossRef]
43. He, X.; Sidi, M.O.; Ahammad, N.A.; Elkotb, M.A.; Elattar, S.; Algelany, A. Artificial neural network joined with lattice boltzmann method to study the effects of mhd on the slip velocity of fmwnt/water nanofluid flow inside a microchannel. *Eng. Anal. Bound. Elem.* **2022**, *143*, 95–108. [CrossRef]
44. Kefayati, G.; Gorji, M.; Sajjadi, H.; Ganji, D.D. Investigation of Prandtl number effect on natural convection MHD in an open cavity by Lattice Boltzmann Method. *Eng. Comput.* **2013**, *30*, 97–116. [CrossRef]
45. Ergun, S. Fluid flow through packed columns. *Chem. Eng. Prog.* **1952**, *48*, 89–94.
46. Mohamad, A. *Lattice Boltzmann Method*; Springer: Berlin/Heidelberg, Germany, 2011; Voulme 70.
47. Fattahi, A.; Hajialigol, N.; Delpisheh, M.; Karimi, N. Lattice-Boltzmann numerical simulation of double-diffusive natural convection and entropy generation in an n-shaped partially heated storage tank. *Eng. Anal. Boundary Elem.* **2023**, *146*, 105–118. [CrossRef]
48. Hasanpour, A.; Farhadi, M.; Sedighi, K.; Ashorynejad, H. Numerical study of Prandtl effect on MHD flow at a lid-driven porous cavity. *Int. J. Num. Meth. Fluids* **2012**, *70*, 886–898. [CrossRef]
49. Guo, Z.; Shu, C. *Lattice Boltzmann Method and Its Application in Engineering*; World Scientific: Singapore, 2013; Volume 3.
50. Hasan, M.F.; Molla, M.M.; Kamrujjaman, M.; Siddiqa, S. Natural convection flow over a vertical permeable circular cone with uniform surface heat flux in temperature-dependent viscosity with three-fold solutions within the boundary layer. *Computation* **2022**, *10*, 60. [CrossRef]
51. Clever, R.; Busse, F. Transition to time-dependent convection. *J. Fluid Mech.* **1974**, *65*, 625–645. [CrossRef]
52. He, X.; Chen, S.; Doolen, G.D. A novel thermal model for the lattice Boltzmann method in incompressible limit. *J. Comput. Phys.* **1998**, *146*, 282–300. [CrossRef]
53. Lera, G.; Pinzolas, M. Neighborhood based Levenberg-Marquardt algorithm for neural network training. *IEEE Trans. Neural Netw.* **2002**, *13*, 1200–1203. [CrossRef] [PubMed]
54. Mailund, T. Manipulating data frames: Dplyr. In *R Data Science Quick Reference*; Springer: Berlin/Heidelberg, Germany, 2019; pp. 109–160.
55. Borchers, H.W. Pracma: Practical numerical math functions. *R Package Version* **2019**, *2*, 519.
56. Chapman, S.J. *Fortran 90/95 for Scientists and Engineers*; McGraw-Hill Higher Education: New York, NY, USA, 2004.
57. R Core Team R. *R: A Language and Environment for Statistical Computing*; R Core Team R: Vienna, Austria, 2013.
58. Seifert, E. OriginPro 9.1: Scientific data analysis and graphing software-software review. *J. Chem. Info. Model.* **2014**, *54*, 1552. [CrossRef]
59. Rudraiah, N.; Barron, R.; Venkatachalappa, M.; Subbaraya, C. Effect of a magnetic field on free convection in a rectangular enclosure. *Int. J. Eng. Sci.* **1995**, *33*, 1075–1084. [CrossRef]
60. Kefayati, G.R. Lattice Boltzmann simulation of natural convection in nanofluid-filled 2D long enclosures at presence of magnetic field. *Theor. Comput. Fluid Dyn.* **2013**, *27*, 865–883. [CrossRef]
61. Sheikholeslami, M.; Gorji-Bandpy, M.; Ganji, D.D. Lattice Boltzmann method for MHD natural convection heat transfer using nanofluid. *Powder Technol.* **2014**, *254*, 82–93. [CrossRef]
62. Sajjadi, H.; Delouei, A.A.; Izadi, M.; Mohebbi, R. Investigation of MHD natural convection in a porous media by double MRT lattice Boltzmann method utilizing MWCNT–Fe3O4/water hybrid nanofluid. *Int. J. Heat Mass Transf.* **2019**, *132*, 1087–1104. [CrossRef]
63. Seta, T.; Takegoshi, E.; Okui, K. Lattice Boltzmann simulation of natural convection in porous media. *Math. Comp. Sim.* **2006**, *72*, 195–200. [CrossRef]
64. Ghasemi, K.; Siavashi, M. Lattice Boltzmann numerical simulation and entropy generation analysis of natural convection of nanofluid in a porous cavity with different linear temperature distributions on side walls. *J. Mol. Liq.* **2017**, *233*, 415–430. [CrossRef]
65. Dixon, A.G.; Partopour, B. Computational fluid dynamics for fixed bed reactor design. *Ann. Rev. Chem. Biomol. Eng.* **2020**, *11*, 109–130. [CrossRef]
66. Lyu, Z.; Asadi, A.; Alarifi, I.M.; Ali, V.; Foong, L.K. Thermal and fluid dynamics performance of MWCNT-water nanofluid based on thermophysical properties: An experimental and theoretical study. *Sci. Rep.* **2020**, *10*, 1–14. [CrossRef]
67. Rahman, A.; Redwan, D.A.; Thohura, S.; Kamrujjaman, M.; Molla, M.M. Natural convection and entropy generation of non-Newtonian nanofluids with different angles of external magnetic field using GPU accelerated MRT-LBM. *Case Stud. Therm. Eng.* **2022**, *30*, 101769. [CrossRef]

axioms

Article

Aggregative Game for Distributed Charging Strategy of PEVs in a Smart Charging Station

Ti Kang , Huaqing Li * and Lifeng Zheng

Chongqing Key Laboratory of Nonlinear Circuits and Intelligent Information Processing, College of Electronic and Information Engineering, Southwest University, Chongqing 400715, China
* Correspondence: huaqingli@swu.edu.cn

Abstract: This paper proposes a charging strategy for plug-in electric vehicles (PEVs) in a smart charging station (SCS) that considers load constraints and time anxieties. Due to the rapidly growing load demand of PEVs and the load capacity investments in infrastructure, PEV charging needs to be subject to overload limits, beyond which failures can occur. The time anxiety is presented to address some of the uncertainties that may arise while charging PEVs. Under an aggregative game framework, this paper constructs a price-driven charging model to minimize costs by choosing the optimal charging strategy. Meanwhile, since the driver information is an aggregated item in the PEV cost function, the drivers' privacy can be protected. Then, a distributed reflected forward–backward (RFB) splitting method is developed to search for the generalized Nash equilibria (GNE) of the game. The convergence of the proposed algorithm and the effectiveness of the charging strategy are verified by the detailed simulation and results.

Keywords: aggregative game; plug-in electric vehicles; distributed charging strategy; time anxiety; load constraints

MSC: 9110

Citation: Kang, T.; Li, H.; Zheng, L. Aggregative Game for Distributed Charging Strategy of PEVs in a Smart Charging Station. *Axioms* **2023**, *12*, 186. https://doi.org/10.3390/axioms12020186

Academic Editor: Leonid Plotnikov

Received: 16 January 2023
Revised: 8 February 2023
Accepted: 9 February 2023
Published: 10 February 2023

1. Introduction

Governments worldwide are promoting plug-in electric vehicles (PEVs) as a clean alternative to conventional gasoline vehicles due to the depletion of fossil fuel resources and environmental pollution, both of which are serious issues. Compared with the conventional gasoline vehicle, the PEV reduces carbon dioxide emissions and the overall operating cost [1]. However, with the rapid growth of PEVs, the total charging load of PEVs entering the electric grid has increased [2], which leads to an overload of charging stations. According to [3], uncoordinated PEV charging lowers the electrical grid's power quality. However, the demand presented in [4] might actually flatten as a result of the coordinated charging of PEVs. To increase the electric grid's operational effectiveness and security, we must formulate effective charging strategies to coordinate and control the charging behavior of PEV drivers.

Since PEVs are capable of storing electrical energy, some PEVs can be used as energy suppliers and transmit power in both directions, to and from a smart charging station (SCS), which can mitigate the effects of overcharging at SCS at peak times. Meanwhile, in order to improve the accuracy of the charging strategy, the driver's time anxiety needs to be taken into account, in addition to the plug-in electric vehicle factor. The authors of Vatanparvar et al. [5] introduced the concept of demand–price elasticity to address the demand–response model based on the drivers' behavior. A robust-index method was developed in [6] to resolve driver behavior uncertainties and minimize violations to comfort in household load scheduling. In [7], a distributed method was proposed and a means of reducing the impact of some uncertain events under a non-cooperative game was proposed.However, some of these driver behavior models were homogeneous and lacked theoretical justification.

In order to improve the validity of the model, four different PEV driver behaviors were proposed to study the effect of time anxiety on plug-in electric vehicle charging. There are two main classes of control architecture in this charging strategy, namely the centralized and distributed approaches.

The authors of Yan et al. [8] proposed a four-stage optimal control method for electric vehicle charging stations (EVCSs) to reduce operating costs and maintain the supply–demand power balance. In [9], a centralized control approach based on multistage droop control was used to operate an island microgrid in the presence of high PEV penetration. Furthermore, the model in [10] utilized a centralized real-time charging scheme based on a convex relaxation method to coordinate PEV charging. However, in the centralized strategy, each driver, as well as the control center, needs to know the complete information, including cost function, local feasible decision sets and affine sharing constraints, which can easily compromise the drivers' privacy. Meanwhile, in many cases, no central node can communicate with all drivers in both directions, and it is not possible to realize large-scale driver communication in a real-time manner. Therefore, a distributed strategy is used to compute the local decisions corresponding to PEVs and generalized Nash equilibria and communicate with the adjacent local decisions using the drivers' local data, which reduces the computational and interaction burden and maintains the privacy of driver information. Based on a decentralized protocol, Li et al. [11] presented a charge control technique for large-scale PEVs under the Newton-type algorithm. The authors of Gan et al. [12] studied a decentralized algorithm to optimize PEV charging during off-peak hours. However, the distributed algorithm described above rarely considers the strategic interaction between multiple PEVs in the charging station.

Therefore, to study communication between multiple PEVs, we then focused on game theory, as this is a powerful tool for analyzing the interactions between multiple decision makers and can improve the model's performance. Since the famous Cournot model was proposed, aggregative games have become an important type of game theory. In the description for the aggregative game in [13], each driver was not subject to a one-to-one interaction, but was subject to a number of aggregations across the charging strategy. Recent studies in [14–17] considered the linear aggregation functions and quadratic cost functions in such games, as well as their interaction with plug-in electric vehicle charging. Based on the literature, a distributed reflected forward–backward splitting method was proposed in [18] to find the generalized Nash equilibria of the aggregative game. Unlike the distributed algorithm presented above, this algorithm exploits the aggregated coupling structure in the cost function, meaning that each agent only needs to exchange and maintain an aggregated estimate, not including the estimate of the even multiplier, reducing the computational and communication burden of the model. In addition, since each PEV only shares its estimate of the total, it does not need to share the complete information, further protecting the privacy of the drivers. The main contributions of this paper are summarized as follows:

(1) A new price-driven charging model combining time anxiety and load constraints is constructed to minimize the cost of an individual PEV driver within the framework of an aggregative game. In particular, as everyone only knows the final summation result, not the specific information, the aggregation game can better protect the privacy of drivers.
(2) Load constraints are proposed to protect the safety of SCS. Then, four PEV driver behaviors are proposed based on different time anxiety states and load constraints. Meanwhile, the effects of time anxiety under four different driver behaviors are compared, and the effects of uncertain occurrence events are reduced by the charging strategy.
(3) A distributed reflected forward–backward algorithm is designed to seek the generalized Nash equilibria of the model. The proposed algorithm seeks its optimal response charging strategy regarding the current load and time anxiety in the electric grid, thus preventing overload in the smart charging station and mitigating the impact of uncertain events that may occur at the PEV charging time. The algorithm obtains an

improvement in significant convergence compared to the numerical values of the FB algorithm [19].

2. System Model and Problem Formulation

As this paper focuses on charging PEVs, the studied system is named SCS, which is considered to be part of the electric grid. In [20–22], SCS is composed of PEVs, which are powered by the electric grid, and the PEV can transmit power in both directions between the electric grid. Meanwhile, each PEV i ($i \in \mathcal{I} = \{1, 2, \cdots, N\}$) is the electric grid user in Figure 1. We define d_t as the basic demand load of the system at time t. Let vector $x_i = \mathrm{col}(x_{i,1}, \ldots, x_{i,T})$ denote the charging curve of PEV i over all the time slots and $x_{i,t}$ define the charging power of PEV i at time t. In the SCS, the driver can query the electrical grid constraint described in Figure 1 to prevent the required cost from exceeding the load. Moreover, as different drivers have different needs, they may have a preference for a particular charging time, as they encounter various forms of uncertainty, i.e., events that may occur at a later charging time; therefore, they may feel an urge to finish charging the PEV earlier. Therefore, each PEV in SCS can be charged flexibly based on the available information, which can meet the demand and reduce the charging cost.

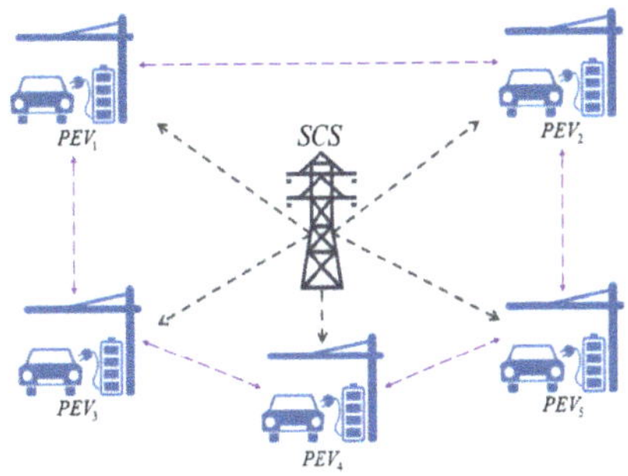

Figure 1. Mechanism of interaction between SCS and PEV combined with the aggregative game.

2.1. *Feasible Charging Coordination Constraint Profiles*

2.1.1. Battery Capacity Constraint for PEV i

We describe the battery dynamics of PEV i using a linear model (1), in which π_i^t represents the state of charge of PEV i at time t. Meanwhile, $\pi_i^{\min}$ denotes the lower limit of battery capacity and $\pi_i^{\max}$ denotes the upper limit of battery capacity for PEV i:

$$\pi_i^{t+1} = \pi_i^t + x_{i,t}, \pi_i^{\min} \leq \pi_i^t \leq \pi_i^{\max} \tag{1}$$

2.1.2. Charging Constraint for PEVs

For each PEV i, the total electrical energy that it obtains needs to meet its charging requirement needs. Based on this constraint, let R_i be the required electrical energy. Therefore, the following equation holds:

$$\sum_{t=1}^{T} x_{i,t} = R_i \tag{2}$$

Moreover, each PEV i can choose to charge and discharge according to its situation, which depends on the PEV battery capacity. Note here that PEVs cannot be selected for charging or discharging at the same time; only one of them can be selected. $\underline{x_i}$ is defined as the minimum discharging power of a PEV i, and $\overline{x_i}$ is defined as the maximum charging power of a PEV i:

$$\underline{x_i} \leq x_{i,t} \leq \overline{x_i} \tag{3}$$

2.1.3. Overload Constraint for Charging PEVs

To avoid overloading the SCS, the charging demand of all the PEVs cannot exceed the maximum electrical energy supplied by the SCS. Let C_{max} denote the total PEV load, and the basic demand load at moment t is set to d_t, which denotes the basic electrical load transmitted through the SCS at time t. As the constraint spatially couples the charging demand of all PEVs using the SCS at time t, constraint (4) is referred to as a joint constraint:

$$\sum_{i=1}^{N} x_{i,t} + d_t \leq C_{\max} \tag{4}$$

2.1.4. Feasible Charging Profiles

We assume that the set $\mathcal{X}_i$ is the feasible charging profile and $\mathcal{X}_i$ is a nonempty, compact convex set. Let $\mathcal{X} = \{(x_1, x_2, \ldots, x_N) \mid x_i \in \mathcal{X}_i, \forall i \in \mathcal{I}\}$ denote the set of charging profiles of all PEVs. Then, $\mathcal{X}$ is also nonempty, compact and convex. Therefore, set $\mathcal{X}_i$ can be written as follows:

$$\mathcal{X}_i = \{x_i \mid (1)\ (2)\ (3)\ (4)\ \} \tag{5}$$

2.2. *Cost Function of PEVs*

We assume that there are N drivers, in which r $(r \leq N)$ start at the charging time and enter a state of anxiety after a period of time. For PEV i, t_i^a is denoted as the arrival time to the SCS and t_i^d as the departure time at which SCS will be left; the discharge is no longer considered after anxiety begins. Considering the aggregative game framework, we defined $\sigma(x) = \frac{1}{N}\sum_{i=1}^{N} x_i$, and the minimization cost function of PEV i under the corresponding energy consumption constraint is defined as follows:

$$\min_{x_i} f_i(x_i, \sigma(x)) = \sum_{t=1}^{T} S_{i,t} \int_0^{x_{i,t}} \rho_t\left(d_t + \mu_{i,t} + \sum_{j\neq i} x_{j,t}\right) d\mu_{i,t} \tag{6}$$

in which $S_{i,t}$ denotes the driver's anxiety influence at time t. Next, the pricing function $\rho_t(L_t)$ is defined

$$\begin{aligned} &\rho_t(L_t) = \alpha_t L_t + \beta_t \\ &\min_{x_i} f_i(x_i, \sigma(x)) = \sum_{t=1}^{T} S_{i,t} \int_0^{x^{i,t}} \left(\alpha_t\left(d_t + \mu_{i,t} + \sum_{j\neq i} x_{j,t}\right) + \beta_t\right) d\mu_{i,t} \end{aligned} \tag{7}$$

where $L_t = d_t + \mu_{i,t} + \sum_{j\neq i} x_{j,t}$ represents the total load at time t. The pricing function is for a nonlinear form that varies with the increasing charging load of PEV i, while the models in [23,24] use ρ_t as a linear function. As a consequence, the linearly decreasing marginal benefits [12,23,24] correspond to the PEV quadratic pricing function. In the meantime, the PEV i increases a load of $\mu_{i,t}$ to the electricity grid at time t; the driver of PEV pays an amount of $\rho_t\left(d_t + \mu_{i,t} + \sum_{j\neq i} x_{j,t}\right)\Delta\mu_{i,t}$ for charging at the time t. When $\Delta\mu_{i,t}$ tends toward zero and $\mu_{i,t}$ changes from zero to $x_{i,t}$, it then evolves into Formula (8):

$$\min_{x_i} f_i(x_i, \sigma(x)) = \left(d + N\sigma(x) - \frac{1}{2}x_i\right)^{\mathrm{T}} S_i \alpha x_i + 1_T^{\mathrm{T}} S_i \beta x_i \tag{8}$$

2.3. *Time Anxiety for Drivers*

The design of the cost function (8) shows that a higher price leads to lower charging power. Furthermore, if the driver has time anxiety about charging and discharging the PEV, the charging power needs to be higher. Based on this conclusion, $S_{i,t}$ must be small to have a high charging power. Following that, at the time of charging, the value of $S_{i,t}$ gradually rises. Meanwhile, the local objective function was defined according to the problem of anxious and non-anxious drivers, so we obtained two different settings.

The first setting is non-anxious drivers, and we defined S as a constant. This means that the driver has a constant anxiety factor at the charging time, which does not change over time. In this model, we denoted $AT(i)$ as the arrival time of PEV i to SCS and denoted $DT(i)$ as the departure time of PEV i from SCS:

$$S_{i,t} = \begin{cases} S, & \mathrm{AT}(i) \le t < DT(i) \\ 0, & \text{otherwise} \end{cases} \tag{9}$$

and the next setting is anxious drivers. We defined $V_{i,t}$ as the ratio of the duration of the anxiety state to the overall anxiety time. Thus, the following equation holds:

$$V_{i,t} = \frac{t - t_i^a}{t_i^d - t_i^a}, t_i^a \le t < t_i^d \tag{10}$$

With the time anxiety studied in this paper, at the time from $AT(i)$ to t_i^a, the time anxiety is at its lowest value $S_{i,\min}$, and from t_i^d to $DT(i)$, it is at its highest value $S_{i,\max}$. This paper will explore the behavior of PEV drivers based on [7] and propose four different behaviors:

(1) Non-time-anxious driver (NTAD): This type of driver reaches an anxious time directly after entering a state of peak anxiety (Figure 2).
(2) Less time-anxious driver (LTAD): This type of driver has anxiety values that rise quickly and then slowly after entering the anxious time. The rise is faster and then slower (Figure 3).
(3) Mid-time-anxious driver (MTAD): This type of driver enters anxious time with anxiety values increasing at a uniform rate (Figure 4).
(4) High-time-anxious driver (HTAD): This type of driver has anxiety values that rise slowly and then quickly after entering anxious time. The rise is slow and then fast (Figure 5).

Thus, we can obtain the following equation:

$$S_{i,t} = \begin{cases} S_{i,\max}, & \text{for NTAD} \\ \ln[V_{i,t}(e-1)+1] \times (S_{i,\max} - S_{i,\min}) + S_{i,\min}, & \text{for LTAD} \\ V_{i,t}(S_{i,\max} - S_{i,\min}) + S_{i,\min}, & \text{for MTAD} \\ \frac{e^{V_{i,t}}-1}{e-1}(S_{i,\max} - S_{i,\min}) + S_{i,\min}, & \text{for HTAD} \end{cases} \tag{11}$$

In Figures 2–5, the value of $S_{i,\max} - S_{i,\min}$ indicates the PEV driver's time anxiety, i.e., the depth of time anxiety. This indicates that the smaller the $S_{i,\min}$, the greater the time anxiety of the PEV driver. Furthermore, a greater time anxiety will increase the willingness of the PEV driver to meet the charging demand earlier and closer to the departure time, thus reducing the amount of time anxiety.

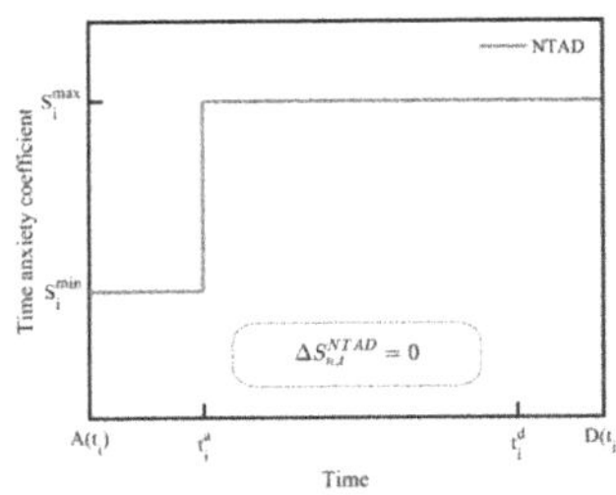

Figure 2. The time anxiety impact of NTAD.

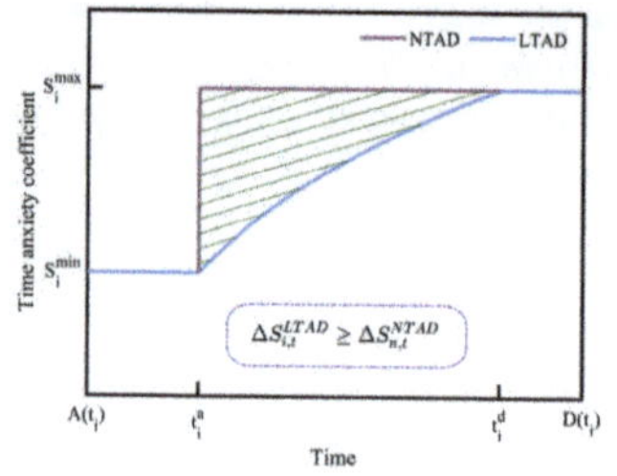

Figure 3. The time anxiety impact of LTAD.

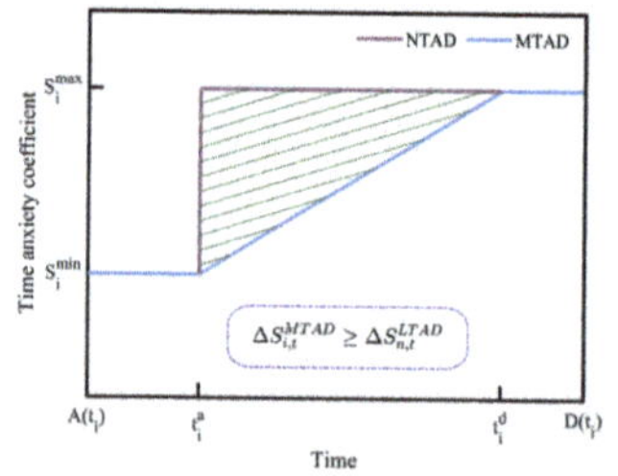

Figure 4. The time anxiety impact of MTAD.

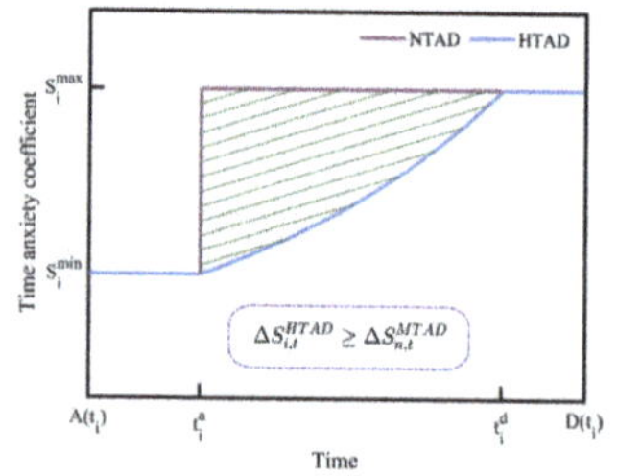

Figure 5. The time anxiety impact of HTAD.

According to the preceding discussions, two factors will influence PEV drivers' time anxiety. The intensity of time anxiety is the first, and the curve's shape is the second. To represent the effect of curve shape, the curve shapes of the four different behavior types regarding time anxiety are shown in Figures 2–5. We chose NATD as the reference, i.e., we chose $S_{i,\max}$ as the reference value. The difference between this and the impact of time anxiety caused by the suggested behavior is referred to as the impact difference. For example, $\Delta S_{i,t}^{\text{LTAD}} = S_{i,t}^{NATD} - S_{i,t}^{\text{LTAD}}$ is defined as the impact difference of the LTAD behavior. Therefore, the following inequality (12) holds in Figures 2–5:

$$\Delta S_{i,t}^{\text{NTAD}} \le \Delta S_{i,t}^{\text{LTAD}} \le \Delta S_{i,t}^{\text{MTAD}} \le \Delta S_{i,t}^{\text{HTAD}} \tag{12}$$

3. Distributed Charging Strategy

In this section, we propose a distributed reflected forward–backward splitting method to find a GNE of the function (8). Then, for the optimal response of each PEV driver, a distributed charging strategy is proposed. We assume that all PEVs in the SCS are selfish and each PEV is only allowed information about local problem data. Centralized control methods are typically unavailable in this situation.

3.1. Game Model

As the PEV charging and discharging problem is a generalized Nash equilibrium [24], we consider a group of agents $\mathcal{I} = \{1, \ldots, N\}$ that seek a GNE of the aggregative game with globally shared affine constraints and the gradient condition of KKT necessary optimality conditions can be then given by (13):

$$\nabla_{x_i} f_i(x_i, \sigma(x)) = (d + N\sigma(x) + x_i)S_i\alpha + 1_T S_i \beta \tag{13}$$

Assumption 1. *For each $i \in \mathcal{I}$, the function in (12) is differentiable and convex, and $\Omega \subset R^n$ is a closed convex set.*

We define R_m as m-dimensional Euclidean space and each agent i chooses its local decision $x_i \in \Omega_i \subset R^n$. We call $\mathbf{x} = \mathrm{col}(x_1, \cdots, x_N) = \mathrm{col}(x_i)_{i \in \mathcal{I}} \in \Omega \subset R^{Nn}$ the decision profile, i.e., the stacked vector $\prod_{i=1}^{N} \Omega_i = \Omega$. The aim of each agent i is to optimize its objective function, $f_i(x_i, \sigma(x)) : \Omega \to R$, within its feasible decision set. Note that $f_i(x_i, \sigma(x))$ is nonlinearly coupled to the decisions of the other agents, but may not be explicitly coupled to the decisions of all other agents. We denote

$$\mathcal{X} := \left\{ x_i \in \Omega_i \mid \sum_{i=1}^{N} A_i x_i \leq \sum_{i=1}^{N} b_i \right\} \tag{14}$$

where $A_i \in R^{m \times n}$, $b_i \in R^m$ are local data from agent i. Thus, we can obtain the following formula:

$$\min_{x_i} f_i(x_i, \sigma(x)) \quad s.t. \sum_{i=1}^{N} A_i x_i \leq \sum_{i=1}^{N} b_i \tag{15}$$

Furthermore, by considering games with affine sharing constraints $Ax \leq b$, and supposing x^* as a GNE of game (12), the optimal solution to the following convex optimization problem is defined as

$$\min_{x_i} f_i(x_i^*, \sigma(x^*)) \quad s.t. x_i \in \Omega_i, A_i x_i \leq b - \sum_{j \neq i, j \in \mathcal{I}} A_j x_j^* \tag{16}$$

where $b = \sum_{i=1}^{N} b_i \in R^m$. The set Ω_i denotes the local decision set of agent i and the matrix A_i defines how agent i is involved in the coupling constraint. However, the constraints of Equation (5) are clearly different from those of Equation (16). Therefore, the matrix A_i and vector b_i are then divided into two submatrices, H_i and W_i, and subvectors, P_i and Q_i. We define these as follows:

$$W_i = \begin{bmatrix} 0_{(i-1)T \times T} \\ -E \\ 0_{(N-i)T \times T} \\ 0_{(i-1)T \times T} \\ E \\ 0_{(N-i)T \times T} \\ I \end{bmatrix}, \quad P_i = \begin{bmatrix} 0_{(i-1)} \\ R_i \\ 0_{(N-i)} \end{bmatrix} \tag{17}$$

$$Q_i = \begin{bmatrix} 0_{(i-1)T} \\ (\pi_i^1 - \pi_i^{\min}) 1_T \\ 0_{(i-1)T} \\ 0_{(i-1)T} \\ (\pi_i^{\max} - \pi_i^1) 1_T \\ 0_{(i-1)T} \\ \frac{C_{\max}}{N} 1_T \end{bmatrix}, \quad H_i = \begin{bmatrix} 0_{(i-1) \times T} \\ 1_{1 \times T} \\ 0_{(N-i) \times T} \end{bmatrix} \tag{18}$$

where 0_T represents the zero vector of T dimension, I represents the unit matrix and $E = \begin{bmatrix} 1 & & & & \\ 1 & 1 & & & \\ \cdot & \cdot & \cdot & & \\ \cdot & \cdot & & \cdot & \\ \cdot & \cdot & & & \cdot \\ 1 & 1 & \cdot & \cdot & \cdot & 1 \end{bmatrix} \in R^{T\times T}$. By definition, the constraints $\sum_{i=1}^{N} W_i x_i \leq \sum_{i=1}^{N} Q_i$, $\sum_{i=1}^{N} H_i x_i = \sum_{i=1}^{N} P_i$ are satisfied. Then, the constraints in Equation (5) are satisfied. Note that $\sum_{i=1}^{N} H_i x_i = \sum_{i=1}^{N} P_i$ is an equation constraint, while Equation (16) contains only the inequality constraint. Therefore, we represent this Equation constraint using two inequality constraints, satisfying both $\sum_{i=1}^{N} H_i x_i \leq \sum_{i=1}^{N} P_i$ and $-\sum_{i=1}^{N} H_i x_i \leq -\sum_{i=1}^{N} P_i$. If using this approach, we need to rewrite $A_i = \begin{bmatrix} W_i \\ H_i \\ -H_i \end{bmatrix}$ and $b_i = \begin{bmatrix} Q_i \\ P_i \\ -P_i \end{bmatrix}$, and the constraints in Equation (5) are converted into the constraints in Equation (16).

Assumption 2. *For each $i \in \mathcal{I}$, and for each $\xi \in$ E, the function $f_i(x_i, \sigma(x), \xi)$ is Lipschitz continuous, convex, and continuously differentiable. For $\sigma(x)$, the Lipschitz constant $\ell(\sigma(x), \xi)$ is integrable in ξ.*

Among all possible generalized Nash equilibria, we are concerned with those solution sets that correspond to the set of solutions to an appropriate variational inequality. For this purpose, let us define the (pseudo) gradient mapping as

$$\mathrm{F}(x) = \mathrm{col}\big((\mathrm{E}[\nabla_{x_i} f_i(x_i, \sigma(x), \xi_i)])_{i\in\mathcal{I}}\big) \tag{19}$$

We define a local Lagrangian function for agent i as $\mathcal{L}(x, z, \lambda) = \langle \mathrm{F}(x^*, Jx^*), x\rangle + \ell_{\Omega}(x) + \lambda^{\mathrm{T}}(Ax - b) + \ell_{R_+^{Nm}}(\lambda) + z^{\mathrm{T}} L_\lambda \lambda$, where $\lambda_i \in R_+^m$ is a dual variable associated with the coupling constraint. When x^* is an optimal solution to (16), the following Karush–Kuhn–Tucker (KKT) conditions are satisfied:

$$\forall i \in \mathcal{I}: \quad \begin{cases} 0 \in \mathrm{F}(x^*, Jx^*) + \mathrm{N}_{\Omega}(x^*) + A^{\mathrm{T}}\lambda^* \\ 0 = L_\lambda \lambda^* \\ 0 \in Ax^* - b - \mathrm{N}_{R_+^{Nm}}(\lambda^*) + L_\lambda z^* \end{cases} \tag{20}$$

In order to ensure that all the preceding signs are $+$ and facilitate the operation, the third formula becomes $0 \in -\boldsymbol{A}\boldsymbol{x}^* + \boldsymbol{b} + L_\lambda \lambda^* + \mathrm{N}_{R_+^{Nm}}(\boldsymbol{\lambda}^*) - \boldsymbol{L}_\lambda \boldsymbol{z}^*$. Since $\hat{\boldsymbol{u}} = Mu$ is orthogonal to Ju, there is no consistent vector in the space of $\hat{\boldsymbol{u}}$ to make $L_u \hat{\boldsymbol{u}} = 0$ when, and only when, $\hat{\boldsymbol{u}} = 0$. Meanwhile, we introduce L_u to implement a distributed estimation. If we use I instead of L_u, we need the central node to pass the average information. Based on this result, the extended KKT condition is as follows:

$$\begin{cases} 0 \in \mathbf{F}(\boldsymbol{x}^*, \boldsymbol{J}^* + \hat{\boldsymbol{u}}^*) + \mathrm{N}_{\Omega}(\boldsymbol{x}^*) + \boldsymbol{A}^{\mathrm{T}}\boldsymbol{\lambda}^* \\ 0 = c\boldsymbol{L}_u \hat{\boldsymbol{u}}^* \\ 0 = \boldsymbol{L}_\lambda \boldsymbol{\lambda}^* \\ 0 \in -\boldsymbol{A}\boldsymbol{x}^* + \boldsymbol{b} + L_\lambda \lambda^* + \mathrm{N}_{R_+^{Nm}}(\boldsymbol{\lambda}^*) - \boldsymbol{L}_\lambda z^* \end{cases} \tag{21}$$

3.2. Distributed Algorithm

We assume that each driver only knows their local data, i.e., $f_i(x_i, \sigma(x))$, Ω_i, A_i and b_i, which contains their own private information. Meanwhile, the shared affine coupling constraints are decomposed such that each driver knows only one local block of the constraint matrix. Note that A_i describes how agent i participates in the coupling constraints (shared global resources), which is also assumed to be privately known by driver i. The globally shared constraint $Ax \leq b$ then couples the set of feasible decisions of the agents, but is not known to any agent. Next, we describe the preprocessing process leading to the distributed iteration proposed in the algorithm. In this, $x_{i,k}$, $u_{i,k}$, $z_{i,k}$ and $\lambda_{i,k}$ are the state variables of agent i at iteration k and τ_i, v_i, α_i are fixed constant step-sizes for driver i. We define the weighted adjacency matrix $W = \left[w_{i,j}\right]_{i,j} \in R^{N\times N}$. The set of neighbors of PEV i is $\mathcal{N}_i^{\lambda} = \left\{j \mid w_{i,j} > 0\right\}$. We define the operators as follows:

$$\overline{\mathrm{A}}: \begin{pmatrix} x \\ u^{\perp} \\ z \\ \lambda \end{pmatrix} \mapsto \begin{pmatrix} \mathrm{F}\left(x, Jx + u^{\perp}\right) \\ cL_u u^{\perp} \\ 0 \\ b \end{pmatrix} + \begin{pmatrix} 0 \\ 0 \\ 0 \\ L_\lambda \lambda \end{pmatrix} \tag{22}$$

$$\overline{\mathrm{B}}: \begin{pmatrix} x \\ u^{\perp} \\ z \\ \lambda \end{pmatrix} \mapsto \begin{pmatrix} \mathrm{N}_\Omega(x) \\ 0 \\ 0 \\ N_{R_+^{\mathrm{Nm}}}(\lambda) \end{pmatrix} + \begin{bmatrix} 0 & 0 & 0 & A^{\mathrm{T}} \\ 0 & 0 & 0 & 0 \\ 0 & 0 & 0 & L_\lambda \\ -A & 0 & -L_\lambda & 0 \end{bmatrix} \begin{pmatrix} x \\ u^{\perp} \\ z \\ \lambda \end{pmatrix} \tag{23}$$

in which $c \in R_+$. Meanwhile, the metric matrix Φ is defined as follows. Note that the matrix Φ is symmetric and positive definite, and the first term of Φ is the antisymmetric matrix in the operator $\overline{\mathrm{B}}$:

$$\begin{aligned} \Phi &= \begin{bmatrix} 0 & 0 & 0 & -A^{\mathrm{T}} \\ 0 & 0 & 0 & 0 \\ 0 & 0 & 0 & -L_\lambda \\ -A & 0 & -L_\lambda & 0 \end{bmatrix} + \begin{bmatrix} \tau^{-1} & 0 & 0 & 0 \\ 0 & 0 & 0 & 0 \\ 0 & 0 & v^{-1} & 0 \\ 0 & 0 & 0 & \alpha^{-1} \end{bmatrix} + \begin{bmatrix} -S_C & -S & 0 & 0 \\ -S & \kappa^{-1}I & 0 & 0 \\ 0 & 0 & 0 & 0 \\ 0 & 0 & 0 & 0 \end{bmatrix} \\ &= \begin{bmatrix} \tau^{-1} - S_C & -S & 0 & -A^{\mathrm{T}} \\ -S & \kappa^{-1}I & 0 & 0 \\ 0 & 0 & v^{-1} & -L_\lambda \\ -A & 0 & -L_\lambda & \alpha^{-1} \end{bmatrix} \end{aligned} \tag{24}$$

Through the RFB algorithm, we can obtain the following formula $\Phi\bar{v}_{k+1} + \overline{\mathrm{B}}\bar{v}_{k+1} = \Phi\bar{v}_k - \overline{\mathrm{A}}v_k$ and $v_{k+1} = 2\bar{v}_{k+1} - \bar{v}_k$, which is in the form of a distributed reflected forward–backward splitting method to find zeros of $(\Phi^{-1}\mathrm{A} + \Phi^{-1}\mathrm{B})$. We substitute the operators into $\Phi\bar{v}_{k+1} + \overline{\mathrm{B}}\bar{v}_{k+1} = \Phi\bar{v}_k - \overline{\mathrm{A}}v_k$, and then, through calculation, we can obtain the following equation:

$$\begin{cases} \mathrm{N}_\Omega(\overline{x}_{k+1}) + \overline{x}_{k+1} = \overline{x}_k + \tau\left(-A^{\mathrm{T}}\overline{\lambda}_k - \mathrm{F}\left(x_k, Jx_k + u_k^{\perp}\right) - cL_k u_k\right) \\ \overline{u}_{k+1}^{\perp} = \overline{u}_k^{\perp} - \kappa c L_k u_k^{\perp} + M(\overline{x}_{k+1} - \overline{x}_k) \\ \overline{z}_{k+1} = \overline{z}_k - vL_\lambda\overline{\lambda}_k \\ \mathrm{N}_{\mathrm{R}_+^{Nm}}\left(\overline{\lambda}_{k+1}\right) + \overline{\lambda}_{k+1} = \overline{\lambda}_k + \alpha\left(A(2\overline{x}_{k+1} - \overline{x}_k) + L_\lambda(2\overline{z}_{k+1} - \overline{z}_k) - b - L_\lambda\lambda_k\right) \\ x_{k+1} = 2\overline{x}_{k+1} - \overline{x}_k \\ u_{k+1}^{\perp} = 2\overline{u}_{k+1}^{\perp} - \overline{u}_k^{\perp} \\ z_{k+1} = 2\overline{z}_{k+1} - \overline{z}_k \\ \lambda_{k+1} = 2\overline{\lambda}_{k+1} - \overline{\lambda}_k \end{cases} \tag{25}$$

its initial condition is $\overline{u}_0 = \overline{x}_0$ and $u_0 = x_0$.

Based on the above conclusions, the algorithm can find the GNE of the game, i.e., the strategy that finds the cost minimization for PEV i. Its convergence was proved in [18]. Meanwhile, by writing the above algorithm in distributed form, we can obtain the following Algorithm 1:

Algorithm 1 : Distributed charging strategy with reflected forward-backward algorithm.

1: **Initialization:** $S_{i,\min} = S_{i,\max} = S$, $x_i \in \Omega_i$, $u_i \in R^n$, $\lambda_i \in R^m$ and $z_i \in R^m$
2: **Task:** slove (16)
3: **For** $k = 1 : k_{\max}$
4: **For** $i = 1 : N$
5: **(1)Receives** $x_{i,k}$ for $j \in \mathcal{N}_i^f$, $u_{j,k}$, $z_{j,k}$ and $\lambda_{j,k}$ for $j \in \mathcal{N}_i^\lambda$ **then updates**
6: $\bar{x}_{i,k+1} = \mathrm{proj}_{\Omega_i}\left(\bar{x}_{i,k} + \tau_i\left(-A_i^{\mathrm{T}}\bar{\lambda}_{i,k} - \nabla_{x_i} f_i(x_{i,k}, u_{i,k}) - c\sum_{j=1}^N \omega_{ij}\left(u_{i,k} - u_{j,k}\right)\right)\right)$
7: $\bar{u}_{i,k+1} = \bar{u}_{i,k} - c\kappa\sum_{j=1}^N \omega_{ij}\left(u_{i,k} - u_{j,k}\right) + \bar{x}_{i,k+1} - \bar{x}_{i,k}$
8: $\bar{z}_{i,k+1} = \bar{z}_{i,k} - v_i\sum_{j=1}^N \omega_{ij}\left(\bar{\lambda}_{i,k} - \bar{\lambda}_{j,k}\right)$
9: $\bar{\lambda}_{i,k+1} = \mathrm{proj}_{R_+^m}\left(\bar{\lambda}_{i,k} + \bar{\alpha}_i A_i(2\bar{x}_{i,k+1} - \bar{x}_{i,k} - \bar{\alpha}_i\left(b_i - \sum_{j=1}^N \omega_{ij}(\lambda_{i,k} - \lambda_{j,k})\right)\right.$
10: $\left. + \bar{\alpha}_i\sum_{j=1}^N \omega_{ij}\left(2\left(\bar{z}_{i,k+1} - \bar{z}_{j,k+1}\right) - \left(\bar{z}_{i,k} - \bar{z}_{j,k}\right)\right)\right)$
11: **(2)Receives** $\bar{x}_{i,k}$, $\bar{u}_{i,k}$, $\bar{z}_{i,k}$, $\bar{\lambda}_{i,k}$ **then updates**
12: $x_{i,k+1} = 2\bar{x}_{i,k+1} - \bar{x}_{i,k}$
13: $u_{i,k+1} = 2\bar{u}_{i,k+1} - \bar{u}_{i,k}$
14: $z_{i,k+1} = 2\bar{z}_{i,k+1} - \bar{z}_{i,k}$
15: $\lambda_{i,k+1} = 2\bar{\lambda}_{i,k+1} - \bar{\lambda}_{i,k}$
16: **end**
17: **end**
18: **if** $\sum_{t=t_i^a}^{t_i^d} x^{i,t} > p_i^c$ **then**
19: $S_{i,\min} = S_{i,\min} - \varepsilon$
20: Go back to step 3
21: **else**
22: **break**
23: **end if**

Step 3 can be explained as follows. We define a step-size ε and set a threshold value p_i^c for time-anxious drivers. Since the initialization $S_{i,\min} = S_{i,\max} = S$, i.e., all drivers are not time-anxious, the algorithm is executed once, and a Nash equilibrium is solved. Then, we determine whether the sum of charging capacity $\sum_{t=t_i^a}^{t_i^d} x_{i,t}$ for PEV i at the time $t \in [t_i^a, t_i^d]$ is greater than the threshold value p_i^c. If $\sum_{t=t_i^a}^{t_i^d} \bar{x}_{i,t} > p_i^c$, let $S_{i,\min} = S_{i,\min} - \varepsilon$. Then, the new $S_{i,\min}$ can be substituted to solve the Nash equilibrium once more. Otherwise, it is straightforward to derive the Nash equilibrium solution.

4. Simulation and Numerical Results

In this section, the performance of the proposed algorithm with load constraints and time anxiety is evaluated by minimizing the charging cost for PEV i in the SCS. For further illustration, we consider the SCS with 10 PEVs in the residential area. We also investigated the charging power distribution under non-time-anxious and time-anxious conditions. The simulation configuration was set up as follows.

4.1. Overload Control for 10 PEVs

In this scenario, depending on the owner's preferences and needs, 10 PEVs arrive at the SCS. The charging needs of each PEV were chosen between 50 and 60 KW, and the charging power was chosen between 10 and 15 KW/h. Table 1 lists the PEV parameters, arrival time (AT), and departure time (DT). Usually, the daily peak charging demand occurs from 12:00 to 17:00, when people go to the SCS to charge their PEVs. Therefore, we define $\alpha_t = 0.3\$/\mathrm{kWh}$ during peak hours (i.e., from 18:00 to 6:00), $\alpha_t = 0.2\$/\mathrm{kWh}$ during off-peak hours (i.e., other times), and the initial electricity price is $\beta_t = 0.3\$/\mathrm{kWh}$. We assume that the maximum power supplied by the SCS at time t is 130 KW, which is obtained from the value of [25].

Table 1. Constraint-related parameters.

PEV i	π_i^1	R_i	π_i^{min}	π_i^{max}	$\overline{x_i}$	$\underline{x_i}$	AT	DT
1	7.5	53	5	75	15	−15	12	17
2	6	56	5	80	10	−10	22	14
3	8	52.5	5	75	10	−10	3	12
4	5.6	51.4	5	70	15	−15	2	10
5	6.7	51	5	65	10	−10	15	21
6	6.5	56	5	75	15	−15	2	8
7	6.1	52.4	5	65	10	−10	4	20
8	9	51	5	80	10	−10	13	24
9	9.2	50.8	5	70	15	−15	10	22
10	7.5	53	5	75	15	−15	18	24

As shown in Figure 6, the red dotted line shows the maximum capacity to support PEV charging at time t, i.e., L_{max} =130 KW. The green and purple solid lines denote the base load and the charging requirement load, respectively, for the 24 h. As shown in Figure 6, the SCS is severely overloaded at peak hours due to uncoordinated PEV charging, which may damage the SCS. Therefore, we present the overload control constraint in the framework of the game (16). After the overload control constraint, the load profile is shown in Figure 7. Compared with Figure 6, as the SCS in Figure 7 holds the charging capacity fixed at peak hours, the charging strategy shifts the excess charging capacity to free time, i.e., off-peak hours. The results show that the charging requirement load is always below the value of L_{max}. Therefore, the strategy ensures the safety of the SCS.

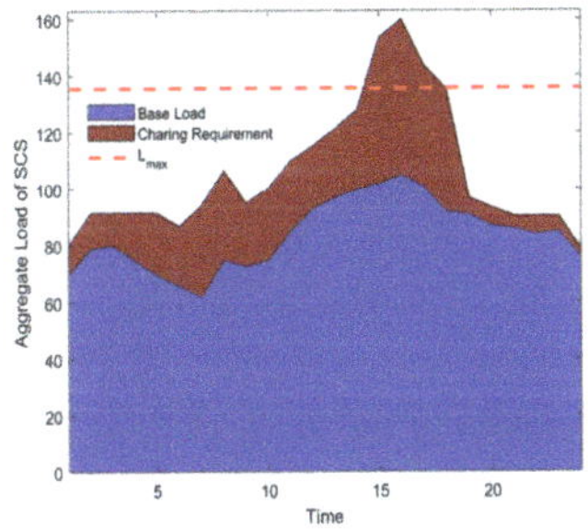

Figure 6. Charging strategies for 10 PEVs without overload control.

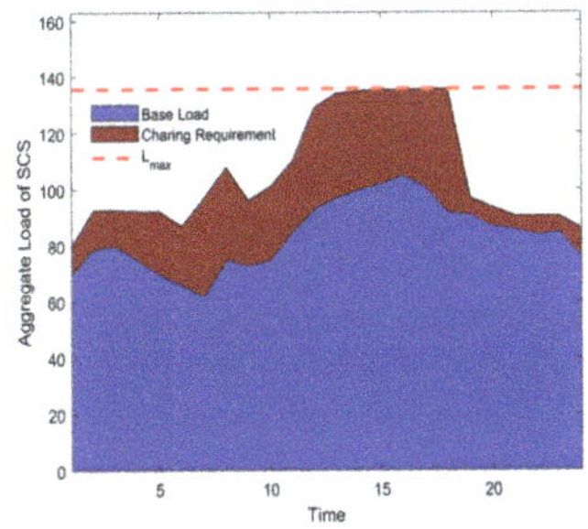

Figure 7. Charging strategies for 10 PEVs with overload control.

4.2. Time Anxiety for PEVs

The simulation provides a further charging strategy considering load constraints. According to the settings in Table 1, the charging and discharging powers of each PEV are shown in Figure 8.

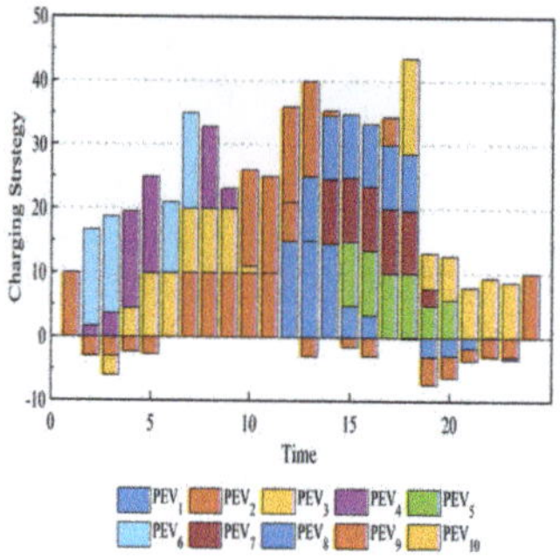

Figure 8. Charging capacity of 10 PEVs in a day.

In these four pictures, the blue line indicates a charging strategy that does not consider time anxiety, while the red-brown line indicates a charging strategy that does consider time anxiety (note that the two are only different charging strategies; the amount that the PEV is charged does not change but is simply shifted). Figure 9 shows the NATD driver's behavior in our simulation. From the discussion in Section 2.3, it follows that the driver has no time anxiety; therefore, the blue and red-brown lines overlap, and we use one of the blue lines to represent the PEV.

Figure 10 shows the LATD driver's behavior in our simulation. From the previous discussion in Section 2.3, it follows that this driver will have some time anxiety, i.e., there may be a delay in the charging of the electric vehicle due to something that occurs during the time anxiety; therefore, a charging strategy that considers time anxiety will move some of the charging within the time interval $[t_i^a, t_i^d]$ of time anxiety in the non-time anxious time interval, effectively avoiding the situation of missed EV charging due to unpredictable circumstances. This is represented in the graph by the transfer of the charge from anxious energy to shifted energy, which can be seen in Figure 10 as a reduction in the charge in the time anxious interval. The PEV driver is satisfied with the current charging method, as it is considered robust.

We simulated the behavior of the MATD driver, and, as can be seen in Figure 11, it shifted more of its charge to the rest of the time interval during the anxiety time interval than in Figure 10. The HATD driver, as can be seen in Figure 12, shifted more of its charge to the rest of the time interval during the anxious period than in Figure 11. This is because PEV drivers of various anxiety levels have a predefined threshold, which is lower if the drivers want to be more robust in response to uncertain events (e.g., HATD has the strongest anxiety). If the current anxiety energy exceeds the threshold, the PEV driver is dissatisfied with the current charge level, which would be insufficient to meet his or her charging needs in an uncertain event. Our algorithm is implemented in an iterative manner until the PEV driver succeeds in bringing his/her anxiety energy below its threshold, as indicated by the red-brown line. Our algorithm is iteratively implemented until the anxiety energy of the electric vehicle driver falls below a particular threshold.

The simulation of these four different driver behaviors leads to the same conclusion as discussed in Section 2.3: the greater the driver's time anxiety, the greater that driver's willingness to meet the charging demand earlier, i.e., more charging is transferred within the interval $[t_i^a, t_i^d]$ of time anxiety.

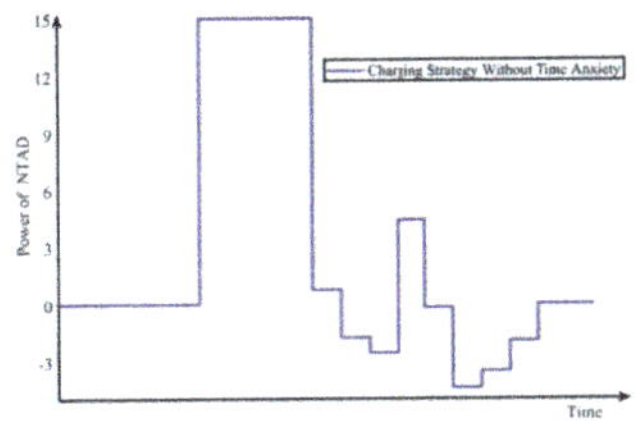

Figure 9. The time anxiety impact of NTAD.

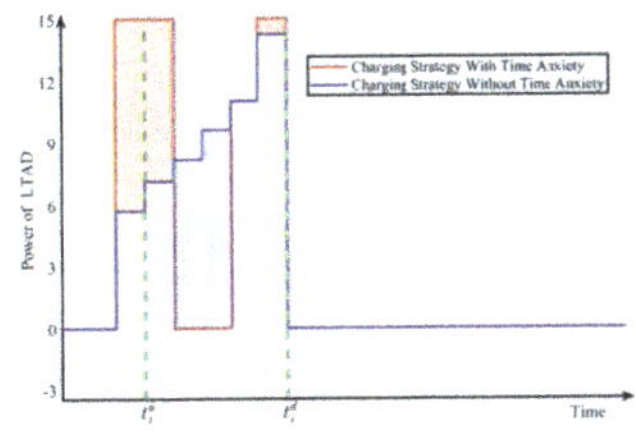

Figure 10. The time anxiety impact of LTAD.

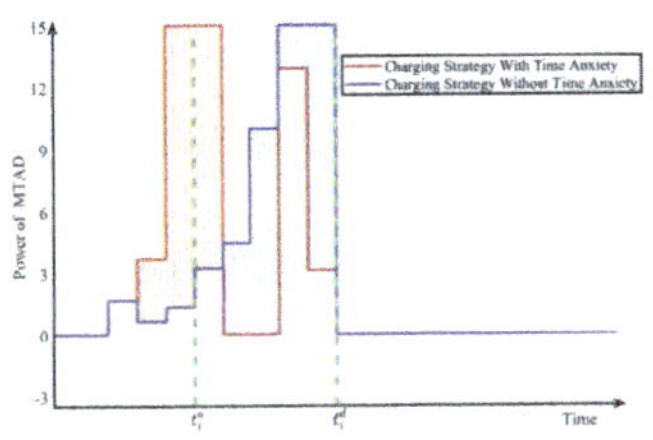

Figure 11. The time anxiety impact of MTAD.

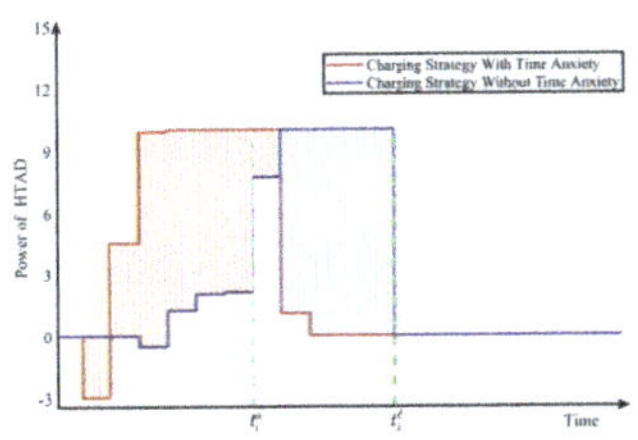

Figure 12. The time anxiety impact of HTAD.

4.3. Convergence Analysis

In this simulation, we provide an iterative process performed by all the PEVs at 18:00. PEVs do not need a central node to be able to bidirectionally communicate with all other PEVs. To preserve privacy, each PEV computes the corresponding decision in a distributed manner using its cost function, feasible set and coupling constraints. At each iteration k, the PEVs update their decisions and their estimates. The iterative process x_k denotes the decision variable with PEV i; u_k denotes the aggregated estimate, which includes the parameters that affect the electricity price; λ_k is used to ensure that the constraints hold; z_k is used as an auxiliary variable to ensure that λ_k is consistent. As shown in Figure 13, the PEVs all converge to their optimal charging strategies during the iterative process. As a result, the proposed distributed RFB algorithm under the aggregative game can solve the charging problem in SCS proficiently.

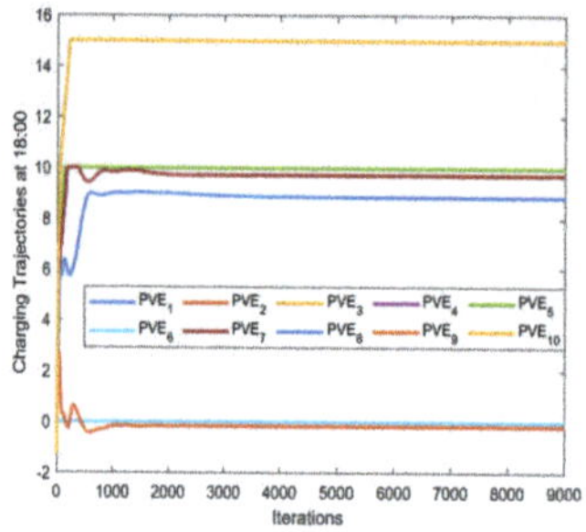

Figure 13. Charging trajectories of PEVs at 18:00.

In addition, we simulated the convergence accuracy of the RFB algorithm and the FB algorithm. Compared to the FB algorithm, the RFB algorithm converged to $10^{-7.1}$ and the FB algorithm converged to $10^{-3.6}$ after 9000 iterations under the same conditions shown in Figure 14. The effectiveness of the algorithm is demonstrated by the fact that it only takes 3000 iterations to converge to $10^{-3.6}$ using the RFB algorithm. The faster convergence indicates that all PEVs are solved to arrive at the optimal strategy faster.

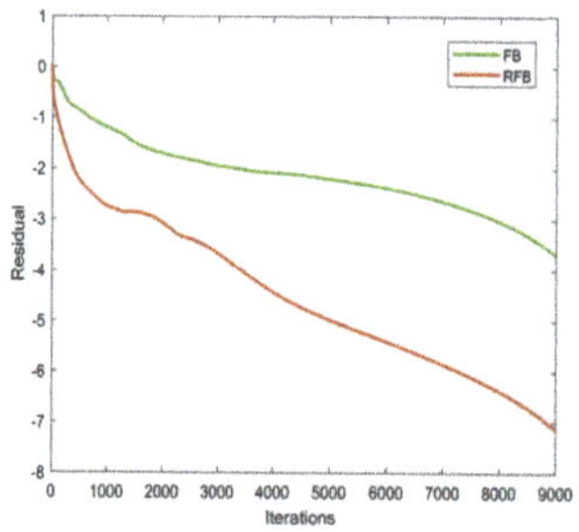

Figure 14. Convergence accuracy.

5. Conclusions

In this paper, a new distributed charging and discharging strategy for PEVs, based on time anxiety and load constraints, was proposed in the framework of the aggregative game. Time anxiety was proposed to mitigate the effects of some uncertain events that may occur during charging. The load constraint was proposed to make the PEV charging more coordinated and protect the safety of the SCS. Detailed case studies were presented, showing that the charging strategy with time anxiety and load constraint considered is more reasonable and reduces the total cost. The distributed reflected forward–backward algorithm was designed to seek the generalized Nash equilibria of the game model. The proposed algorithm achieved optimal driver response in a theoretically fast, distributed manner and protected the driver's privacy. The effectiveness of the proposed algorithm was verified by example simulations. In the future, we will focus on the aggregative game for charging stations.

Author Contributions: Conceptualization, T.K. and H.L.; methodology, T.K. and L.Z.; software, L.Z. and H.L.; validation, H.L.; formal analysis, T.K. and L.Z.; investigation, H.L.; resources, T.K. and H.L.; data curation, T.K. and L.Z.; writing—original draft preparation, T.K; writing—review and editing, H.L. and L.Z.; visualization, T.K. and H.L. and L.Z.; supervision, T.K. and H.L.; project administration, H.L.; funding acquisition, H.L. All authors have read and agreed to the published version of the manuscript.

Funding: The work described in this paper is supported in part by the National Natural Science Foundation of China (Grant No. 62173278) and in part by the Science and Technology Research Program of Chongqing Municipal Education Commission (Grant No. KJQN202100228).

Data Availability Statement: Not applicable.

Conflicts of Interest: The authors declare no conflict of interest.

References

1. Poornesh, K.; Nivya, K.P.; Sireesha, K. A comparative study on electric vehicle and internal combustion engine vehicles. In *Proceedings of the 2020 International Conference on Smart Electronics and Communication (ICOSEC), Trichy, India, 10–12 September 2020*; IEEE: Piscataway, NJ, USA, 2020; pp. 1179–1183.
2. Qian, K.; Zhou, C.; Allan, M.; Yuan, Y. Modeling of load demand due to EV battery charging in distribution systems. *IEEE Trans. Power Syst.* **2010**, *26*, 802–810. [CrossRef]
3. Fischer, D.; Harbrecht, A.; Surmann, A.; McKenna, R. Electric vehicles' impacts on residential electric local profiles—A stochastic modelling approach considering socio-economic, behavioural and spatial factors. *Appl. Energy* **2019**, *233*, 644–658. [CrossRef]
4. Bilh, A.; Naik, K.; El-Shatshat, R. A novel online charging algorithm for electric vehicles under stochastic net-load. *IEEE Trans. Smart Grid* **2016**, *9*, 1787–1799. [CrossRef]
5. Vatanparvar, K.; Faezi, S.; Burago, I.; Levorato, M.; Al Faruque, M.A. Extended Range Electric Vehicle With Driving Behavior Estimation in Energy Management. *IEEE Trans. Smart Grid* **2019**, *10*, 2959–2968. [CrossRef]
6. Wang, C.; Zhou, Y.; Wu, J.; Wang, J.; Zhang, Y.; Wang, D. Robust-index method for household load scheduling considering uncertainties of customer behavior. *IEEE Trans. Smart Grid* **2015**, *6*, 1806–1818. [CrossRef]
7. Alsabbagh, A.; Wu, B.; Ma, C. Distributed electric vehicles charging management considering time anxiety and customer behaviors. *IEEE Trans. Ind. Inform.* **2020**, *17*, 2422–2431. [CrossRef]
8. Yan, Q.; Zhang, B.; Kezunovic, M. Optimized operational cost reduction for an EV charging station integrated with battery energy storage and PV generation. *IEEE Trans. Smart Grid* **2018**, *10*, 2096–2106. [CrossRef]
9. Abdelaziz, M.M.A.; Shaaban, M.F.; Farag, H.E.; El-Saadany, E.F. A multistage centralized control scheme for islanded microgrids with PEVs. *IEEE Trans. Sustain. Energy* **2014**, *5*, 927–937. [CrossRef]
10. Yao, L.; Lim, W.H.; Tsai, T.S. A Real-Time Charging Scheme for Demand Response in Electric Vehicle Parking Station. *IEEE Trans. Smart Grid* **2017**, *8*, 52–62. [CrossRef]
11. Li, J.; Li, C.; Xu, Y.; Dong, Z.Y.; Wong, K.P.; Huang, T. Noncooperative Game-Based Distributed Charging Control for Plug-In Electric Vehicles in Distribution Networks. *IEEE Trans. Ind. Inform.* **2018**, *14*, 301–310. [CrossRef]
12. Gan, L.; Topcu, U.; Low, S.H. Optimal decentralized protocol for electric vehicle charging. *IEEE Trans. Power Syst.* **2013**, *28*, 940–951. [CrossRef]
13. Paccagnan, D.; Gentile, B.; Parise, F.; Kamgarpour, M.; Lygeros, J. Nash and wardrop equilibria in aggregative games with coupling constraints, *IEEE Trans. Autom. Control* **2018**, *64*, 1373–1388.
14. Savadkoohi, S.J.S.; Kebriaei, H.; Aminifar, F. Aggregative Game for Charging Coordination of PEVs in a Network of Parking Lots. In *Proceedings of the 2019 Smart Grid Conference (SGC), Tehran, Iran, 18–19 December 2019*; IEEE: Piscataway, NJ, USA, 2019; pp. 1–6.
15. Li, C.; Liu, C.; Deng, K.; Yu, X.; Huang, T. Data-Driven Charging Strategy of PEVs Under Transformer Aging Risk. *IEEE Trans. Control. Syst. Technol.* **2018**, *26*, 1386–1399. [CrossRef]
16. Alsabbagh, A.; Yin, H.; Ma, C. Distributed Electric Vehicles Charging Management with Social Contribution Concept. *IEEE Trans. Ind. Inform.* **2020**, *16*, 3483–3492. [CrossRef]
17. Gadjov, D.; Pavel, L. Distributed GNE seeking over networks in aggregative games with coupled constraints via forward-backward operator splitting. In *Proceedings of the 2019 IEEE 58th Conference on Decision and Control (CDC)*, Nice, France, 11–13 December 2019; pp. 5020–5025.
18. Cevher, V.; Vũ, B.C. A reflected forward-backward splitting method for monotone inclusions involving Lipschitzian operators. *Set-Valued Var. Anal.* **2021**, *29*, 163–174.
19. Franci, B.; Staudigl, M.; Grammatico, S. Distributed forward-backward (half) forward algorithms for generalized Nash equilibrium seeking. In Proceedings of the 2020 European Control Conference (ECC), Saint Petersburg, Russia, 12–15 May 2020; pp. 1274–1279.
20. Wan, Y.; Qin, J.; Li, F.; Yu, X.; Kang, Y. Game theoretic-based distributed charging strategy for PEVs in a smart charging station. *IEEE Trans. Smart Grid* **2020**, *12*, 538–547. [CrossRef]
21. Jasim, A.M.; Jasim, B.H.; Neagu, B.C.; Alhasnawi, B.N. Efficient Optimization Algorithm-Based Demand-Side Management Program for Smart Grid Residential Load. *Axioms* **2023**, *12*, 33. [CrossRef]
22. Liu, Y.; Deng, R.; Liang, H. A stochastic game approach for PEV charging station operation in smart grid. *IEEE Trans. Ind. Informatics* **2017**, *14*, 969–979. [CrossRef]
23. Ma, Z.; Zou, S.; Liu, X. A distributed charging coordination for large-scale plug-in electric vehicles considering battery degradation cost. *IEEE Trans. Control. Syst. Technol.* **2015**, *23*, 2044–2052. [CrossRef]

24. Ghavami, A.; Kar, K.; Gupta, A. Decentralized Charging of Plug-in Electric Vehicles With Distribution Feeder Overload Control. *IEEE Trans. Autom. Control.* **2016**, *61*, 3527–3532. [CrossRef]
25. Pourakbari-Kasmaei, M.; Contreras, J.; Mantovani, J.R.S. A demand power factor-based approach for finding the maximum loading point. *Electr. Power Syst. Res.* **2017**, *151*, 283–295. [CrossRef]

axioms

Article

A Fast Calculation Method for Sensitivity Analysis Using Matrix Decomposition Technique

Qiuwei Yang [1,2,*] and Xi Peng [1,2]

1 School of Civil and Transportation Engineering, Ningbo University of Technology, Ningbo 315211, China
2 Engineering Research Center of Industrial Construction in Civil Engineering of Zhejiang, Ningbo University of Technology, Ningbo 315211, China
* Correspondence: yangqiuwei@nbut.edu.cn or yangqiuwei79@gmail.com

Abstract: The sensitivity reanalysis technique is an important tool for selecting the search direction in structural optimization design. Based on the decomposition perturbation of the flexibility matrix, a fast and exact structural displacement sensitivity reanalysis method is proposed in this work. For this purpose, the direct formulas for computing the first-order and second-order sensitivities of structural displacements are derived. The algorithm can be applied to a variety of the modifications in optimal design, including the low-rank modifications, high-rank modifications, small modifications and large modifications. Two numerical examples are given to verify the effectiveness of the proposed approach. The results show that the presented algorithm is exact and effective. Compared with the existing two reanalysis methods, this method has obvious advantages in calculation accuracy and efficiency. This new algorithm is very useful for calculating displacement sensitivity in engineering problems such as structure optimization, model correction and defect detection.

Keywords: sensitivity reanalysis; flexibility matrix; disassembly perturbation; structural displacement; exact method

check for updates

Citation: Yang, Q.; Peng, X. A Fast Calculation Method for Sensitivity Analysis Using Matrix Decomposition Technique. *Axioms* **2023**, *12*, 179. https://doi.org/10.3390/axioms12020179

Academic Editor: Leonid Plotnikov

Received: 28 December 2022
Revised: 6 February 2023
Accepted: 7 February 2023
Published: 9 February 2023

1. Introduction

Sensitivity analysis is often used in structural optimization design, vibration control, and damage identification. In general, sensitivity refers to the first derivative of structural response parameters to its physical parameters [1,2]. In engineering design, it is often necessary to modify the structure repeatedly. As a result, the computational cost for sensitivity analysis will be very expensive. To reduce the computational burden, reanalysis and sensitivity reanalysis techniques have been studied continuously in the past decades [3–8]. Sensitivity reanalysis uses the original response of the structure and its sensitivity to find the response sensitivity coefficients of the modified structure, whose calculation cost is far lower than the cost required for the complete analysis. For a structure under a given load vector y, the displacement vector x in the initial design can be computed by the static equilibrium equation as

$$K \cdot x = y \tag{1}$$

in which K is the structural stiffness matrix of $n \times n$ dimension in the initial finite element model (FEM). From Equation (1), the displacement x and its sensitivity $\frac{\partial x}{\partial p_i}$ of the initial design can be calculated from the complete analysis as

$$x = K^{-1} \cdot y = F \cdot y \tag{2}$$

$$\frac{\partial x}{\partial p_i} = -K^{-1}\frac{\partial K}{\partial p_i} \cdot x = -F\frac{\partial K}{\partial p_i}F \cdot y \tag{3}$$

where p_i is a design variable such as geometry size, elastic modulus, and so on. The matrix F is called the structural flexibility matrix, that is, $F = K^{-1}$. Correspondingly, the static balance equation of the modified structure can be expressed as

$$K_d \cdot x_d = y \tag{4}$$

$$K_d = K + \Delta K \tag{5}$$

in which K_d is the modified stiffness matrix, ΔK is the stiffness change caused by the optimal design, and x_d is the modified displacement vector. From Equation (4), x_d and its sensitivity $\frac{\partial x_d}{\partial p_i}$ can also be computed by the complete analysis as

$$x_d = K_d^{-1} \cdot y = F_d \cdot y \tag{6}$$

$$\frac{\partial x_d}{\partial p_i} = -K_d^{-1} \frac{\partial K_d}{\partial p_i} \cdot x_d = -F_d \frac{\partial K_d}{\partial p_i} F_d \cdot y \tag{7}$$

in which F_d is the modified flexibility matrix, i.e., $F_d = K_d^{-1}$. As mentioned earlier, when the half-bandwidth of the stiffness matrix is large, the complete analysis based on Equations (6) and (7) is very inefficient and time-consuming. For solving this problem, many reanalysis algorithms have been presented to calculate x_d and its sensitivity $\frac{\partial x_d}{\partial p_i}$ more effectively. The existing sensitivity reanalysis methods can be divided into two types: finite-difference method [9–12] and direct (analytic) method [13–16]. Most of the existing reanalysis methods can only obtain the approximate solution of displacement sensitivity. Moreover, these methods may be inefficient for large modifications or high-rank modifications. The high-rank modification refers to the design changes in many components of the structure. In view of this, an exact sensitivity reanalysis approach using flexibility disassembly perturbation (FDP) [17–19] is developed in this work for computing the displacement sensitivity. The presented algorithm is accurate and efficient, and it can be used for many types of modifications in design, such as the low-rank, high-rank, small and large modifications. Numerical examples show that the results obtained by the presented sensitivity reanalysis algorithm are the same as those obtained by the complete analysis. In addition, this approach has higher computing efficiency than the existing sensitivity reanalysis methods.

2. Sensitivity Reanalysis Using FDP

Reference [19] presented a static reanalysis method using the FDP technique for quickly and exactly calculating the displacement vector after structural modification. In addition to the displacement vector, the displacement sensitivity is another quantity that needs to be repeatedly calculated in structural optimization design, which indicates the direction of optimization design. So, in this work, FDP is used again to exactly compute the displacement sensitivity after structural modification. The research content of this work can be seen as an extension of reference [19]. From Equation (7), the modified displacement sensitivity $\frac{\partial x_d}{\partial p_i}$ can be easily calculated by the modified flexibility matrix F_d. Thus, the reanalysis problem of displacement sensitivity can be transformed into the reanalysis problem of structural flexibility matrix after modification. According to references [17–19], the modified flexibility matrix can be fast computed using FDP. The core idea of FDP is to decompose the flexibility matrix into a connected matrix reflecting the topological relationship between the degrees of freedom (DOFs) and the diagonal matrix reflecting the material and geometric information. The formulas of FDP are briefly derived as follows. According to the FEM theory, structural stiffness matrix K is the sum of all elementary stiffness matrices K_i $(i = 1 \sim N)$, that is

$$K = \sum_{i=1}^{N} K_i \tag{8}$$

in which N is the number of all elements in FEM. Performing the spectral decomposition on K_i yields

$$K_i = [c_i^1, \cdots, c_i^r] \begin{bmatrix} p_i^1 & & \\ & \ddots & \\ & & p_i^r \end{bmatrix} [c_i^1, \cdots, c_i^r]^T \tag{9}$$

In Equation (9), the non-zero eigenvalues $p_i^1, \cdots, p_i^r$ are purely functions of the material and geometric properties such as elastic modulus E, cross-sectional area A and moment of inertia I. The eigenvectors $c_i^1, \cdots, c_i^r$ reflect the topological relationship between degrees of freedom. For instance, the spectral decomposition on a plane beam element gives [20]:

$$[p_i] = \begin{bmatrix} \frac{2EA}{L} & 0 & 0 \\ 0 & \frac{2EI}{L} & 0 \\ 0 & 0 & \frac{6EI(L^2+4)}{L^3} \end{bmatrix} \tag{10}$$

$$[c_i] = \begin{bmatrix} \frac{1}{\sqrt{2}} & 0 & 0 \\ 0 & 0 & \frac{\sqrt{2}}{\sqrt{L^2+4}} \\ 0 & \frac{-1}{\sqrt{2}} & \frac{L}{\sqrt{2}\sqrt{L^2+4}} \\ \frac{-1}{\sqrt{2}} & 0 & 0 \\ 0 & 0 & \frac{-\sqrt{2}}{\sqrt{L^2+4}} \\ 0 & \frac{1}{\sqrt{2}} & \frac{L}{\sqrt{2}\sqrt{L^2+4}} \end{bmatrix} \tag{11}$$

in which L denotes the beam element length. Thus, $p_i^1, \cdots, p_i^r$ are also called the elementary stiffness coefficients and $c_i^1, \cdots, c_i^r$ are called the topological connection vectors. From Equations (8) and (9), the stiffness disassembly formula can be obtained as

$$K = CPC^T \tag{12}$$

$$C = [C_1^1, \cdots, c_1^r, c_2^1, \cdots, c_2^r, \cdots, c_N^r] \tag{13}$$

$$P = \begin{bmatrix} p_1^1 & & & & \\ & \ddots & & & \\ & & p_1^r & & \\ & & & \ddots & \\ & & & & p_N^r \end{bmatrix} \tag{14}$$

in which C is a $n \times rN$ dimension matrix, and P is a $rN \times rN$ dimension matrix. C is a full-rank matrix with $rank(C_{n\times rN}) = n$ because of $rank(K_{n\times n}) = n$. For the statically determinate system, C is a square matrix of $n = rN$. For the statically indeterminate system, C is a rectangular matrix of $n < rN$. Commonly, structural modifications such as the section correction or material correction only lead to the change of stiffness coefficients $p_i^1, \cdots, p_i^r$. This means that only P is changed in the structural modifications. As a result, the disassembly of the stiffness matrix K_d after modification can be derived as

$$K_d = CP_dC^T \tag{15}$$

$$P_d = \begin{bmatrix} p_1^1(1+\alpha_1^1) & & & & \\ & \ddots & & & \\ & & p_1^r(1+\alpha_1^r) & & \\ & & & \ddots & \\ & & & & p_N^r(1+\alpha_N^r) \end{bmatrix} \tag{16}$$

where α_i^j $(i = 1 \sim N, j = 1 \sim r)$ denotes the modification ratio of the stiffness parameter p_i^j. As stated before, C is a full-rank square matrix for the statically determinate system. Thus, the flexibility matrix F_d can be fast computed from Equation (15) by $F_d = K_d^{-1}$ as

$$F_d = DQ_dD^T \tag{17}$$

$$D = (C^{-1})^T \tag{18}$$

$$Q_d = P_d^{-1} = \begin{bmatrix} \frac{1}{p_1^1(1+\alpha_1^1)} & & & & \\ & \ddots & & & \\ & & \frac{1}{p_1^r(1+\alpha_1^r)} & & \\ & & & \ddots & \\ & & & & \frac{1}{p_N^r(1+\alpha_N^r)} \end{bmatrix} \tag{19}$$

It should be pointed out that the computational burden of the flexibility matrix reanalysis is only focused on the diagonal matrix Q_d, which only requires simple division operation when the modification ratios α_i^j are given. The computation of the matrix D should be attributed to the initial analysis, since D is unchanged in each modification. For the statically indeterminate structure, the flexibility disassembly as in Equation (17) is nonexistent, since C is a rectangular matrix with $n < rN$. In this case, the flexible disassembly can be realized by converting the statically indeterminate system into a statically determinate substructure and the redundant constraints. Correspondingly, the stiffness disassembly of the statically indeterminate system can be expressed from Equation (15) by

$$K_d = CP_dC^T = C'P_d'(C')^T + C''P_d''(C'')^T \tag{20}$$

where C' and P_d' are associated with the statically determinate substructure, while C'' and P_d'' are associated with the redundant constraints. The dimensions of C' and P_d' are both $n \times n$. The dimensions of C'' and P_d'' are $n \times (rN - n)$ and $(rN - n) \times (rN - n)$, respectively. From Equation (20), the flexibility disassembly can be derived by $F_d = K_d^{-1}$ with the help of Sherman–Morrison–Woodbury formulas [21,22] as

$$F_d = D'Q_d'(D')^T - D'Q_d'(D')^T C''P_d''\left[I_e + (C'')^T D'Q_d'(D')^T C''P_d''\right]^{-1}(C'')^T D'Q_d'(D')^T \tag{21}$$

$$D' = ((C')^{-1})^T, \tag{22}$$

$$Q_d' = (P_d')^{-1} \tag{23}$$

where I_e is the identity matrix, while Q_d' and P_d'' are the corrections corresponding to the statically determinate subsystem and the redundant constraints. Equation (21) is the flexibility reanalysis formula for the statically indeterminate system with the given Q_d' and P_d''.

According to the above theory and derivation, the modified displacement sensitivity $\frac{\partial x_d}{\partial p_i}$ can be fast computed using Equation (7) with F_d determined by Equation (17) or (21). It is clear that Equation (17) is an exceptional case of Equation (21). The step-by-step summary for the proposed sensitivity reanalysis approach is as follows. Step 1: Perform the stiffness disassembly of the initial structure using Equations (8)–(14) to obtain the matrices C, or C' and C''. Step 2: Compute the matrix D or D' by Equation (18) or (22). Step 3: Calculate the modified flexibility matrix F_d by Equation (17) or (21) with the given modifications Q_d, or Q_d' and P_d''. Step 4: Compute the displacement sensitivity $\frac{\partial x_d}{\partial p_i}$ of the modified structure using Equation (7). Note that the calculations in steps 1 and 2 should be attributed to the initial analysis. The computational burden of the sensitivity reanalysis algorithm is the focus of steps 3 and 4. Another virtue of this algorithm is that it can be readily extended to calculate the second-order sensitivity of static displacement. Differentiating Equation (4) with respect to p_i twice and rearranging gives the second-order sensitivity $\frac{\partial^2 x_d}{\partial p_i^2}$ as

$$\frac{\partial^2 x_d}{\partial p_i^2} = -F_d \frac{\partial^2 K_d}{\partial p_i^2} F_d y - 2F_d \frac{\partial K_d}{\partial p_i} \cdot \frac{\partial x_d}{\partial p_i} \tag{24}$$

Apparently, the second-order sensitivity of static displacement can also be fast calculated by Equation (24) using the proposed method for the modified structure.

3. Numerical Examples

3.1. Statically Determinate Structure

As presented in Figure 1, a statically determinate system of a 23-bar truss is used firstly to demonstrate the proposed approach. The values of the concentrated loads applied to the structure shown in Figure 1 are $f_1 = f_2 = f_3 = f_4 = f_5 = 10$ kN. Assuming the change rate of cross-sectional area is the correction factor α_i, Table 1 gives several modification cases including the low-rank, high-rank, small and large corrections. Tables 2 and 3 present the first-order sensitivity $\frac{\partial x_d}{\partial p_{10}}$ and second-order sensitivity $\frac{\partial^2 x_d}{\partial p_{10}^2}$ using the proposed method and complete analysis for these modification cases. It is found from Tables 2 and 3 that the reanalysis results of the presented algorithm are the same as the complete analysis results. This shows that the proposed method is an exact algorithm for displacement sensitivity reanalysis.

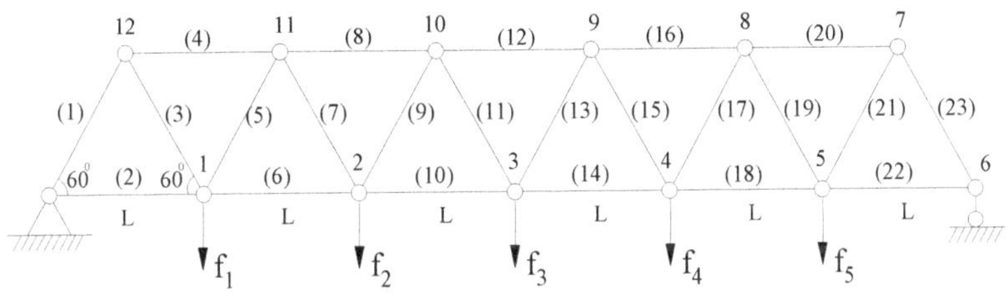

Figure 1. An initial structure of a 23-bar truss. Material parameters: Elastic modulus is 200 GPa, density is 7800 kg/m^3, L = 1 m, and initial cross-sectional area of each bar is 175.9 mm^2.

3.2. Statically Indeterminate Structure

As presented in Figure 2, a statically indeterminate system of a 275-bar truss is used to conduct the comparison study on the computation efficiency between this method and two existing sensitivity reanalysis approaches. The first existing technique is the combined approximate (CA) method proposed by Kirsch in reference [10]. The second existing technique is the method proposed by Zuo et al. in reference [16], which combines Taylor series expansion and the CA method. Table 4 gives three types of corrections for this example. As shown in Figure 2, the modified bars of the three types of corrections are: bars 1~10, bars 1~93 (the first story), and all bars (1~275) of the system, respectively. For each correction, 200 modifications are performed, and the total calculation times of displacement sensitivities $\frac{\partial x_d}{\partial p_8}$ using the complete analysis, the CA method, Zuo's method, and the proposed method are given in Table 5. Note that the correction coefficient α_i increases with the modification number z ($z = 1 \sim 150$). This means that the early stage corresponds to small modifications and the later stage corresponds to large modifications. Tables 6–11 show the displacement sensitivity data of some DOFs for each correction scenario with $z = 1$, $z = 2$, $z = 10$ and $z = 15$, respectively. From Table 5, one can see that the presented algorithm has the highest calculation efficiency among the four sensitivity reanalysis methods. For type 1 (10 bars are modified), the calculation times of the four methods are: t_1 = 0.262 s (the complete analysis), t_2 = 0.166 s (CA method), t_3 = 0.161 s (Zuo's method) and t_4 = 0.083 s (the presented algorithm), respectively. For type 2 (93 bars are modified), the calculation times of the four methods are: t_1 = 0.254 s (the complete analysis), t_2 = 0.191 s (CA method), t_3 = 0.174 s (Zuo's method) and t_4 = 0.097 s (the presented algorithm), respectively. For the third type (all bars are modified), the calculation times of the four methods are: t_1 = 0.292 s (the complete analysis), t_2 = 0.232 s (CA method), t_3 = 0.217 s (Zuo's method) and t_4 = 0.140 s (the presented algorithm), respectively. Overall, the calculation time of the presented algorithm is about 30~40% of that of the complete analysis method, and it is about 50~60% of that of CA or Zuo's method. This means that whether the number of correction bars is small or large, the presented algorithm always has the high computation efficiency. According to Tables 6–11, it can be seen that the results achieved by the presented approach and the complete analysis method are exactly the same. One can also find that the results obtained by CA and Zuo's methods have some errors compared with the exact results. These results show that the presented approach is an exact algorithm for displacement sensitivity reanalysis, and the CA and Zuo's methods are approximate methods.

Table 1. Different correction cases of a 23-bar truss.

The Correction Coefficient α_i	Scenario 1: Low-Rank Correction	Scenario 2: High-Rank Small Correction	Scenario 3: High-Rank Large Correction
α_1	0	0.15	4.87
α_2	0	0.17	4.07
α_3	0	−0.08	−4.22
α_4	0	0.15	3.32
α_5	0.21	0.19	−1.93
α_6	0	−0.09	−1.15
α_7	0	−0.10	−0.88
α_8	0	0.14	−0.53
α_9	0.44	−0.02	−1.40
α_{10}	0	0.19	−4.66
α_{11}	0	−0.18	0.32
α_{12}	0	0.12	1.81
α_{13}	0	0.06	−1.32
α_{14}	−0.32	−0.16	3.08
α_{15}	0	0.17	−0.87
α_{16}	0	−0.08	1.16
α_{17}	0	0.13	0.54
α_{18}	0	0.09	−1.76
α_{19}	0	−0.13	−0.03
α_{20}	0	−0.10	4.27
α_{21}	0	0.05	4.35
α_{22}	0	0.10	−3.18
α_{23}	0	0.08	4.06

Table 2. The first-order sensitivities of displacements for modified structures ($\times 10^{-3}$).

DOF Number	Scenario 1: Low-Rank Correction		Scenario 2: High-Rank Small Correction		Scenario 3: High-Rank Large Correction	
	The Complete Analysis	The Proposed Reanalysis Algorithm	The Complete Analysis	The Proposed Reanalysis Algorithm	The Complete Analysis	The Proposed Reanalysis Algorithm
1	0.000	0.000	0.000	0.000	0.000	0.000
2	0.940	0.940	0.664	0.664	0.070	0.070
3	0.000	0.000	0.000	0.000	0.000	0.000
4	1.879	1.879	1.327	1.327	0.140	0.140
5	−1.395	−1.395	−0.985	−0.985	−0.104	−0.104
6	2.013	2.013	1.422	1.422	0.150	0.150
7	−1.395	−1.395	−0.985	−0.985	−0.104	−0.104
8	1.342	1.342	0.948	0.948	0.100	0.100
9	−1.395	−1.395	−0.985	−0.985	−0.104	−0.104
10	0.671	0.671	0.474	0.474	0.050	0.050
11	−1.395	−1.395	−0.985	−0.985	−0.104	−0.104
12	−0.814	−0.814	−0.575	−0.575	−0.061	−0.061
13	0.336	0.336	0.237	0.237	0.025	0.025
14	−0.814	−0.814	−0.575	−0.575	−0.061	−0.061
15	1.007	1.007	0.711	0.711	0.075	0.075
16	−0.814	−0.814	−0.575	−0.575	−0.061	−0.061
17	1.678	1.678	1.185	1.185	0.125	0.125
18	−0.814	−0.814	−0.575	−0.575	−0.061	−0.061
19	2.349	2.349	1.659	1.659	0.175	0.175
20	−0.814	−0.814	−0.575	−0.575	−0.061	−0.061
21	1.409	1.409	0.995	0.995	0.105	0.105
22	−0.814	−0.814	−0.575	−0.575	−0.061	−0.061
23	0.470	0.470	0.332	0.332	0.035	0.035

Table 3. The second-order sensitivities of displacements for modified structures ($\times 10^{-3}$).

DOF Number	Scenario 1: Low-Rank Correction		Scenario 2: High-Rank Small Correction		Scenario 3: High-Rank Large Correction	
	The Complete Analysis	The Proposed Reanalysis Algorithm	The Complete Analysis	The Proposed Reanalysis Algorithm	The Complete Analysis	The Proposed Reanalysis Algorithm
1	0.000	0.000	0.000	0.000	0.000	0.000
2	−1.879	−1.879	−1.115	−1.115	0.038	0.038
3	0.000	0.000	0.000	0.000	0.000	0.000
4	−3.758	−3.758	−2.230	−2.230	0.077	0.077
5	2.790	2.790	1.656	1.656	−0.057	−0.057
6	−4.027	−4.027	−2.390	−2.390	0.082	0.082
7	2.790	2.790	1.656	1.656	−0.057	−0.057
8	−2.685	−2.685	−1.593	−1.593	0.055	0.055
9	2.790	2.790	1.656	1.656	−0.057	−0.057
10	−1.342	−1.342	−0.797	−0.797	0.027	0.027
11	2.790	2.790	1.656	1.656	−0.057	−0.057
12	1.627	1.627	0.966	0.966	−0.033	−0.033
13	−0.671	−0.671	−0.398	−0.398	0.014	0.014
14	1.627	1.627	0.966	0.966	−0.033	−0.033
15	−2.013	−2.013	−1.195	−1.195	0.041	0.041
16	1.627	1.627	0.966	0.966	−0.033	−0.033
17	−3.356	−3.356	−1.991	−1.991	0.068	0.068

Table 3. *Cont.*

DOF Number	Scenario 1: Low-Rank Correction		Scenario 2: High-Rank Small Correction		Scenario 3: High-Rank Large Correction	
	The Complete Analysis	The Proposed Reanalysis Algorithm	The Complete Analysis	The Proposed Reanalysis Algorithm	The Complete Analysis	The Proposed Reanalysis Algorithm
18	1.627	1.627	0.966	0.966	−0.033	−0.033
19	−4.698	−4.698	−2.788	−2.788	0.096	0.096
20	1.627	1.627	0.966	0.966	−0.033	−0.033
21	−2.819	−2.819	−1.673	−1.673	0.057	0.057
22	1.627	1.627	0.966	0.966	−0.033	−0.033
23	−0.940	−0.940	−0.558	−0.558	0.019	0.019

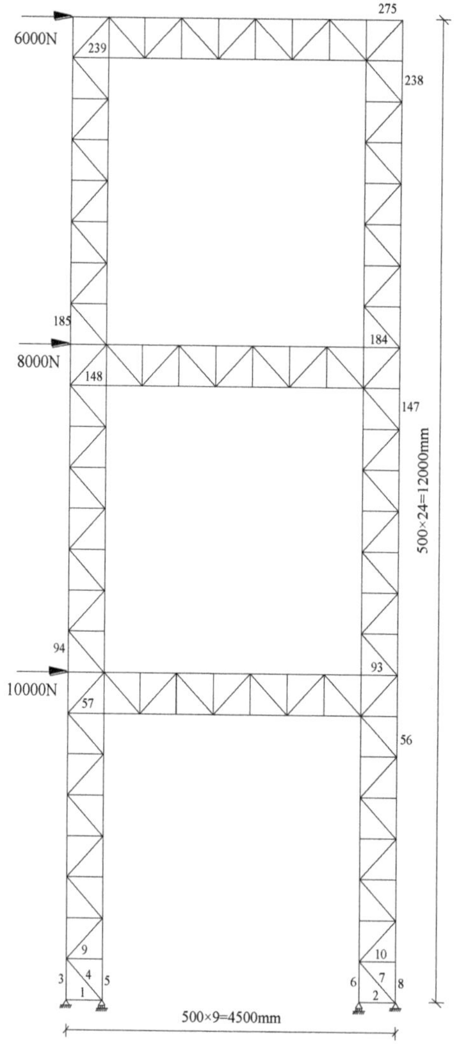

Figure 2. An initial structure of a 275-bar truss. Material parameters: Elastic modulus is 200 GPa, density is 7800 kg/m^3, L = 0.5 m, and initial cross-sectional area of each bar is 314 mm^2.

Table 4. Types of corrections in the 275-bar truss system.

Type of Correction	Modified Bars	Correction Coefficients α_i^z (i is the Bar Number, z is the Modification Number, z = 1~150)
Type 1	Bars 1~10 as shown in Figure 2	$\alpha_i^z = \frac{z}{20}, i = 1 \sim 10$
Type 2	Bars 1~93 of the first story as shown in Figure 2	$\alpha_i^z = \begin{cases} \frac{z}{40}, i = 1 \sim 56 \\ \frac{z}{50}, i = 57 \sim 93 \end{cases}$
Type 3	All bars (1~275) in Figure 2	First story: $\alpha_i^z = \begin{cases} \frac{z}{40}, i = 1 \sim 56 \\ \frac{z}{50}, i = 57 \sim 93 \end{cases}$ Second story: $\alpha_i^z = \begin{cases} \frac{z}{60}, i = 94 \sim 147 \\ \frac{z}{75}, i = 148 \sim 184 \end{cases}$ Third story: $\alpha_i^z = \begin{cases} \frac{z}{80}, i = 185 \sim 238 \\ \frac{z}{100}, i = 239 \sim 275 \end{cases}$

Table 5. Computation times of the four algorithms for the three types of modifications.

Type of Modification	The Complete Analysis t_1	CA Method t_2	Zuo's Method t_3	The Proposed Method t_4
Type 1 (10 elements are revised)	t_1 = 0.262 s	t_2 = 0.166 s	t_3 = 0.161 s	t_4 = 0.083 s
		$(t_1 - t_2)/t_1$ = 36.6%	$(t_1 - t_3)/t_1$ = 38.5%	$(t_1 - t_4)/t_1$ = 68.3%
			$(t_2 - t_3)/t_2$ = 3.0%	$(t_2 - t_4)/t_2$ = 50.0%
				$(t_3 - t_4)/t_3$ = 48.4%
Type 2 (93 elements are revised)	t_1 = 0.254 s	t_2 = 0.191 s	t_3 = 0.174 s	t_4 = 0.097 s
		$(t_1 - t_2)/t_1$ = 24.8%	$(t_1 - t_3)/t_1$ = 31.5%	$(t_1 - t_4)/t_1$ = 61.8%
			$(t_2 - t_3)/t_2$ = 8.9%	$(t_2 - t_4)/t_2$ = 49.2%
				$(t_3 - t_4)/t_3$ = 44.3%
Type 3 (all elements are revised)	t_1 = 0.292 s	t_2 = 0.232 s	t_3 = 0.217 s	t_4 = 0.140 s
		$(t_1 - t_2)/t_1$ = 20.5%	$(t_1 - t_3)/t_1$ = 25.7%	$(t_1 - t_4)/t_1$ = 52.1%
			$(t_2 - t_3)/t_2$ = 6.5%	$(t_2 - t_4)/t_2$ = 39.7%
				$(t_3 - t_4)/t_3$ = 35.5%

Table 6. Displacement sensitivities for modification type 1 when $z = 1$ and $z = 2$ ($\times 10^{-5}$).

DOF Number	The Complete Analysis		CA Method		Zuo's Method		The Proposed Method	
	$z = 1$	$z = 2$	$z = 1$	$z = 2$	$z = 1$	$z = 2$	$z = 1$	$z = 2$
10	1.659	1.518	1.657	1.516	1.663	1.531	1.659	1.518
11	−0.551	−0.496	−0.551	−0.495	−0.552	−0.500	−0.551	−0.496
12	−0.187	−0.169	−0.187	−0.169	−0.188	−0.171	−0.187	−0.169
13	−0.551	−0.496	−0.551	−0.495	−0.552	−0.500	−0.551	−0.496
14	0.169	0.154	0.168	0.154	0.169	0.155	0.169	0.154
15	−0.979	−0.908	−0.979	−0.909	−0.981	−0.915	−0.979	−0.908
16	0.344	0.310	0.344	0.309	0.345	0.313	0.344	0.310
17	−0.979	−0.908	−0.979	−0.909	−0.981	−0.915	−0.979	−0.908
18	1.492	1.364	1.491	1.362	1.496	1.376	1.492	1.364
19	−1.047	−0.949	−1.046	−0.948	−1.049	−0.957	−1.047	−0.949
20	−0.260	−0.237	−0.260	−0.237	−0.260	−0.239	−0.260	−0.237
21	−1.047	−0.949	−1.046	−0.948	−1.049	−0.957	−1.047	−0.949

Table 7. Displacement sensitivities for modification type 1 when $z = 10$ and $z = 15$ ($\times 10^{-5}$).

DOF Number	The Complete Analysis		CA Method		Zuo's Method		The Proposed Method	
	$z = 10$	$z = 15$	$z = 10$	$z = 15$	$z = 10$	$z = 15$	$z = 10$	$z = 15$
10	0.837	0.622	0.835	0.619	0.971	0.813	0.837	0.622
11	−0.244	−0.171	−0.240	−0.165	−0.283	−0.224	−0.244	−0.171
12	−0.087	−0.062	−0.086	−0.061	−0.101	−0.081	−0.087	−0.062
13	−0.244	−0.171	−0.240	−0.165	−0.283	−0.224	−0.244	−0.171
14	0.083	0.060	0.084	0.062	0.096	0.079	0.083	0.060
15	−0.540	−0.413	−0.563	−0.447	−0.626	−0.540	−0.540	−0.413
16	0.155	0.110	0.143	0.093	0.180	0.144	0.155	0.110
17	−0.540	−0.413	−0.563	−0.447	−0.626	−0.540	−0.540	−0.413
18	0.747	0.553	0.742	0.547	0.866	0.723	0.747	0.553
19	−0.492	−0.355	−0.490	−0.352	−0.571	−0.464	−0.492	−0.355
20	−0.128	−0.094	−0.129	−0.095	−0.149	−0.123	−0.128	−0.094
21	−0.492	−0.355	−0.490	−0.352	−0.571	−0.464	−0.492	−0.355

Table 8. Displacement sensitivities for modification type 2 when $z = 1$ and $z = 2$ ($\times 10^{-5}$).

DOF Number	The Complete Analysis		CA Method		Zuo's Method		The Proposed Method	
	$z = 1$	$z = 2$	$z = 1$	$z = 2$	$z = 1$	$z = 2$	$z = 1$	$z = 2$
10	1.738	1.660	1.736	1.658	1.739	1.664	1.738	1.660
11	−0.587	−0.561	−0.587	−0.560	−0.588	−0.562	−0.587	−0.561
12	−0.198	−0.190	−0.198	−0.189	−0.198	−0.190	−0.198	−0.190
13	−0.587	−0.561	−0.587	−0.560	−0.588	−0.562	−0.587	−0.561
14	0.177	0.170	0.177	0.169	0.177	0.170	0.177	0.170
15	−1.012	−0.967	−1.010	−0.966	−1.012	−0.969	−1.012	−0.967
16	0.366	0.350	0.366	0.349	0.367	0.350	0.366	0.350
17	−1.012	−0.967	−1.010	−0.966	−1.012	−0.969	−1.012	−0.967
18	1.565	1.496	1.563	1.494	1.566	1.498	1.565	1.496
19	−1.107	−1.058	−1.106	−1.056	−1.108	−1.060	−1.107	−1.058
20	−0.273	−0.261	−0.273	−0.261	−0.273	−0.262	−0.273	−0.261
21	−1.107	−1.058	−1.106	−1.056	−1.108	−1.060	−1.107	−1.058

Table 9. Displacement sensitivities for modification type 2 when $z = 10$ and $z = 15$ ($\times 10^{-5}$).

DOF Number	The Complete Analysis		CA Method		Zuo's Method		The Proposed Method	
	$z = 10$	$z = 15$	$z = 10$	$z = 15$	$z = 10$	$z = 15$	$z = 10$	$z = 15$
10	1.193	0.996	1.181	0.976	1.242	1.079	1.193	0.996
11	−0.403	−0.336	−0.399	−0.329	−0.420	−0.364	−0.403	−0.336
12	−0.137	−0.114	−0.136	−0.113	−0.142	−0.124	−0.137	−0.114
13	−0.403	−0.336	−0.399	−0.329	−0.420	−0.364	−0.403	−0.336
14	0.122	0.102	0.121	0.100	0.128	0.111	0.122	0.102
15	−0.698	−0.584	−0.692	−0.574	−0.727	−0.633	−0.698	−0.584

Table 9. *Cont.*

DOF Number	The Complete Analysis		CA Method		Zuo's Method		The Proposed Method	
	$z = 10$	$z = 15$	$z = 10$	$z = 15$	$z = 10$	$z = 15$	$z = 10$	$z = 15$
16	0.250	0.208	0.247	0.203	0.260	0.225	0.250	0.208
17	−0.698	−0.584	−0.692	−0.574	−0.727	−0.633	−0.698	−0.584
18	1.075	0.898	1.065	0.880	1.120	0.973	1.075	0.898
19	−0.761	−0.635	−0.753	−0.623	−0.792	−0.688	−0.761	−0.635
20	−0.188	−0.157	−0.187	−0.155	−0.196	−0.171	−0.188	−0.157
21	−0.761	−0.635	−0.753	−0.623	−0.792	−0.688	−0.761	−0.635

Table 10. Displacement sensitivities for modification type 3 when $z = 1$ and $z = 2$ ($\times 10^{-5}$).

DOF Number	The Complete Analysis		CA Method		Zuo's Method		The Proposed Method	
	$z = 1$	$z = 2$	$z = 1$	$z = 2$	$z = 1$	$z = 2$	$z = 1$	$z = 2$
10	1.736	1.656	1.734	1.655	1.737	1.660	1.736	1.656
11	−0.587	−0.560	−0.586	−0.559	−0.587	−0.561	−0.587	−0.560
12	−0.198	−0.189	−0.198	−0.189	−0.198	−0.190	−0.198	−0.189
13	−0.587	−0.560	−0.586	−0.559	−0.587	−0.561	−0.587	−0.560
14	0.177	0.169	0.177	0.169	0.177	0.169	0.177	0.169
15	−1.010	−0.964	−1.009	−0.963	−1.011	−0.966	−1.010	−0.964
16	0.366	0.349	0.366	0.349	0.366	0.350	0.366	0.349
17	−1.010	−0.964	−1.009	−0.963	−1.011	−0.966	−1.010	−0.964
18	1.563	1.492	1.562	1.490	1.564	1.495	1.563	1.492
19	−1.106	−1.055	−1.105	−1.054	−1.106	−1.058	−1.106	−1.055
20	−0.273	−0.260	−0.272	−0.260	−0.273	−0.261	−0.273	−0.260
21	−1.106	−1.055	−1.105	−1.054	−1.106	−1.058	−1.106	−1.055

Table 11. Displacement sensitivities for modification type 3 when $z = 10$ and $z = 15$ ($\times 10^{-5}$).

DOF Number	The Complete Analysis		CA Method		Zuo's Method		The Proposed Method	
	$z = 10$	$z = 15$	$z = 10$	$z = 15$	$z = 10$	$z = 15$	$z = 10$	$z = 15$
10	1.181	0.981	1.173	0.969	1.234	1.071	1.181	0.981
11	−0.399	−0.332	−0.397	−0.328	−0.417	−0.362	−0.399	−0.332
12	−0.135	−0.112	−0.134	−0.111	−0.141	−0.123	−0.135	−0.112
13	−0.399	−0.332	−0.397	−0.328	−0.417	−0.362	−0.399	−0.332
14	0.121	0.101	0.120	0.099	0.126	0.110	0.121	0.101
15	−0.689	−0.573	−0.685	−0.567	−0.720	−0.626	−0.689	−0.573
16	0.248	0.206	0.246	0.203	0.259	0.224	0.248	0.206
17	−0.689	−0.573	−0.685	−0.567	−0.720	−0.626	−0.689	−0.573
18	1.064	0.884	1.057	0.874	1.111	0.965	1.064	0.884
19	−0.753	−0.626	−0.749	−0.619	−0.787	−0.683	−0.753	−0.626
20	−0.186	−0.155	−0.185	−0.153	−0.194	−0.169	−0.186	−0.155
21	−0.753	−0.626	−0.749	−0.619	−0.787	−0.683	−0.753	−0.626

4. Conclusions

In this paper, an exact algorithm for the reanalysis of static displacement sensitivity based on flexibility disassembly perturbation is proposed. The presented algorithm is exact and efficient, and it can be used in many types of corrections in structural optimal design, including the low-rank, high-rank, small and large corrections. Numerical examples show that the presented approach can achieve the same results as the complete analysis method with less computational time. Compared with CA and Zuo's techniques, this algorithm has obvious advantages in computational efficiency and accuracy. It has been shown that the proposed algorithm has great application potential in structural optimization design based on gradient.

Author Contributions: Methodology, Q.Y.; Software, X.P. All authors have read and agreed to the published version of the manuscript.

Funding: This research was funded by the Ningbo natural science foundation (202003N4169), Zhejiang public welfare project (LGF22E080021), Natural Science Foundation of China (52008215), Zhejiang Province Natural Science Foundation (LQ20E080013), and the project of Ningbo science and technology innovation 2025 (2019B10076).

Data Availability Statement: The data generated and/or analyzed during the current study are not publicly available for legal/ethical reasons but are available from the corresponding author on reasonable request.

Conflicts of Interest: The authors declare no conflict of interest.

References

1. Lin, R.; Mottershead, J.; Ng, T. A state-of-the-art review on theory and engineering applications of eigenvalue and eigenvector derivatives. *Mech. Syst. Signal Process.* **2020**, *138*, 106536. [CrossRef]
2. Yang, Q.; Peng, X. An Exact Method for Calculating the Eigenvector Sensitivities. *Appl. Sci.* **2020**, *10*, 2577. [CrossRef]
3. Kirsch, U. *Reanalysis of Structures*; Springer: Dordrecht, The Netherlands, 2008.
4. Yang, Q.W.; Peng, X. A highly efficient method for structural model reduction. *Int. J. Numer. Methods Eng.* **2023**, *124*, 513–533. [CrossRef]
5. Cheikh, M.; Loredo, A. Static reanalysis of discrete elastic structures with reflexive inverse. *Appl. Math. Model.* **2002**, *26*, 877–891. [CrossRef]
6. Chen, S.H.; Yang, Z.J. A universal method for structural static reanalysis of topological modifications. *Int. J. Numer. Methods Eng.* **2004**, *61*, 673–686. [CrossRef]
7. Wu, B.; Li, Z. Static reanalysis of structures with added degrees of freedom. *Commun. Numer. Methods Eng.* **2006**, *22*, 269–281. [CrossRef]
8. Zuo, W.; Yu, Z.; Zhao, S.; Zhang, W. A hybrid Fox and Kirsch's reduced basis method for structural static reanalysis. *Struct. Multidiscip. Optim.* **2012**, *46*, 261–272. [CrossRef]
9. Adelman, H.M.; Haftka, R.T. Sensitivity analysis of discrete structural systems. *AIAA J.* **1986**, *24*, 823–832. [CrossRef]
10. Kirsch, U. Reanalysis and sensitivity reanalysis by combined approximations. *Struct. Multidiscip. Optim.* **2009**, *40*, 1–15. [CrossRef]
11. Kirsch, U.; Bogomolni, M.; Sheinman, I. Efficient structural optimization using reanalysis and sensitivity reanalysis. *Eng. Comput.* **2007**, *23*, 229–239.
12. Zuo, W.; Huang, K.; Bai, J.; Guo, G. Sensitivity reanalysis of vibration problem using combined approximations method. *Struct. Multidiscip. Optim.* **2017**, *55*, 1399–1405. [CrossRef]
13. Chen, W.; Zuo, W. Component sensitivity analysis of conceptual vehicle body for lightweight design under static and dynamic stiffness demands. *Int. J. Veh. Des.* **2014**, *66*, 107–123. [CrossRef]
14. Thomas, H.; Zhou, M. Issues of commercial optimization software development. *Struct. Multidiscip. Optim.* **2002**, *23*, 97–110. [CrossRef]
15. Liu, J.; Wang, H. Fast sensitivity reanalysis methods assisted by Independent Coefficients and Indirect Factorization Updating strategies. *Adv. Eng. Softw.* **2018**, *119*, 93–102. [CrossRef]
16. Zuo, W.; Bai, J.; Yu, J. Sensitivity reanalysis of static displacement using Taylor series expansion and combined approximate method. *Struct. Multidiscip. Optim.* **2016**, *53*, 953–959.
17. Yang, Q. A new damage identification method based on structural flexibility disassembly. *J. Vib. Control* **2011**, *17*, 1000–1008. [CrossRef]
18. Yang, Q.; Sun, B. Structural damage localization and quantification using static test data. *Struct. Health Monit.* **2011**, *10*, 381–389. [CrossRef]

19. Yang, Q.W. Fast and Exact Algorithm for Structural Static Reanalysis Based on Flexibility Disassembly Perturbation. *AIAA J.* **2019**, *57*, 3599–3607. [CrossRef]
20. Di, W.; Law, S. Eigen-parameter decomposition of element matrices for structural damage detection. *Eng. Struct.* **2007**, *29*, 519–528. [CrossRef]
21. Akgün, M.A.; Garcelon, J.H.; Haftka, R.T. Fast exact linear and non-linear structural reanalysis and the Sherman-Morrison-Woodbury formulas. *Int. J. Numer. Methods Eng.* **2001**, *50*, 1587–1606. [CrossRef]
22. Ren, J.; Zhang, Q. Structural Reanalysis Based on FRFs Using Sherman–Morrison–Woodbury Formula. *Shock Vib.* **2020**, *3*, 8212730.

www.ingramcontent.com/pod-product-compliance
Lightning Source LLC
Chambersburg PA
CBHW061816210926
51850CB00006B/719